# Practical VoIP Using VOCAL

# Practical VoIP Using VOCAL

*Luan Dang, Cullen Jennings, and David Kelly*

**O'REILLY**®

Beijing · Cambridge · Farnham · Köln · Paris · Sebastopol · Taipei · Tokyo

**Practical VoIP Using VOCAL**

by Luan Dang, Cullen Jennings, and David Kelly

Published by O'Reilly & Associates, Inc., 1005 Gravenstein Highway North,
Sebastopol, CA 95472.

O'Reilly & Associates books may be purchased for educational, business, or sales promotional use. Online editions are also available for most titles (*safari.oreilly.com*). For more information, contact our corporate/institutional sales department: (800) 998-9938 or *corporate@oreilly.com*.

| | |
|---|---|
| **Editor:** | Jim Sumser |
| **Production Editor:** | Jane Ellin |
| **Cover Designer:** | Ellie Volckhausen |
| **Interior Designer:** | David Futato |

**Printing History:**

| | |
|---|---|
| July 2002: | First Edition. |

**Library of Congress Cataloging-in-Publication Data**

Dang, Luan
   Practical VoIP Using VOCAL/Luan Dang, Cullen Jennings, & David Kelly.
     p. cm.
   Includes index.
   ISBN 0-596-00078-2
    1. Internet telephony  I. Jennings, Cullen  II. Kelly, David  III. Title.

  TK5105.8865 .D35 2002
  621.382'12--dc2120
                                             2002023323

[C]

# Table of Contents

# Preface

This book is the product of a Silicon Valley success story called Vovida Networks, a high-technology start-up founded by Alan Knitowski and Luan Dang. Vovida recruited an international team of talented developers to build an open source (*http://www.opensource.org*) phone system that became known as the Vovida Open Communication Application Library (VOCAL). Vovida's aim was to enable a community of developers to produce new Voice over IP (VoIP) applications that would challenge the established companies, dispel popular beliefs about phone systems, and contribute to a new world of interactivity and interoperability. All of the technolology that Vovida built is open source. You can build a complete phone system from it, and this book describes how to acquire, compile, install, and work with the applications as an end user, and how to work with the code as a developer.

During the time that Vovida was building its team and developing its applications, the industry was going through an exciting period of growth and innovation. The Telecommunications Act of 1996 (*http://www.fcc.gov/telecom.html*) had opened up a previously closed industry to "...promote competition and reduce regulation in order to secure lower prices and higher quality services for American telecommunications consumers and encourage the rapid deployment of new telecommunications technologies." At the same time, many venture capital firms had relaxed their guidelines about lending and initial public offerings (IPOs). This mixture of new law and new money helped establish more than 1000 new telecom service providers, along with many new software and hardware manufacturers. A strong stock market also permitted established organizations to expand their capabilities by acquiring innovative start-ups.

At the height of this prosperity, Vovida Networks was acquired by Cisco Systems, which enabled VOCAL and a new web site called *Vovida.org* (*http://www.vovida.org*) to continue growing and expanding their communities. VOCAL is currently being developed not only at Cisco, but also inside the laboratories, cubicles, and basements of community members scattered throughout the world. *Vovida.org* is a communications community site dedicated to providing a forum for open source software used in datacom and telecom environments. From April 2001 to January 2002, the number of downloads of VOCAL source code* grew from 697 to 1293 per month, almost a 100% increase in 10 months, demonstrating the steady momentum of the open source movement in Voice over IP.

Whether it is through the mailing lists, from meeting developers at different trade shows, or from talking directly to people who have worked with VOCAL in their corporate labs, we hear a consistent story coming from the community: those who have built their projects on top of VOCAL components, rather than from scratch, have significantly reduced their development costs and time to market.

We want you to share in our success by creating the next killer application, hosting a friends and family phone system at home, or finding another creative outlet that satisfies your curiosity about VoIP. We look forward to receiving your messages on our mailing lists as we hope to find out more about your experiences with running VOCAL.

## How to Use This Book

This book has been written to appeal to three types of VOCAL users:

*Hobbyist*
> One who is interested in using VOCAL as a small phone system, who may make some minor code changes but has no interest in turning it into a commercially viable product

*System administrator*
> One who has been tasked to maintain a VOCAL-based phone system but who has no interest in VoIP software development

*Professional developer*
> One who is using VOCAL to test commercial applications or who is building new commercial applications on top of the VOCAL source code

Here are some suggestions about how each type of user can get the most out of this book:

*Hobbyist*
> Use the instructions in Chapter 2 to install VOCAL onto a single host and to test your installation, and then refer to Chapters 4, 5, and 6 for information about working with users, dial plans, and servers. Refer to Chapter 8 to learn about the SIP stack architecture and Chapter 9 to learn about the base code. Afterward, refer to the chapters from the second half of the book for more information about the other VOCAL modules.

---

* There is no correlation between number of downloads and active VOCAL systems. Any given person could download one copy of VOCAL over another or could run multiple systems from a single download. However, the growth in the monthly number of downloads is a strong indication of the growth of the community.

*System administrator*

Refer to Chapters 2 and 3 for information about installing VOCAL and configuring IP phones and gateways, and then refer to Chapters 4, 5, and 6 for information about provisioning users, dial plans, and servers.

*Developer*

Refer to Chapter 2 about installing VOCAL; refer to Chapters 4, 5, and 6 about provisioning users, dial plans, and servers; and then refer to the chapters in the rest of the book that best suit your needs.

## How This Book Is Organized

We have organized the book so that you can get started and make something happen on your home or test PC quickly and easily. We're hoping that your excitement from making calls on your own phone system will inspire you to look further into the system's complexity.

In order to make our delivery of this material as straightforward as possible, we have refrained from discussing the technical details of the signaling methods and architectural models throughout the first six chapters. If you want the theory first and the practical second, turn to Chapter 7 and read through the rest of the book before returning to Chapter 2. We think that most readers will prefer getting their hands dirty first.

Chapter 1, *VOCAL: Say, What?*

Gives you a brief overview of what this book, the software, and Voice over IP are all about.

Chapter 2, *Setting Up a Phone System at Home*

Provides instructions about getting a phone system working at home on a single Linux host and then adding new phone devices to your system.

Chapter 3, *Setting Up an Internal Trial System*

Provides instructions about setting up an internal system that can support dozens of users working in a professional environment. This chapter also includes configuration instructions for gateways and how to deploy the system onto a distributed network of hosts.

Chapter 4, *Provisioning Users*

Provides information about adding users to the system and the different options available for features and other end-user parameters.

Chapter 5, *Configuring System Parameters and Dial Plans*

Provides information about setting up global system values such as dial plans and the multicast address used for heartbeats.

Chapter 6, *Provisioning Servers*

Provides information about setting up individual server types, configuring their IP addresses, and adding new servers to the system.

Chapter 7, *Session Initiation Protocol and Related Protocols*

A general overview of SIP, SDP, and the different message types used by these protocols. Includes some illustrated call flows and a line-by-line analysis of the basic message content.

Chapter 8, *Vovida SIP Stack*

A specific overview of Vovida's implementation of a SIP stack including class diagrams, discussions about data structures, some insight into how the stack was developed, and what the engineers were working on as this book was being written.

Chapter 9, *Base Code*

A short but important chapter about the base code that is common to most of the VOCAL servers.

Chapter 10, *VOCAL User Agent*

Discusses the SIP user agent (UA) that comes with VOCAL. This UA is useful for testing and demonstrating how the software works, but it was never intended as a practical softphone for end users. This chapter discusses the data structures and some basic call flows.

Chapter 11, *SIP Proxy: Marshal Server*

The Marshal server is our name for a SIP-edge proxy server that provides authentication and security for VOCAL. This chapter looks at the data structures that make the Marshal server work and includes additional, general information about authentication, security, and working with firewalls.

Chapter 12, *Redirect Server*

The VOCAL Redirect server performs the duties described in the standard (RFC 2543) and also provides registration and location services. This chapter looks at the data structures and provides additional information about routing, ENUM, and Telephony Routing over IP (TRIP).

Chapter 13, *CPL Feature Server*

The Call Processing Language (CPL) Feature server is an implementation of the SIP proxy that provides basic features such as call forwarding. This chapter provides information about CPL and the data structures that make up the server.

Chapter 14, *Unified Voice Mail Server*

As a trade show demo, we wrote a voice mail server. Despite its basic functionality, it works for a small user population and has become a popular module within the user community. This chapter explains the data structures and a few of the solutions that we implemented to make this service work.

Chapter 15, *MGCP Translator*

The MGCP translator allows VOCAL to talk to MGCP gateways, which are normally attached to analog phone sets. This chapter discusses the MGCP protocol stack, the translator, and call flows through the state machine.

Chapter 16, *H.323 Translator*

The H.323 translator allows H.323 endpoints such as NetMeeting or H.323 gateways to be used with VOCAL. This chapter provides a simplified look at the data structures with some basic call flows.

Chapter 17, *System Monitoring*

We took the UC Davis SNMP stack and adopted it into VOCAL to provide network monitoring. This chapter discusses how this was accomplished and how you can add a new Management Information Base (MIB) to your system.

Chapter 18, *Quality of Service and Billing*

Advanced topics. Quality of Service (QoS) and using the Open Settlement Protocol (OSP) are still in a state of development. This chapter also discusses how we built a Remote Authentication Dial-In User Service (RADIUS) stack to talk to billing servers.

Chapter 19, *Provisioning*

Provides an analysis of the code behind the provisioning system that is shown in Chapters 4, 5, and 6. It then discusses the new provisioning system that we plan to merge with VOCAL.

Appendix A, *VOCAL SIP UA Configuration File*

An annotated configuration file that explains all the available settings for the VOCAL SIP UA.

Appendix B, *Testing Tools*

# Conventions Used in This Book

The following formatting conventions are used throughout this book:

- *Italic* is used for commands, filenames, directories, functions, threads, and operators.
- `Constant width` is used for IP addresses, branches in a SIP message flow, and equations.
- `Constant width italic` is used for replaceable text.
- **`Constant width bold`** is used for user input.

 This icon designates a note, which is an important aside to the nearby text.

This icon designates a warning relating to the nearby text.

## How to Contact Us

Please address comments and questions concerning this book to the publisher:

O'Reilly & Associates, Inc.
1005 Gravenstein Highway North
Sebastopol, CA 95472
(800) 998-9938 (in the United States or Canada)
(707) 829-0515 (international or local)
(707) 829-0104 (fax)

There is a web page for this book, which lists errata, examples, or any additional information. You can access this page at:

*http://www.oreilly.com/catalog/voip*

To comment or ask technical questions about this book, send email to:

*bookquestions@oreilly.com*

For more information about O'Reilly's books, conferences, Resource Centers, and the O'Reilly Network, see the O'Reilly web site at:

*http://www.oreilly.com*

## Acknowledgments

First of all, we would like to thank the members of the VOCAL development team who helped us bring this book to you:

- Alan Knitowski co-founded Vovida Networks, Inc, along with Luan Dang, and served as its CEO from inception until its acquisition by Cisco Systems. VOCAL and the Vovida.org website were derived from Alan's vision of providing developers with open source tools to build SIP-based Voice over IP systems worldwide.

- Julie Chan coauthored the user guides, which provided useful material for most of the chapters throughout the book.

- Enlai Chu, Doug Hawk, Raghavan Kripakaran, Kim Le, Tom Maneri, Barbara Samson, Mahesh Shankar, and Oscar Thomas helped us with the material that became Chapters 2 through 6.

- Amit Bhadoria, Amit Choksi, Sunitha Kumar, Bryan Ogawa, and Surendra Prajapat provided material about the SIP stack for Chapter 8.

- Eoin Canny and Chok Lam provided material about the base code for Chapter 9.
- Chok Lam and Mitch Zollinger provided material about the User Agent for Chapter 10.
- Charles Eckel provided material about Marshal servers for Chapter 11.
- Marc Luettchau provided material about the Redirect server for Chapter 12.
- Kenny Hom provided material about the CPL Feature server for Chapter 13.
- Vincent Rubiere provided material about the Voice Mail server for Chapter 14.
- Mason Kudo, Hong Liu, Hsin-shi Lo, and Mallik Medala provided material about the MGCP translator for Chapter 15.
- Ian Cahoon provided material about the H.323 translator for Chapter 16.
- Neetha Kumar, Tom Maneri, and Mai Vu provided material about system management for Chapter 17.
- Wendy Breu, Francisco Hong, Wenqing Jin, and Surendra Prajapat provided the material about Quality of Service and billing for Chapter 18.
- David Bryan, Matt Naish, Barbara Samson, Monika Smoczynska, Cheung Tam, and Mai Vu provided material about the new Provisioning server for Chapter 19.
- Rohan Mahy provided a technical review of the first draft.

We would also like to salute the other employees of Vovida Networks who contributed long hours to the development of VOCAL and *Vovida.org*: Teju Ajani, Steve Anderson, Shervin Bakhtiari, Kim Boortz, Laura Brannon, Vicki Bryan, Dion Campisi II, Riva Canton, Renee Chen, Jeff Gao, Amy Goerges, James Gosnell, Don Gulcher, Grace Huang, Taroon Kapur, Mike Lehmann, Jack Liou, Kevin McDermott, Eddie Mendonca, Par Merat, Sean O'Neil, Krista Plasky, Deepali Rashingkar, Jennifer Sheng, Larry Trovinger, Rick Tywoniak, Krishan Veer, Doug Wadkins, Quinn Weaver, Han Yu, Tina Zhang, Jenny Zhao, and Vlad Zubarev.

We would also like to thank the many individual contributors to *Vovida.org*. To those who have helped us by answering mailing list questions, discovering bugs, testing new releases, and contributing patches, we are eternally grateful.

Thank you to everyone who has contributed to VOCAL and to *Vovida.org*, especially our wives, Linh, Lyndsay, and Mireya, who provided us with moral support while we spent our quality family time writing this book. We would also like to tip our hats to the editors, illustrators, and the rest of the staff at O'Reilly & Associates who helped us bring this book from concept to completion. Thank you for your fine effort.

# CHAPTER 1

# VOCAL: Say, What?

We wrote this book to find a new set of motivated individuals to take our software, twist it into something new, and create their own important breakthroughs. Are you driven by the desire to discover how things work and the challenge to make them work better? If so, this book is for you.

## What's This All About?

This book, along with the *Vovida.org* web site (*http://www.vovida.org*), provides tools and knowledge to help you build a phone system, find out how it works, and then tinker with it to make it better. The source code and suggested operating system, Linux, are both open source. You can look "under the hood" down to the base code and protocol stack levels and discover not only how the system works, but also how common problems are being worked out in the development environment. We're hoping that you will be inspired to take this system to another level by implementing a feature or functionality that nobody has ever thought about before.

Maybe you have already devoted your professional career to the telecom field and have suffered through the glacial pace of change that characterizes the Public Switched Telephone Network (PSTN). Here is your opportunity to try out an idea that you can demonstrate after a few days of coding. Maybe you work as a software developer in another sector and have a new idea that would normally be discredited by telecom professionals. Here is your opportunity to prove them wrong. Maybe you have a general interest in technology and thought it would be cool to play around with Voice over IP (VoIP). Our invitation to you, to tinker with this system, is as sincere and open as it is to the professionals.

The software described in this book, the Vovida Open Communication Application Library (VOCAL), is not just another Internet Protocol (IP) softphone that connects to others over the Internet, but a full-blown system complete with call control, operations service support, and features. This system is scalable and limited only by the quantity and quality of the computing hardware and network resources at your disposal.

VOCAL is the software that enables a core network to support VoIP. On a scaled-down, single host in our lab, VOCAL has successfully supported more than 40 calls per second (cps), which translates into supporting about 72,000* users, depending on the available memory and the users' calling frequency. On a distributed network of 31 computers, this same software, downloaded from our community web site, *http://www.vovida.org*, can support up to 500 cps,† which roughly translates into 900,000 users. If you consider that the type of computer used to achieve these results can be purchased for less than $500 per unit (at a retail store in the United States at the end of 2001), you can see that, even with some additional fixed and variable business costs, the per-user cost of running a phone system with VOCAL is pretty attractive.

## Voice over IP

For years, software has been available for making "free" long-distance calls between workstations over the Internet. The early versions of this software provided poor quality, but users were willing to suffer packet loss, jitter, and latency in return for bypassing normal long-distance toll charges. Today, users can choose from a large variety of VoIP software packages. Improvements in bandwidth and the processing speeds of home PCs have enabled practical conversations through VoIP devices.

In North America and Europe, the novelty of "free" long-distance calling has withered away with the reduction of toll charges from the major telecom carriers. For example, in the United States, cell phone providers are presently bundling free long-distance calling with their regular services. By itself, a perceived advantage in long-distance charges will not be enough to convince individuals and organizations in these countries to replace their traditional phone systems with IP-based systems. The current belief being expounded by the pundits is that before IP phone systems become widely accepted in developed markets, they are going to require advanced, integrated features of the type that are practically impossible to implement in traditional private branch exchanges (PBXs) and central offices (COs).

In other regions of the globe, the intersection of costs and features will play an important role in the adoption of VoIP. Some areas suffer from both crippling poverty and extravagant import duties imposed on communication equipment. In these

---

* Using a common set of assumptions whereby an average call lasts for 3 minutes and 10% of the users in a subscriber base are making calls at the same time, if the system provides 40 calls per second (cps):

    40 cps * 60 seconds per minute * 3 minutes per average call = 7,200 active calls

  Assuming that 10% of the users are making a call at the same time during a busy hour, this means:

    7,200 / 0.10 = 72,000 users

† Refer to the VOCAL home page on *Vovida.org* for documentation and an executable, *500cps.cxx*, which you can download and run on your host.

countries, there are service providers who will do everything possible to bring low-cost systems into their market space. In others areas, landlines may be scarce, but capital and technical expertise are available to build new phone systems. Many of these new ventures are looking toward VoIP as a flexible solution that enables deploying phone services to millions of subscribers quickly.

The nature of VoIP, also known as *packet telephony*, permits the type of advanced features that will win over new users. As anyone who uses the World Wide Web knows, packets running over an IP network can deliver text, pictures, and audio and video content. The PC is becoming a redundant tool for Internet access with the advent of personal digital assistants (PDAs), cell phones, and other portable devices that provide access to email and other web content. Unlike the PSTN, the Internet is decentralized and permits smart endpoints. Someday, the concept of making a phone call may become obsolete by the concept of simply being in touch with people through a variety of smart IP-based devices.

Advances in packet telephony could also lead to new forms of virtual offices that would seem alien to our current telecommuting practices. Our descendants could know an enhanced mode of long-distance communication in which body language, along with the other 90% of communication that is lost over audio-only devices, is transmitted and received intact. This could make face-to-face meetings a rare novelty. What this might do to city planning, traffic jams, and, indeed, our lifestyles is a worthy subject for another book.

Where do you get the ideas to build this new world? Here's a hint: you don't get them from the "martini lunch" crowd down at the golf club. You get them from insightful individuals who are inspired to think outside of the box. These are the pioneers: people like Linus Torvalds who wrote a new operating system called Linux that can run on home PCs and compete with Microsoft Windows for market share, and Richard Stallman who created the recursively named GNU's Not Unix (GNU) open source license, also known as a *copyleft* agreement, that defines the legal basis for many open source projects. Already, VoIP has benefited from the insightful contributions of legions of developers, and we're hoping that we can recruit you to help us take this development further.

## Open Source

*Open source* is a large topic, covered in a number of books, such as O'Reilly's *Open Sources: Voices from the Open Source Revolution*. The concept behind open source software has been around a lot longer than the notion of proprietary intellectual property. One widely touted example is the common law, where the most brilliant lawyers in the country can make revolutionary arguments to the Supreme Court without retaining the rights to their content. Once these arguments have been presented in court, they become part of the public record and can be viewed and used by anyone.

Of the thousands of open source projects scattered throughout the Internet, a large number are end-user applications, utilities, and other small packages intended to run on a single host. Some of these projects began life as castoffs set adrift by software companies that were either unable or unwilling to continue supporting them. Others originated as *skunkworks* or *bootleg projects*, put together by professional developers and students in their spare time, which then grew in popularity through word of mouth. Some of these projects, such as the Linux operating system, the Apache web server, and VOCAL, acquired funding and support from the biggest companies in the industry. Today, if you were to visit the LinuxWorld (*http://www.linuxworld.com*) and ApacheCon (*http://www.apachecon.com*) trade shows, you would discover a whole world of open source development and its derivative products.

Anyone who has a late-model computer and access to the Internet can download most open source projects and try them out at home. What you'll find with many of these projects is that they appear unfinished and not anything like the sleek, shrink-wrapped products that populate the shelves at your local computer software store. Indeed, most of these projects were never intended for home use. There is no 1-800 help line and no salesperson will call, mostly because there aren't any salespeople on staff to follow up on the downloads. There are exceptions, such as Red Hat Linux (*http://www.redhat.com*), that provide user support, but as a rule, open source projects were intended to express ideas, to implement standards, and to provide building blocks for others to transform the code into hobby projects or commercial-grade solutions.

Many have asked what motivates developers to contribute to the open source movement. According to a study completed by Karim Lakhani and Eric von Hipple (*http://web.mit.edu/evhippel/www/opensource.PDF*), it is the satisfaction gained by solving problems along with signaling a sense of relevance to one's peers that keeps these projects flourishing. Another important reason is that the code has been worked on and tested by hundreds of developers. These projects are robust and feature-rich, because those who have new ideas and the inclination to implement them have been welcomed with open arms. Developers who contribute to the open source VoIP space find themselves acting as catalysts for new ideas on the bleeding edge of the technology curve.

## VOCAL

The *V* in VOCAL refers to the name of our Silicon Valley start-up, Vovida Networks, which existed as an independent company from its first day early in 1999 until its acquisition, by Cisco Systems, in November 2000. Vovida was a made-up name that combined the first two letters of the words *voice*, *video*, and *data* into a unique expression representing our aspiration toward building communication products and working with new technologies. Vovida was a good name for a startup company because it was distinctive, easy to say, and easy to remember. Vovida also sounds like the well-known cheese, "Velveeta," which has led to endless cheese jokes. In fact, all of our projects have been code named after different cheeses.

VOCAL has an Open Source Initiative (OSI, *http://www.opensource.org*)–approved open source license that is similar to the license that governs the Berkeley Software Development (BSD) flavor of the Unix operating system. Unlike the GNU Public License, which requires code changes to remain open source, VOCAL's license makes the code free: free of cost, free of obligation, and free of the copyright restrictions that govern proprietary software. If you want to take our code, change it, and build the changes into your own proprietary project, go for it! There are no port fees, no royalties due upon your production release, and no obligation to contribute your enhancements back to the common code base. The VOCAL code is open source with no strings attached.

The *Vovida.org* web site hosts VOCAL along with a variety of open source protocol stacks, applications, and other software projects that are available for downloading. By writing this book, we're hoping to encourage more people to check out our software and improve their understanding of VoIP technology.

## The business model

Whenever we discuss our work with friends, neighbors, and industry colleagues, the same question always arises: "What's your business model: how do you make your money?" Currently, we are paid employees of Cisco Systems, Inc., which supports our work as part of a corporate strategy to accelerate the adoption of VoIP. *Vovida. org* is helping Cisco achieve that objective by providing open source software based on open communication standards.

One of the legacies of the Internet is its list of open standards expressed as requests for comments (RFCs) on the Internet Engineering Task Force (IETF)'s web site, *http://www.ietf.org*. To quote the IETF's definition offered in RFC 2026 (*http:// www.ietf.org/rfc/rfc2026.txt*):

> In general, an Internet Standard is a specification that is stable and well-understood, is technically competent, has multiple, independent, and interoperable implementations with substantial operational experience, enjoys significant public support, and is recognizably useful in some or all parts of the Internet.

An example of a well-known Internet Standard is the Internet Protocol (RFC 791), which, combined with the Transmission Control Protocol (TCP, RFC 793), has become the *de facto* means for transmitting data between diverse systems. While some might argue that it is not as secure as other protocols, TCP/IP has gained a large install base through its simplicity and adherence to an open set of standards.

In general, open standards have enabled interoperability between equipment manufactured by partners and rivals to the benefit of all users. At Vovida, we incorporated open standards in communication both with outside devices and between the call control components found within the system. In addition, by open sourcing the code, we have enabled a community of developers to contribute their enhancements to the code base and potentially to the standard itself. Even though the developers in the community are not obliged to contribute anything back to the code base, a growing number have been sending us bug fixes, patches, applications and protocol stacks for the benefit of everyone. This has made working on VOCAL and *Vovida.org* truly gratifying.

We have set out to provide software that could be used by outside developers for a wide variety of implementations. Just as interoperability fueled the growth of TCP/IP, we required interoperability to fuel the growth of our community. Our success depended on building an interoperable system that had the ability to support not only equipment from a wide variety of vendors, but also, as you will see, equipment based upon several different communication protocols.

## Gaining acceptance

How does a new company begin an open source project so that it can quickly gain a wide following? The first thing to consider is the machine type and operating system. When IBM introduced its PC in the early 1980s with off-the-shelf components, it set a standard for bringing the adoption curve forward so that the mainstream felt comfortable with grasping the technology while it was still relatively new. Nobody was going to be interested in a project that ran only on a platform that required a significant investment; we chose to work with standard, run-of-the-mill Intel-based PCs.

As for the operating system, our development team had two simple design goals from the outset of our project. First, we wanted a high-quality core platform, one that was reliable, scalable, and stable. Second, we needed it to be inexpensive to allow as many outside developers as possible to join our community. With this in mind, we immediately ruled out Microsoft Windows NT and all other "flavors" of Unix due to their prohibitive costs. We could have chosen FreeBSD, and while it would have met our core needs, we saw the increasing popularity of Linux as an attractive benefit to our project and, therefore, selected Linux as our initial development platform.

Using Linux as our development and product platform has been rewarding. Linux has proven itself to be an extremely stable platform for many mission-critical applications. We expect that many new VoIP applications will be built on top of Linux in coming years, since Linux not only provides an inexpensive, reliable, and stable platform for development, but also provides an excellent development environment with freely available tools.

As VOCAL has grown, it has been ported to additional platforms. In 2000, we ported VOCAL to Sun Solaris to support an early customer trial, and recently, some members of the community have reported success running VOCAL on FreeBSD. Version 1.3.0 of VOCAL also provides a Session Initiation Protocol (SIP) stack that runs on Microsoft Windows 2000.

The following sections briefly describe the VoIP signaling protocols and the architecture of VOCAL. If you prefer, you can skip ahead to Chapter 2 and begin your installation without reading the rest of this chapter. However, if you would like to know more about how the system operates, please continue reading.

# System Architecture

When we started designing VOCAL, we had three primary goals in mind:

- Build a distributed architecture.
- Build a system that was scalable.
- Ensure no single point of failure.

A distributed architecture suited our aim to open source VOCAL as it provided components that developers from the community could build upon or build into their projects. Scaling the system meant assigning one type of server, which became known as the Marshal server, with the task of being a single point of contact for the subscribers and enabling duplicates of this server to be added to the system as the subscriber population grew. Our original idea was to achieve load balancing by assigning each additional Marshal server with a specific population of subscribers. Our original plan also called for a multihost system with redundancy for all call control servers to avoid a single point of failure.

## Data Types

In its most basic form, a Voice over IP system is a set of data combined with the capacity to process calls. There is persistent data, such as the provisioning databases of users and server configurations, as well as the dynamic data that is potentially different for each call. The call control servers handle the following types of data:

*Registration*
> When a user connects to her service provider, the system needs to add her address to a list of active endpoints.

*Security*
> The system requires a perimeter to allow qualified users in and keep intruders out. Security is multifaceted and includes, for example:

*Authentication*
> > The system needs to ensure that the connecting users are who they say they are, that the contents of the message have not been modified, and that no one else could have sent the same message.

*Call admission*
> > The system must determine the types of calling that qualified subscribers are permitted to use.

*Routing*

One server needs to know who the call is for, in terms of whether the called party is a local subscriber, where to send off-network calls, and how to route the call through the system with respect to features and final destination.

*Features*

Phone users are used to working with a variety of features, including voice mail, call forwarding, and call blocking. VoIP has long been touted as a possible source for advanced features that are impossible to implement in the PSTN.

*Billing*

Although this feature is not important for a test or hobby system, commercial VoIP applications that require billing are growing in size and population.

*Policy*

How do you work with other VoIP networks? How do you share billing and allow access to calls coming from known or unknown systems? These issues are handled by the Policy server.

Having set forth our goals and a generic architecture, the next phase in our planning was evaluating the different protocol stacks available to us.

## VoIP Protocol Stacks

In 1997, the only VoIP protocol with any following was H.323, a specification created by the International Telecommunications Union (ITU) for the transport of call signaling over networks. By 1999, when we started Vovida Networks, there were two new options: Media Gateway Control Protocol (MGCP) and SIP.

Then, as now, each protocol had its own set of advantages. H.323, being the first widely available VoIP protocol, enjoyed a head start as developers implemented it as toll-bypass systems as well as PC-to-phone and video-conferencing applications. The best-known H.323 application was Microsoft Netmeeting. MGCP is well suited for centralized systems that work with dumb endpoints, such as analog phones. The most celebrated use of MGCP is for high-capacity gateways designed to work with traditional telecom equipment. There is also momentum building for a replacement to MGCP called MEGACO/H.248. SIP is an easy-to-use protocol that enables developers to push the intelligence to the edge of the networks, implement a distributed architecture, and create advanced features.

We chose to base VOCAL on SIP because it suited our needs for rapid development, and we liked its similarities to Hypertext Transfer Protocol (HTTP, RFC 2616) and Simple Mail Transfer Protocol (SMTP, RFC 2821). At the same time, we provided translating endpoints to help us include H.323 and MGCP developers in our community. Chapters 7 and 8 discuss specific details about SIP and our implementation. The MGCP and H.323 translators are discussed in Chapters 15 and 16, respectively.

Looking at the different organizations present at recent trade shows such as Voice on the Net (VON), we have seen more and more implementations of VoIP using SIP. One example is Microsoft announcing its decision to drop its H.323-based Netmeeting product in favor of Messenger, a SIP application that integrates voice, video, application sharing, and instant messaging and runs on Microsoft's operating system. Also, 3G Wireless, the new cellular phone standard from the ITU, has chosen SIP as its VoIP protocol.

Having chosen SIP, let's look at how the standard describes the roles that different server types play within call processing and then how we implemented our requirements into a SIP-based system.

## SIP Architecture Components

RFC 3261 describes the components that are required to develop a SIP-based network. In many implementations, some of these components are combined into the same software modules. As you might suspect, there are also many different ways to achieve the same results. Some implementations may duplicate some components to enable more options for interoperability with other systems.

### SIP user agents

RFC 3261 defines the telephony devices as user agents (UAs), which are combinations of user agent clients (UACs) and user agent servers (UASs). The UAC is the only entity on a SIP-based network that is permitted to create an original request. The UAS is one of many server types that are capable of receiving requests and sending back responses. Normally, UAs are discussed without any distinction made between their UAC and UAS components.

SIP UAs can be implemented in hardware such as IP phone sets and gateways or in software such as softphones running on the user's computer. It is possible for two user agents to make SIP calls to each other with no other software components. When we start talking about message flows, we'll look at examples that include just two IP phones. Later in the chapter, we will look at the more complex configurations that involve other system components.

### SIP servers

Even though the UA contains a server component, when most developers talk about SIP servers, they are referring to server roles usually played by centralized hosts on a distributed network. Here is a description of the four types of SIP servers that are discussed in the RFC:

*Location server*
 Used by a Redirect server or a Proxy to obtain information about a called party's possible location.

*Proxy server*

> Also referred to as a Proxy. Is an intermediary program that acts as both a server and a client for the purpose of making requests on behalf of other clients. Requests are serviced internally or transferred to other servers. A proxy interprets and, if necessary, rewrites a request message before forwarding it.

*Redirect server*

> An entity that accepts a SIP request, maps the address into zero or more new addresses, and returns these addresses to the client. Unlike a Proxy, it cannot accept calls but can generate SIP responses that instruct the UAC to contact another SIP entity.

*Registrar server*

> A server that accepts REGISTER requests. A registrar is typically colocated with a Proxy or Redirect server and may offer location services. The Registrar saves information about where a party can be found.

In VOCAL, the SIP Location, Redirect, and Registrar servers are combined together into a single server called the VOCAL Redirect server. SIP servers can provide a security function by authenticating users before permitting their messages to flow through the network. Frequently, all four server types are included in one implementation. Proxies can also provide features such as Call Forward No Answer (CFNA).

### SIP messages

Although the messages are an integral part of the protocol, it is not necessary to understand them before working through the instructions offered in Chapter 2. See Chapter 7 for information about messages including definitions, call flows, and descriptions of the message headers and bodies.

## VOCAL Servers

VOCAL contains two types of Proxy servers, the Marshal and the Feature servers. Also, VOCAL has implemented both the SIP Redirect and Location servers into the VOCAL Redirect server. Figure 1-1 shows a simplified view of VOCAL's architecture and how it connects to different types of endpoints.

Here's a look at the servers that are included in VOCAL:

*Marshal servers*

> The Marshal servers are front-line devices that receive all incoming signals, authenticate the users, and forward authenticated signals to the Redirect server. The Marshal servers also receive and forward signals from other servers within the VOCAL network. See Chapter 11 for more information. Four types of Marshal servers are found within VOCAL:
>
> *Gateway Marshal server*
>
> > Works with signals coming from and going to the PSTN gateway.

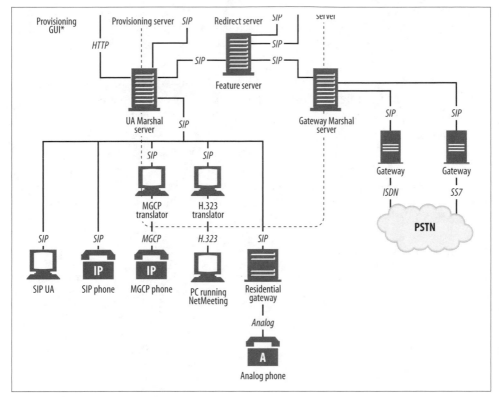

*Figure 1-1. Simplified system overview*

*Internetwork Marshal server*
>   Works with signals coming from and going a known SIP Proxy server on another IP network.

*User Agent Marshal server*
>   Works with signals either coming from or going to IP phones that are connected to the network.

*Conference Bridge Marshal server*
>   (Not shown in Figure 1-1.) Designed to work with third-party conferencing servers. This Marshal works with calls that are destined for an ad hoc conference call. At this time, we are not aware of any open source conference bridges available on the Web. Perhaps someone in our development community will write one!

*GUI*
>   The graphical user interface screens have been built as Java applets, and you need to install the Java Runtime Environment (JRE) before you install the software. If you plan to make any changes to the Java code, then you will need to install the Java Development Kit (JDK) as well. See Chapters 4 through 6 for more information.

*Feature servers*

The Feature servers provide enhanced telephony features such as call forwarding and call blocking. See Chapter 13 for more information.

*Redirect server*

This server keeps track of the users who are registered on the network and provides routing information to help incoming and outgoing calls arrive at their intended destinations. See Chapter 12 for more information.

*Other system components*

Besides the previous components, there are others, such as the Call Detail Record server, which captures user information for billing or accounting applications, and the Policy server, which enables calling between IP networks owned by separate entities. For more information about billing, policy, and related protocols, see Chapter 18.

### Peripheral equipment: endpoints

These items are separate pieces of hardware that talk SIP and must be connected to the VOCAL system to permit calling. For more information about SIP endpoints, see Chapter 10.

*IP phone*

Internet Protocol phone. This can be any kind of IP telephony device including a softphone or an IP phone.

*PSTN gateway*

Public Switched Telephone Network gateway. This is a device that translates the SIP-based signals into signals that can be understood by traditional phone systems.

*Residential gateway*

This includes equipment from a variety of manufacturers. It takes the analog signal from a phone and converts it into a digital VoIP signal. The VOCAL system can connect to SIP-based, MGCP-based, or H.323-based residential gateways.

As you can see, all the VOCAL components are revisited in detail throughout this book. As stated several times, we don't consider VOCAL to be finished and we continue to look to the community to help us upgrade its components.

## Writing the Code

Several engineers have asked us why we chose to write the bulk of the code in C++ as opposed to C or Java. We chose C++ because that language seemed to have a good mix of modern object-oriented design support while still having the performance characteristics that traditional C language telecom people felt comfortable with using. We could have chosen to write the entire application in Java, and the reason that we did not has nothing to do with any perceived speed advantage that C++

code may have over Java code: traditional telecom developers feel more comfortable with the performance of C/C++ than they do with Java. We did, however, use Java to write the Provisioning GUI because of its portability.

## Where's This Going?

We believe that, as a result of our efforts, interoperability will be taken to new, higher levels, and the total cost of ownership of communication systems and solutions will be greatly reduced. In this end state, consumers will be the big winners, as they will be provided with the capability of reaching anyone, anywhere, anytime in the world in more ways, and for less money, than ever before. While this may not lead to immediate changes in the United States, where the investment into the PSTN over the past 100 years is widely estimated at $1 trillion, VoIP will contribute to providing communication services to areas of the globe where access to the PSTN is difficult and expensive. Third-world economies whose modernization has leapfrogged over the industrial revolution will continue leaping toward wireless systems that can provide services to far-flung customers at a fraction of the cost of traditional landline networks. Marshal McLuhan once spoke of a global village; will the new wave of communication technologies reduce our world to a global *house*?

Enabling this type of interoperability extends beyond working with other vendors; we are looking for opportunities for a broader range of implementations of our software. If you need assistance or if you want to participate in our community, you can subscribe to the mailing lists available on *Vovida.org*. While we may not be able to re-create every possible scenario in our lab, we are more than willing to help anyone who is experiencing trouble to the best of our ability.

## What's in This for You?

We're providing you with a functioning open source application that has been available on *Vovida.org* since March 2001. This code has been tested in our lab and in the labs of our community members all over the world. The Voice over IP community needs fresh ideas and new features. The ball is in your court.

The first thing that we suggest you do is go to our community web site, *http://www.vovida.org*, read through the site's resources, and sign up for the mailing lists. These lists give you access to the developers, and even though they may not be able to offer individual support like a help desk, they can point you toward information that may help you achieve more success with your system.

Besides VOCAL, the Vovida site also offers the following open source protocol stacks, each with its own mailing list:

- Session Initiation Protocol (SIP)
- Media Gateway Control Protocol (MGCP)

- Real-time Transport Protocol (RTP)
- Common Open Policy Service (COPS)
- Telephony Routing over IP (TRIP)
- Remote Authentication Dial-In User Service (RADIUS)
- Real-time Streaming Protocol (RTSP)

Before you load the software on your PC, you should also flip through this book to get a better idea about what the VOCAL system is all about. Then, when you're ready, install VOCAL and start playing with it.

*Good luck! Hope to see you on our mailing lists soon.*

# CHAPTER 2

# Setting Up a Phone System at Home

This chapter provides instructions about installing a phone system on a PC running the Linux operating system. If you prefer to install VOCAL on a machine running Sun Solaris, those instructions are available from the VOCAL home page at *Vovida.org* (*http://www.vovida.org*). From these instructions, you will be able to install VOCAL, configure a couple of IP telephony devices, and make calls to other phones over an IP network. Your IP telephony devices can be softphones running on PCs, analog phones plugged into translating devices, IP telephone sets, or any combination of these. If you don't have any IP phone devices, VOCAL provides a softphone that is useful for testing purposes, and most of the instructions in this chapter are directed toward using this device. Once you have your single-host system set up, you can use it as a hobby box at home, a test machine in a lab, and/or a demonstration machine at a trade show.

VOCAL has been designed to run on either a single PC or a multihost network. This can create some confusion in the networking terminology used to describe the system. As VOCAL is a distributed system, the term *server* is used to describe software modules, and the term *host* is used to describe the machines where those servers reside. In large networks, it is common for one type of server to be duplicated onto many hosts for redundancy. See Chapter 3 for more information about loading VOCAL onto larger systems.

This is not a user guide for Linux. If you are unfamiliar with Linux, you should look through some of the user guides and Frequently Asked Questions (FAQs) available on the Web. Some places to begin looking are *http://www.linux.org* and *http://www.linuxdot.org/nlm*.

You don't need to be a Linux guru to run VOCAL. What you need is some familiarity with Unix, some knowledge about commonly used shells, such as bash, and basic experience with installing and running software applications in the Linux environment. If you intend to contribute code to the VOCAL project or use the VOCAL code to build your own application, you should be familiar with the Concurrent Versioning System (CVS, *http://www.cvshome.org*), object-oriented programming, the

GNU compiler, and the basic routines and procedures involved in professional software development.

 We used the bash shell to enter all the command examples shown later.

Most of our software modules have been built with the object-oriented languages C++ and Java. (One exception is the MGCP translator, which was written as a functional C language project.) We would expect contributors to have experience with those languages and the concepts that are derived from the object-oriented world. End users can make calls through VOCAL without needing to know anything about how the code was written.

# Hardware Requirements

The requirements in the following sections are suggested configurations to help you achieve satisfying results from your installation. VOCAL may run on machines with less capacity, but slow processors or lack of available memory will hinder the system's performance.

## Host Machine Requirements

Your machine can be a basic off-the-shelf unit from your local PC retailer. In order to run VOCAL, it should have at least the following attributes:

- 400 MHz, Intel Pentium II PC processor
- 128 MB of RAM
- 1 GB of hard disk space

## Peripheral Requirements

To run a basic test system that doesn't make external calls, you don't need a gateway or an IP phone. VOCAL comes with all the components required to test the signaling between User Agents as calls are set up or torn down.

Later in this chapter, when we discuss adding users, you will need some or all of the following additional equipment depending on how much functionality you desire:

- Two or more IP phone devices. They can be PCs loaded with softphones or IP phones.
- A standard sound card.
- Ethernet cable and hub to connect the IP phone devices to the host machine.

# Software Requirements

Before the VOCAL system can be installed onto your machine, you must have the following software installed and running:

- One of the following versions of Linux, Red Hat:
  - Version 6.2; kernel: 2.2.14; compiler: g++ 2.91.66
  - Version 7.1; kernel: 2.4; compiler: g++ 2.96
  - Version 7.2; kernel: 2.4.9; compiler: g++ 2.96
- Apache server (we tested with the version that came with Red Hat 6.2, Version 1.3.19-5)
- Java Runtime Environment (JRE) Version 1.3.1_01 or higher
- Any browser that supports the JRE version that is currently being offered by Sun Microsystems (see *http://java.sun.com*)

 We tested with JRE Version 1.3.1_01. New versions of this software come out on a regular basis. Check with *Vovida.org* to see which versions are being supported today.

This section tells you how to acquire and install these products if you don't already have them. If you are familiar with Linux and Apache and have JRE Version 1.3.1_01 or later loaded on your machine, skip ahead to "Verifying Networking Requirements."

## Versions

This material is being written in the fall of 2001 with respect to the current versions of VOCAL, Red Hat, and the JRE. By the time you read this, newer versions of these packages will be available, and if you are working with a later version of VOCAL than 1.3.0, different version numbers of the supporting software may be required in some of the following commands. We also have no control over the availability of either JRE 1.3.1_01 or Red Hat 7.1. As always, check the *Vovida.org* web site for information about the most current versions of VOCAL and the supporting software packages.

## Secure Shell

If you load the File server package of Linux Red Hat, you will load and enable the Secure Shell (SSH) server. This is useful for distributing the UA and other components of VOCAL to other hosts. For the tests described in this chapter, SSH is not required for every host, but it's nice to have.

## Linux Red Hat

Linux Red Hat is a commercially sold compilation of files used to load a Linux operating system. Go to *http://www.redhat.com* for information about downloading or acquiring a CD-ROM of their package. Versions 1.2.0 and 1.3.0 of VOCAL will work on Red Hat Version 6.2. However, at the time of this writing, we were doing most of our testing on Red Hat Version 7.1. As new versions of Red Hat are released, we will likely start using them in our lab. We suggest that you check *Vovida.org* to see what version of Red Hat is currently supported by VOCAL.

VOCAL will run on most other distributions of Linux, but it may not install automatically on anything other than the Linux Red Hat versions previously mentioned.

### Options

When we loaded Red Hat, we accepted the default options except for these:

- Server system
- Automatically partition
- No firewall
- All packages (X Windows and web server are required)

### Partitions

We found that, with enough disk space on the host, the automatic partition works for us. Make sure you have lots of swap space; we recommend you at least double your RAM.

To see how your disk is partitioned, type:

```
cfdisk /dev/hda
```

## Apache Server

The Apache server is used to transmit data between the Provisioning server and the other VOCAL server modules. If this server is not running, you will not be able to add any users or edit any of the server settings. If you don't already have the Apache server loaded on your machine, you can acquire it from your Linux Red Hat CD-ROM. Check the *readme* files and other documentation for instructions about how to use it.

## Verifying Networking Requirements

Before downloading and compiling VOCAL, it would be wise to verify that your network settings and domain name service (DNS) are set properly.

## Verifying your INET address and multicast

This test is intended to make sure that your machine has a valid IP address, as opposed to a loopback address, and that the multicast is up and running.

From root, type:

```
ifconfig -a eth0
```

The following appears:

```
eth0      Link encap:Ethernet  HWaddr 00:03:47:9C:2E:BA
          inet addr:192.168.0.3  Bcast: 192.168.0.3  Mask:255.255.255.0
          UP BROADCAST RUNNING MULTICAST  MTU:1500  Metric:1
          RX packets:91383 errors:0 dropped:0 overruns:0 frame:0
          TX packets:53 errors:0 dropped:0 overruns:0 carrier:0
          collisions:0 txqueuelen:100
          Interrupt:11 Base address:0xdf00
```

If your inet addr: is 127.0.0.1, which is a loopback address, you will have trouble running VOCAL and need to change it to something that resembles the address in the previous example.

If the third line does not show:

```
UP BROADCAST RUNNING MULTICAST
```

you need to change this as well. Consult your Linux manual for instructions about how to fix these problems.

## Verifying DNS

This test is intended to make sure that your DNS server is set up properly.

From root, type:

```
cat /etc/resolv.conf
```

The following appears:

```
search <domain name>
nameserver 171.69.2.133
```

If no nameserver appears, you need to troubleshoot your DNS server.

## Testing DNS

This test is just a simple ping request to a well-known domain on the Internet.

Type:

```
ping www.yahoo.com
```

The following appears:

```
PING www.yahoo.akadns.net (216.115.102.77) from 128.107.140.170 : 56(84) bytes of
data.
64 bytes from w5.snv.yahoo.com (216.115.102.77): icmp_seq=0 ttl=237 time=11.6 ms
```

```
64 bytes from w5.snv.yahoo.com (216.115.102.77): icmp_seq=1 ttl=237 time=11.5 ms
64 bytes from w5.snv.yahoo.com (216.115.102.77): icmp_seq=2 ttl=237 time=14.9 ms
64 bytes from w5.snv.yahoo.com (216.115.102.77): icmp_seq=3 ttl=237 time=11.7 ms
...
--- www.yahoo.akadns.net ping statistics ---
14 packets transmitted, 14 packets received, 0% packet loss
round-trip min/avg/max = 10.9/11.6/14.9 ms
```

This is not a big deal as far as running VOCAL is concerned; however, if you can't connect to the Internet, you can't download VOCAL and the JRE.

### Verifying your hostname

From root, type:

```
hostname
```

The following appears:

```
<hostname>
```

You will need to know this name after installing VOCAL, when you are ready to call up Provisioning.

### Verifying your hosts file

To check your /etc/hosts file, type:

```
cat /etc/hosts
```

This file should contain a mapping for the loopback address and your IP address, as shown here:

```
127.0.0.1        localhost.localdomain localhost
<ip address>     <hostname>
```

If a hostname other than *localhost* appears in the first line, edit your file to match the preceding example.

 Don't remove the loopback address (127.0.0.1); otherwise, other programs that require network functionality will fail.

If your /etc/hosts file has your hostname listed as an alias for the loopback address (127.0.0.1), VOCAL will not work. Make sure that the address associated with your hostname is your Ethernet IP address.

This is incorrect:

```
127.0.0.1     localhost.localdomain localhost hostname
```

This is correct:

```
127.0.0.1     localhost.localdomain localhost
<ip-addr>     hostname
```

## Java Runtime Environment

At the time of this writing, the Java Runtime Environment (JRE) was available from *http://java.sun.com*. Version 1.3.0 of VOCAL requires Version 1.3.1_01 or higher of the JRE.

Download the Red Hat RPM Shell Script version to a temporary directory and install it before compiling VOCAL.

# Acquiring VOCAL Software

VOCAL software is available on the *Vovida.org* web site as a Linux source file, *vocal-1.3.0.tar.gz*. The tarball is more than 21 MB in size and, depending on your connection speed, downloading this file could require a few minutes, a few hours, or, hopefully not, a few days. To find this file on *Vovida.org*, go to the VOCAL home page and look for the link under Source Code.

While we were completing the final draft of this book, Version 1.4.0 of VOCAL was released on Vovida.org. Version 1.4.0 gives you the option of installing VOCAL without a Java requirement: using a simple HTTP Provisioning interface to add users instead of the GUI described in Chapter 4. There are also RPM files available that enable you to bypass the compiling stage of the installation.

Go to *Vovida.org* for more information about installing Version 1.4.0 along with more commentary about some of the other changes found in that release. The majority of the effort that went into Version 1.4.0 was directed towards simplifying the installation process. No significant changes were made to the servers.

# Installing and Deploying VOCAL

Once you have finished downloading the tarball, you need to untar it, recompile VOCAL, and run the deploy script before you can test it.

   All of these steps should be done as root.

## Untarring VOCAL

Extract the tarball, *vocal-1.3.0.tar.gz*, into a temporary directory on your machine.

To untar the tarball, as root, type:

```
tar -xvzf vocal-1.3.0.tar.gz
```

After untarring this file, we suggest deleting *vocal-1.3.0.tar.gz* to free up some space.

## Compiling VOCAL

The source code comes uncompiled, because if we offered it as binaries, the file size (267.8 MB) would be too large for most of our community to download over the public Internet. By the way, if you have both the JRE RPM file and the VOCAL tarball on your machine, you might want to move them to */tmp* or delete them to free up some disk space for compiling.

### Optimizing the code

In the instructions, we recommend running *make all* with the *CODE_OPTIMIZE=1* option. Running the *make* command with this option passes the *-O* option to the C++ compiler, which is shorthand for "enable optimization." This works on almost every C or C++ compiler that has a command-line interface.

The *gcc* compiler supports the *-O, -O0, -O1, -O2,* and *-O3* options for different levels of optimization. If you were to compile VOCAL with *CODE_OPTIMIZE=2,* the command would pass the *-O2* option, which is a GNU C++-specific variant of *-O* that means "optimize more than *-O*." Refer to the documentation for your compiler for more information about its optimization options.

We don't support compiling VOCAL with *CODE_OPTIMIZE=2,* although some people in the community like to compile it this way.

By default, *make all* compiles with debug turned on. If you choose to run *make all* and *make allinone* without *CODE_OPTIMIZE=1,* the command will not pass *-O* or *-O2* to the compiler and therefore will not optimize the code. Nonoptimized code runs slower than optimized code, but it usually compiles more quickly and generally produces more straightforward code, which is helpful when debugging. In addition, we enable debugging symbols by passing *-g,* which allows the use of a symbolic debugger with VOCAL.

In summary, a system running with debug enabled can be slow but more useful for testing new code. Compiling the code with the optimizing option speeds up the system and enables more calls per second.

 Both the *make all* and *make allinone* commands must be run with the same options. For example, if you run *make CODE_OPTIMIZE=1 all,* you must run *make CODE_OPTIMIZE=1 allinone;* otherwise, VOCAL will not install correctly. Similarly, if you run *make all* without any options, you must run *make allinone* without any options.

### Source code information

The files that control *CODE_OPTIMIZE=1* as well as other options you can pass to *make* are in the */usr/local/vocal-1.3.0/build* directory. The files *Makefile.osarch* and *Makefile.tools* in that directory control which options are available—for example,

*CODE_OPTIMIZE=1*—and how they operate. More information about these files can be found in the *readme* file for that directory.

### Instructions for compiling VOCAL

To compile VOCAL, as root, type:

```
cd vocal-1.3.0
```

 These instructions assume that you are using VOCAL Version 1.3.0. If you are using a later version, substitute your version number in the commands.

Press Enter, then type:

```
make CODE_OPTIMIZE=1 all
```

This process may take as long as 4 hours to complete. There is no interaction in this process; it returns the command-line prompt when it is finished. It is possible that the script will come to a grinding halt if it runs out of disk space. If that happens, try to free up at least a gigabyte and try again. You can run this script over and over again without doing any damage to the software.

## Deploying VOCAL

This next set of instructions will deploy VOCAL onto a single server and create two provisioned users for testing. This routine takes about 10 minutes to complete; however, the first interactive prompts appear within the first 2 minutes, so don't walk away.

If you are redeploying, make sure that you are in the *vocal-1.3.0* directory, which contains the source code. Don't confuse this directory with */usr/local/vocal*, which contains the binaries.

To deploy VOCAL onto one machine, as root, type:

```
make CODE_OPTIMIZE=1 allinone
```

The script copies a large number of files to new directories, then runs the *allinone-configure* script, which brings up the following warning:

```
WARNING WARNING WARNING WARNING WARNING WARNING WARNING WARNING WARNING

The following may destroy any configuration that you currently have on your system.
If you would like to exit, press Control-C now.

WARNING WARNING WARNING WARNING WARNING WARNING WARNING WARNING WARNING

Welcome to the VOCAL all-in-one configuration system. This program is intended to
configure a small example system which has all of the servers running on one box,
known as the "all-in-one" system.
```

> This all-in-one system is NOT intended as a production system, but as a simple example to get users started using VOCAL.
>
> This configuration WILL destroy any currently configured system on this machine. If this is not acceptable, please quit by pressing Control-C now.

If you are not comfortable with loading VOCAL onto your machine or if you're not supposed to be fooling around with the family computer, now's a good time to bail out by pressing Ctrl-C. If you *are* comfortable about installing VOCAL onto your machine, let's get on with it.

The following prompts appear. Press Enter to accept the defaults for all of these:

```
Host IP Address [<your IP address>]:
Remote Contact hostname or address (this should NOT be loopback or 127.0.0.1) [<your
IP address>]:
Multicast Heartbeat IP Address [224.0.0.100]:
Multicast Heartbeat Port [9000]:
Log Level [LOG_ERR]:
User to run as [nobody]:
HTML directory to install .jar and .html files into [/usr/local/vocal/html]:
```

This next prompt offers two options; read both before continuing.

> Provisioning your VOCAL system requires the ability to view the contents of /usr/local/vocal/html from the web. There are two ways to do this, review both options before answering the next prompt.
>
> Option 1:
>
> Step 1: Answer y to the next prompt. This will let this script attempt to add the following to your Apache httpd.conf file:
>
> ```
> Alias /vocal/ "/usr/local/vocal/html/"
> <Directory "/usr/local/vocal/html">
>     AllowOverride None
>     Order allow,deny
>     Allow from all
> </Directory>
> ```
>
> Adding this script creates an alias from the following URL:
>
> http://<hostname>/vocal/
>
> which points to /usr/local/vocal/html.
>
> Step 2: After this script has completed running, restart your copy of Apache (httpd) for the change to take effect.
>
> Option 2: Answer n to the next prompt. Then, manually copy the directory /usr/local/vocal/html to

```
your web server's HTML directory.  You
should not need to restart your copy of
Apache after the script has completed running.
```

```
Would you like this script to attempt Option 1, Step 1 (y), or would you like to
perform Option 2 manually (n)? (If y, you must restart Apache after this script has
completed running.) [y]:
```

Press Enter to accept the default [y].

Accept the defaults to the next prompt by pressing Enter:

```
Directory where Apache's httpd.conf is located [/etc/httpd/conf]:
```

The following prompt requires a different answer than the default:

```
Path to Java VM (if none, automated provisioning will not work)
 (please include name of interpreter, e.g., /usr/java/bin/java)
[none]:
```

Type the location of your Java binary file. When we tested with *jre1.3.1_01*, the location was:

**/usr/java/jre1.3.1_01/bin/java**

 Later versions of the JRE may load the Java binary into a different location. If you are using the JDK, you will need to type in its location rather than the location of the JRE. If you are not using JDK, make sure the version numbering in the command matches the JRE version that you are using.

Press Enter, and a confirmation of your configuration appears:

```
******************************

Configuration:

Host IP Address:                172.19.174.207
Remote Contact Address:         172.19.174.207
Multicast Heartbeat IP Address: 224.0.0.100
Multicast Heartbeat Port:       9000
Log Level:                      LOG_ERR
User to run as:                 nobody
HTML directory:                 /usr/local/vocal/html
Add alias to:                   /etc/httpd/conf/httpd.conf
Java Runtime:                   /usr/java/jre1.3.1_01/bin/java


******************************

Continue [n]:
```

This was the configuration for our test machine; your configuration may be different.

Type **y**.

Press Enter, and the following appears:

```
**********************************************************************

Beginning the VOCAL configuration process.  This may take a few seconds.

setting umask to 0022 -- users other than root must be able to run VOCAL in its
default configuration.

fixing permissions...
creating uavm config files...
creating UA config files for 1000 and 1001...
Stopping VOCAL...
Creating and filling provisioning directory...
Creating and filling HTML directory...
  Adding alias to httpd.conf directory...
Creating users...
Starting VOCAL...
Creating configuration files...
Restarting VOCAL...


**********************************************************************
Configuration complete!

  To configure your VOCAL system:

  * Go to

    http://<hostname>/vocal/

    and select Provisioning.

[root@<hostname> vocal1.3.0]#
```

The deployment is complete.

If you are redeploying, the alias may have already been added to your Apache configuration file, and the lines:

```
It appears that the alias has already been added to httpd.conf.
Skipping...
```

will appear in the script. This did not create any problems in our testing.

Next, if you chose the default option from the *make allinone* script for adding the alias script to *httpd.conf*, you need to restart the Apache server.

To restart the Apache server, as root, type:

**/etc/rc.d/init.d/httpd restart**

Press Enter, and the following appears:

```
Shutting down httpd:                                        [  OK  ]
Starting httpd:                                             [  OK  ]
```

If Shutting down httpd returns [FAILURE], no problem; that means that Apache was not running. As long as Starting httpd returns [OK], you're OK.

---

# Testing Your Installation

The SIP User Agent may not be the most exciting application that you've ever run on your machine, but it is satisfying to see the call signaling working. You won't be able to talk or hear any audio during this test; this is only meant to prove that your installation is good and ready for more users. Once you get past this stage and start making real calls with a phone or a sound card, you will experience some of the excitement that we've experienced at Vovida.

## Launching Two User Agents

The *allinone* deploy script has provisioned two User Agents (UAs)—1000 and 1001—for testing. From your Linux desktop, launch two terminal panels, and then size the panels so that they do not overlap each other. After submitting registration commands, these panels will become your two test UAs.

The configuration of the UAs is controlled by their respective configuration files. For these test units, the deploy script has created configuration files with names that contain the local numbers; for example, the file that controls *ua1000* is called *ua1000.cfg*. See Appendix A for more information about editing these files.

### Registering ua1000

In the first panel, type:

```
cd /usr/local/vocal/bin
```

Press Enter, then type:

```
./ua -f ua1000.cfg
```

The following appears:

```
Ready
01/09/22 14:46:51 Registration OK
Ready
```

This means that *ua1000* has successfully registered with the Registration server. In VOCAL, the Redirect server acts as the SIP Registration server. For a diagram of how this signaling is processed, see Chapter 8. For an explanation of how the Redirect server was built, see Chapter 12.

If you have either a sound card or an Internet phone card from Quicknet (*http://www.quicknet.com*) installed in your host, you can run the UA with the *-s* option for sound (`./ua -s -f ua1000.cfg`) and make basic voice calls using headphones and a microphone. We suggest initially testing VOCAL without sound to prove that your installation is working properly. After successfully running these signaling tests, you should make voice calls with either a sound card or a Quicknet card.

When you use the -s option, the output on your screen will be appear slightly differently than the way it appears in this book.

### Registering ua1001

In the second panel, type:

```
cd /usr/local/vocal/bin
```

Press Enter, then type:

```
./ua -f ua1001.cfg
```

The following appears:

```
Ready
01/09/22 14:47:21 Registration OK
Ready
```

*ua1001* is now registered with VOCAL.

If the line 01/09/22 14:47:21 Registration OK does not appear, VOCAL may not be running. Type the following commands to restart VOCAL:

```
/usr/local/vocal/bin/vocalstart restart
```

Try the Registration instructions again. If your registration continuously fails, redo the *allinone* installation, making sure that all the prerequisites listed at the beginning of this chapter are in place.

## Making a Call

Now that the extensions are registered, we can call one from the other. In this test, *ua1000* is the calling party, and *ua1001* is the called party. You can also reverse roles between the locals if you want.

Operating the User Agent is very simple: type **a** to take it off-hook, use any of the numbers on your keyboard to dial, and type **z** to hang up.

### Dialing

From the first panel, *ua1000*, type:

```
a1001
```

As you type, the following appears:

```
a
Dialing %%%    Start dial tone %%%
11%%%   Stop dial tones %%%
00%%%   Stop dial tones %%%
00%%%   Stop dial tones %%%
11
%%%    Stop dial tones %%%
```

```
%%%   Stop dial tones %%%

01/09/22 14:49:04Calling sip:1001@<host name>:5060;user=phone

%%%   Start local ringback   %%%
```

## Ringing

On the second panel, *ua1001*, the following appears:

```
01/09/22 14:49:04 Call From: UserAgent-1<sip:1000@172.19.174.207:5000;user=phone>

%%%   Start ringing   %%%
```

## Answering

Typing **a** from the second panel, *ua1001*, takes the UA off-hook.

The following appears:

```
a%%%   Stop ringing %%%
%%%   Establishing audio   %%%
%%%   Listening on port: 10168
%%%   Sending to host: 172.19.174.64
%%%   Sending to port: 10027
```

On *ua1000*, the following appears at roughly the same time:

```
%%%   Stop local ringback   %%%
%%%   Establishing audio   %%%
%%%   Listening on port: 10027
%%%   Sending to host: 172.19.174.207
%%%   Sending to port: 10169
```

At this point, the audio channel is open. During this test, there is no audio, so you will have to use your imagination. The good news is that if you have been seeing similar results to these examples, your VOCAL system has been installed correctly.

## Hanging up

Either side can hang up first by typing **z**. For this example, we will have *ua1000* end the call.

From *ua1000*, type **z**.

The following appears:

```
z
01/09/22 14:51:08 Bye To: 1001<sip:1001@<hostname>:5060;user=phone>;tag=326fa403
Ready
%%%   Stopping audio   %%%
```

At roughly the same time, the following appears on *ua1001*:

```
01/09/22 14:51:08 Call end From: UserAgent-1<sip:1000@172.19.174.207:5000;user=phone>

%%%   Stopping audio   %%%
```

From *ua1001*, hang up by typing **z**:

```
z
Ready
```

The first UA to hang up, whether it originally initiated or answered the call, displays:

```
Ready
%%%   Stopping audio   %%%
```

This is different than the lines displayed by the other UA as it hangs up, which are:

```
z
Ready
```

Even though both are displaying different text, both are ready for the next call.

That completes the first test. This chapter will now lead you into slightly more complicated territory including editing the UA configuration files, hooking up other endpoints, and making audible calls over a LAN.

# Accessing Provisioning

First, you must ensure that the Java plug-ins are in the correct location for use by provisioning. Set the environment variable for the *NPX_PLUGIN_PATH* to the location of *ns4*. With JRE 1.3.1_01, this path was:

```
/usr/java/jre1.3.1_01/plugin/i386/ns4
```

Finally, launch Netscape (or your favorite browser, providing it supports JRE 1.3.1_01 or later).

Your browser appears. If you are using Netscape, you can check the status of your plug-ins by typing the following into the location field:

```
about:plugins
```

You should see an obvious reference to JRE at the top of the listing.

You will need your browser to provision more users and servers to your system. To test your access to provisioning, type the following URL into the browser's location field:

```
http://<hostname>/vocal/
```

A simple, text web page appears with the following title and menu:

```
VOCAL Configuration for <hostname>
Choose one:
· Provisioning
· System Status
· User Configuration
```

If this appears, you can reduce or close your browser for now. We have made it possible to test your VOCAL installation without having to learn about the provisioning screens. If you would like to learn about the Provisioning server, go to Chapter 4.

---

For the time being, you can test your installation with the two test users that the deploy script has provisioned for you from the command-line interface.

# Installing and Running a UA from Separate Hosts

This is not as complex as you might think. All you have to do is copy some files from your *allinone* machine to another host. Besides the UA files, you will need the sound files from the */vocal1.3.0/sip/ua/Tone* directory to enable dial tone and ring-back. In our lab, we used cheap headsets and microphones. The UAs by themselves support only the free G.711 codec.

 These instructions show you how to distribute the UAs over additional hosts. See Chapter 3 for information about distributing VOCAL's servers over several hosts.

## Local Area Network

The simplest setup involves two Linux boxes running the same version of Linux Red Hat. To hear your call, you need a basic sound card and some form of microphone and speaker combination for both machines.

1. On the remote machine, from */usr/local*, type **mkdir vocal** and from *vocal*, **mkdir bin**.

2. From */usr/local/vocal/bin*, copy the UA and *ua1000.cfg* files to the */usr/local/vocal/bin* directory on the other machine.

3. From */usr/local/vocal1.3.0/sip/ua*, copy the complete *Tone* directory to the same directory on the other machine. Remember, */usr/local/vocal* contains the binaries; */usr/local/vocal1.3.0* contains the source code.

4. Following the instructions from the earlier section, "Testing Your Installation," register both users and make calls.

## Wide Area Network

If you have a broadband Internet connection with a fixed IP address, you can send the *ua* and *Tone* files to someone else who also has broadband and is running the same version of Linux Red Hat. As this involves making a call over the Internet, the remote user will have to make a small change to his *ua1001.cfg* file.

Using a favorite text editor, make this change for the remote user:

```
Registration
Register_On            bool        False
Register_From          string      <your public IP address>
```

```
Register_To          string        <your public IP address>
Register_Expire      int           60000
Register_Action      string        proxy
```

After changing the IP addresses in those fields, follow these instructions:

1. On the remote machine, from */usr/local*, type **mkdir vocal** and from *vocal*, **mkdir bin**.

2. From */usr/local/vocal/bin*, copy the UA and *ua1000.cfg* files to the */usr/local/vocal/bin* directory on the other machine.

3. From */usr/local/vocal1.3.0/sip/ua*, copy the complete *Tone* directory to the same directory on the other machine.

4. Following the instructions from the earlier section "Testing Your Installation," register both users and make calls.

# Configuring Software UAs

As far as your network is concerned, adding a User Agent is much the same as adding any other Ethernet device. You need to configure an IP address, the net mask, and the address of the Domain Name Server (DNS). Normally, you would not have a DNS server at home; you would use the address of your ISP's DNS server.

You will also have to configure a default proxy—use the IP address of the UA Marshal server. As for your username, local numbers such as 1000 and 1001 are valid. Passwords are unnecessary unless you are testing digest authentication.

## Using the UA

The following provides more detailed information about running two UAs in test mode.

### Configuring two UAs

The *ua.cfg* file is stored in */usr/local/vocal1.3.0/sip/ua/SampleConfigFiles*; see Appendix A for a completely annotated copy of the file. Configure two SIP UAs, both with identical settings except for the following:

For *ua1000*:

```
Device_Name          string        /dev/phone0
User_Name            string        1000
Display_Name         string        Phone
Pass_Word            string        test
Local_SIP_Port       string        6060
```

Save the *ua.cfg* file for *ua1000* as *ua1000.cfg*.

For *ua1001*:

| | | |
|---|---|---|
| Device_Name | string | /dev/**phone1** |
| User_Name | string | **1001** |
| Display_Name | string | Phone |
| Pass_Word | string | test |
| Local_SIP_Port | string | **7060** |

Save the *ua.cfg* file for *ua*1001 as *ua1001.cfg*.

If you want to create more user agents, you will need to provision the new users; this is discussed later in this chapter.

### Launching two UAs

The options for running the UAs are as follows:

```
Options:
  -d              Run in daemon mode( not used )
  -f <config-file> Configuration file (default is ua.cfg)
  -h              Prints "ua [-dhqr] [-v[<log-level>]] [-f <config-file>]"
  -q              Use Quicknet card
  -r              Run with retransmission of SIP messages off (not used,
                  currently retransmission is always off)
  -s              Work with sound card
  -v[<log-level>]  Set the lowest log level
        Levels:
            - LOG_EMERG
            - LOG_ALERT
            - LOG_CRIT
            - LOG_ERR
            - LOG_WARNING (default)
            - LOG_NOTICE
            - LOG_INFO
            - LOG_DEBUG
            - LOG_DEBUG_STACK
Note that there is no ' ' between 'v' and the level.
```

A typical command to launch a UA is:

```
./ua -vLOG_WARNING -f ua.cfg
```

For *ua1000*, one possible command is:

```
./ua -vLOG_WARNING -f ua1000.cfg
```

For *ua1001*, one possible command is:

```
./ua -vLOG_WARNING -f ua1001.cfg
```

### Working with the UAs

We have already demonstrated how to use the UA earlier in this chapter. You can also use the UA to communicate with other IP phones.

## Setting Up Other IP Phones

Many different IP phones are available commercially from manufacturers such as Cisco, 3COM, Nortel, Pingtel, and others that provide sales, service, and technical assistance. IP phones are niche products and tend to be expensive. This will remain true until large-scale manufacturing companies see a worldwide market potential for selling millions of sets a year and start mass production. Many experts do not expect this "commoditization" of IP phone sets to happen for a few years.

There are also a number of free softphones that you can download from the Internet. Some of these products are open source, others are demos of proprietary software. One of our favorite sites for downloading components is *http://www.sipcenter.com*. Here is a sampling of some of the free User Agents that we have worked with:

*kphone (http://www.div8.net/kphone)*

> Billy Biggs, who, at the time of writing, is a student at the University of Waterloo in Ontario, Canada, developed the kphone. The kphone is open source; however, it uses the GNU public license.

*SIPHON (http://siphon.sourceforge.net )*

> The SIPHON phone was developed by a small group of engineers in Montreal who were funded by Vovida Networks to build an open source UA based on the Vovida SIP stack.

*Sippo (http://sippo.hotsip.com/download/index.html)*

> Sippo is a SIP User Agent, offered by HotSIP in Stockholm. Sippo is primarily used for instant messaging and voice calling through servers hosted by HotSIP.

Each phone set comes with its own unique instructions. Here are some generic items to keep in mind to make these phones work on your LAN:

- The IP address is your choice.
- The Gateway IP address is the address of your *allinone* system. On your host, type `ifconfig -a etho` to find your address.
- Before your phone can register with VOCAL, you need to provision a new user for it into the system (see the following section). You can do the provisioning before or after you configure your phone. The Registration event will pick up the IP address from the phone.
- If you happen to have an upscale IP phone, like the Cisco 7960 IP phone, you will need to have a Trivial File Transfer Procotol (TFTP) server running to provide a configuration file for the phone based on its MAC address.

Most of these phone sets are expensive for home use, but you may have access to some units from your employer or another source. Chapter 3 discusses distributing VOCAL servers over larger networks.

## Provisioning New Users

As you have seen, the deploy script automatically provisions two test user IDs. If you want to add more devices to your system, you will have to provision more users. We suggest that you provision only the usernames and leave the other fields with their default settings for the time being. As you become more familiar with VOCAL, you may choose to enable features, authentication, and other options for each user. The screens shown throughout these instructions are explained thoroughly in Chapter 4.

1. From the web browser, go to *http://<host_name>/vocal* and select Provisioning. The screen shown in Figure 2-1 appears.

*Figure 2-1. Provisioning login screen*

2. Select Administrator, and type **vovida** as both your login ID and password. A dialog box appears notifying you about creating a new user ID and password. Maintaining user IDs and passwords is discussed in Chapter 4.

3. Click OK.

4. Click Login. The User Configuration screen appears.

5. Two test UAs are listed in the table. These were created when you ran the *allin-one* deploy scipt. Select Show Admin Data, then right-click over the middle of the screen. A menu, as shown in Figure 2-2, appears.

6. Select New. The Edit User screen, a long screen with many fields, appears.

7. Fill in a name, such as **4000**, and scroll down to the bottom of the screen. Click Add to submit this user to the system.

   For now, you can ignore the rest of the fields. If you are curious, see Chapter 4 for a complete description of all fields on this screen.

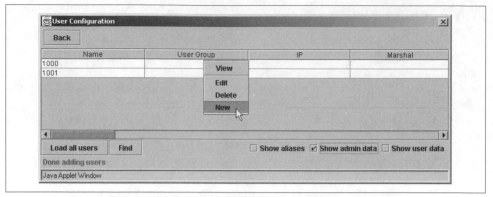

*Figure 2-2. Right-click menu appearing over the middle of the User Configuration screen*

8. Repeat Steps 5 through 7 to add another username, such as **4001**.

9. When you are finished adding two users, your User Configuration screen looks like Figure 2-3.

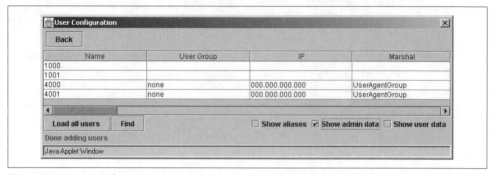

*Figure 2-3. User Configuration screen showing two new users*

Having provisioned two new users, you can use the phone sets to call each other or the original test users. Calling out to the PSTN will require working with the dial plans (see Chapter 5) and plugging into a gateway device (see Chapter 3). We have heard of some small, residential SIP/PSTN gateways being available on the market, but we do not currently have any in our lab. A software residential gateway that works with the Vovida SIP stack and line cards from Quicknet Technologies (*http:// www.quicknet.com*) was contributed by Tata Technologies of India to *Vovida.org*. Look for the SIPRG link in the *Vovida.org* menubar.

## Starting, Restarting, and Stopping VOCAL

This section describes how to start, restart, or stop VOCAL on a deployed system.

# Syntax for Restarting, Starting, and Stopping

There are three methods for starting, stopping, and restarting servers:

- Using the SNMP GUI (as explained in Chapter 17)
- Using *vocalstart*
- Using direct command-line instructions

## Using vocalstart

The syntax for restarting, starting, or stopping is:

```
/usr/local/vocal/bin/vocalstart [command] [server type or port number]
```

where:

- *[command]* can be restart, start, or stop.
- (Optional) *[server type]* can be rs, ms, fs, js, cdr, net, vm, pd, or hs:

  rs = Redirect server
  ms = Marshal server
  fs = Feature server
  js = JTAPI Feature server
  cdr = Call Detail Record server
  net = Network Management Station
  vm = Voice Mail server
  pd = Policy server
  hs = Heartbeat server

- (Optional) *[port number]* is the port number used by the server.

Some examples of this command in use:

1. This starts all VOCAL servers:

   ```
   ./vocalstart start
   ```

2. This stops just the Redirect server:

   ```
   ./vocalstart stop rs
   ```

3. This stops just the User Agent Marshal server:

   ```
   ./vocalstart stop 5060
   ```

4. 5060 is the default IP port number for SIP, and it is normally assigned to the UA Marshal server. You can check the *vocal.conf* file (*/usr/local/vocal/etc/vocal.conf*) for the port numbers of the other servers.

## Using the command line

Many developers would rather type their own options than use the *vocalstart* script. If you use a direct command, you must start the Provisioning server before starting the other VOCAL servers.

The syntax for starting the Provisioning server is:

```
./pserver [Option]...
```

```
Where [OPTION] is one of:
  -v[<log_level>]                    set verbosity
  -r <psRootFileSystem>              set file system root
  -f <logfile>                       set log file
  -b <redundant server host:port>   set redundant ("backup")
                                     server hostname and port
  -p <local port>                    set port to listen for
                                     incoming requests
  -d                                 disable daemon mode (run in
                                     foreground)
  -u <number of users (1000)>        number of users
  -n <number of bins>                number of hash bins
                                     (overrides number users)
  -t <number of threads (25)>        number of request proccess
                                     threads
  -M <multicast host (224.0.0.100)>  multicast host
  -P <port (9000)>                   multicast port
```

The option *log_level* can use one of the following options:

```
-v[<log-level>]    Set the lowest log level
                   Levels:
                     - LOG_EMERG
                     - LOG_ALERT
                     - LOG_CRIT
                     - LOG_ERR
                     - LOG_WARNING (default)
                     - LOG_NOTICE
                     - LOG_INFO
                     - LOG_DEBUG
                     - LOG_DEBUG_STACK
                   Note that there is no ' ' between 'v' and the level.
```

Therefore, you can start the *pserver* without any options, or you can choose a debug level, a log file location, and its local port. To start the Provisioning server with these options, from */usr/local/vocal/bin*, type:

```
./pserver -vLOG_DEBUG_STACK -f /usr/local/vocal/log/ps-6005.log -p 6005
```

The local port of the *pserver* is now set to 6005, and this will be useful information for starting up the other servers.

The syntax for starting the other VOCAL servers is:

```
./<executable> [Options] <pserverA:port> <pserverB:port> <executable's port>
```

For example, to start the User Agent Marshal server (local port 5060), from */usr/local/vocal/bin*, type:

```
./ms -vLOG_DEBUG_STACK -f /usr/local/vocal/log/ms-5060.log 192.168.20.200:6005 192.168.20.200:
6005 5060
```

In this example, we have chosen the debug stack log level and identified the log file as *ms-5060.log*, which resides in the directory */usr/local/vocal/bin*. There is only one Provisioning server, and its address, 192.168.20.200:6005, is repeated, because if you have a redundant Provisioning server, you can access both through this command. The last item, 5060, repeats the local port for this Marshal server.

For the Redirect server, from the same directory, you can type:

```
./rs -vLOG_DEBUG_STACK -f /usr/local/vocal/log/ms-5070.log 192.168.20.200:6005 192.168.20.200:
6005 5070
```

You can look up the local port number for the Redirect server in the *vocal.conf* file. This file resides in */usr/local/vocal/etc*.

You can run the same command for all server executables that reside in */usr/local/vocal/bin*. Remember to start the Provisioning server first, or nothing will work.

For fun, you could stop *vocal*, by running *./vocalstart stop*, then start up the Provisioning server, UA Marshal server, and Redirect server, then return to "Testing Your Installation," from earlier in this chapter, and run the two user agents.

This completes the instructions for installing VOCAL onto a PC and performing some minor tasks with it. If your intention is to play around with VOCAL on a standalone PC without ever taking it into a professional environment, you can skip past Chapter 3.

Chapter 3 shows you how to take the same binaries and make them work over a distributed network and how to configure and connect to a SIP/PSTN gateway.

# CHAPTER 3
# Setting Up an Internal Trial System

Since the summer of 2000, we have used our internal trial system as an office phone system. When we were working at Vovida Networks, just about everyone had a Cisco 7960 IP phone at their desk and had stored their analog phone away in a drawer for emergency use only. We liked to refer to our use of our development project for practical purposes as "eating our own dog food." This chapter speaks from our experience: we've been there, and we want to take you with us.

The instructions in Chapter 2 covered setting up a single-host phone system and running IP phones. This chapter covers how to run different servers from different hosts after setting up the *allinone* system.

There are three phases to setting up an internal trial system:

*Configuring a gateway to permit calling to and from the PSTN*
> We will discuss configuring dial peers and show the complete configuration listing for a Cisco 5300 gateway for our setup as it was at Vovida Networks.

*Deploying VOCAL onto a distributed, redundant network*
> We will give you an example of how to distribute VOCAL over a small system. You can extrapolate those instructions to match the system size and server distribution suggested later in the "Distributed Hosting Guidelines" section.

*Provisioning your users*
> Provisioning users was demonstrated in Chapter 2 and is discussed in greater detail in Chapter 4.

Before we talk about expanding your deployment, we need to make a few remarks about some basic concepts of telephony.

## Interfacing with the PSTN

While we use VoIP phone sets and systems in our offices, out in the real world most people still use the traditional phone system. Therefore, before we can order pizza or call our moms, we have to make our systems talk to the Public Switched Telephone

Network (PSTN). Fortunately, many different gateways are available in today's market that translate IP packets into signals that the PSTN understands. The tricky part is configuring the gateway to make sure that the meaning of the messages doesn't get lost in the translation. This chapter shows you how to configure a Cisco gateway.

We are familiar with the PSTN as it exists in California; however, despite the international following that VOCAL has received, we have not had the opportunity to test with phone systems from other regions of the world. For those who make up our international community of developers, we hope that you can take our examples of configuring equipment for use with the PSTN and make the necessary adjustments to match the requirements of your local phone system. We would also welcome additions to *Vovida.org* Faq-o-matic and mailing lists discussing IP/PSTN connectivity issues from different parts of the globe.

## Gateways

Gateways link the PSTN to IP networks. They can take a VoIP call from a network interface and put it out onto a PSTN interface such as analog, Integrated Services Digital Network (ISDN), or Signaling System 7 (SS7). Gateways also enable incoming calls from the PSTN to be sent to the IP network.

## Analog Signaling

When we talk about analog signaling, we're referring to a "dumb" phone set sending dual-tone multifrequency (DTMF) tones to a central office for interpretation and routing. There are two flavors of analog signaling: Foreign Exchange Office (FXO) and Foreign Exchange Station (FXS). The FXO port is offered on standard analog phone sets and provides an interface to the central office to permit calling out to the PSTN. The FXS port provides voltage, dial tone, and ringing and connects to a phone to enable it to make and receive calls but does not connect to the central office. For example, a Cisco ATA 186 Analog Telephone Adaptor has two FXS ports that you can plug phones into and an Ethernet port that you can connect to the Internet.

The distinction between FXO and FXS will be important to note in our discussion about configuring the gateway.

## Channel Associated Signaling

Channel Associated Signaling (CAS) is a type of signaling used on T1 lines, whereby the control bit is taken from the voice and data channel. This control method has led some to refer to CAS as Robbed Bit Signaling (RBS). CAS is an alternative to Common Channel Signaling (CCS), in which a group of voice and data channels share a separate channel that is used only for control signals.

## ISDN Primary Rate Interface

Primary Rate Interface (PRI) is a type of signaling used in ISDN lines. There are two standards for bundling phone lines into cables, the North American T-carrier and the European E-carrier standards.

The T-carrier standard includes the T1 designation, which is a popular cable standard used by Internet Service Providers. A T1 line contains 24 channels with one control channel and 23 bearer channels, together carrying a maximum data transmission rate of 1.544 MBps. The T3 line, which contains 672 lines or 21 T1s, is commonly used for backbone connections over the Internet or large IP networks. It is also common to obtain fractional T1 line service from telephone service providers, with portions of T1 lines divided between different installations.

The E-carrier standard includes an E1 line, analogous to the T1 line, except that it bundles 32 lines together and achieves a maximum transmission rate of 2.048 MBps.

# Setting Up a Redundant System

At its peak, our old office had about 60 people using the internal trial as their phone system. In terms of capacity, we could have handled all the office phone traffic through a single host. There were only five or six people who might have been considered power phone users. The rest could have been considered light users. Therefore, if we had been running a single host, capacity would not have been an issue.

The issue would have been availability. Although Linux is a stable operating system, to ensure system availability, we needed to build redundancy into the system.

## What Is Redundancy?

A truly redundant network has no single point of failure within its system. For phone systems, the goal is to achieve what is known as carrier class reliability, in which the system is up and running 99.999% of the time. Another term used to describe this is *five nines* reliability. The difference between a system being 99.999% or 99% reliable translates into a potential downtime of just over 5 minutes compared to 3.5 days over an average year.

Availability is measured by:

```
(Total User Time - Total User Outage Time) x 100 / Total User Time
```

In this equation, `Total User Time` is equal to the total number of end users multiplied by the time that transpires during the reporting period.

Some organizations prefer to measure availability in Defects Per Million (DPM), which is measured by:

```
Total User Outage Time x 1,000,000 / Total User Time
```

In this equation, `Total User Outage Time` is equal to the number of end users multiplied by the time that transpires during the reporting period.

Another benefit of a redundant system is that it is usually easier to scale into a larger configuration than a nonredundant system is. System architects like to plan for redundancy early in their development because redundancy is difficult to add later when the system is operating. As a system grows to match a growing demand for capacity, it can be made more valuable if adding hosts also increases the system's availability.

Redundant systems can also be spread over a wide geographic area. Many businesses make arrangements for mirrored sites far away from their offices to permit access to their data in case of a disaster. How long would your business remain solvent if an earthquake prevented access to your site for more than a week? It would be difficult but not impossible to continue your operations if your data was stored at a mirrored site that was not affected by a local disaster.

There are two different types of redundancy schemes: primary and backup and load sharing.

### Primary and backup

A primary machine has all processes loaded and running on it. A second machine waits in standby mode. The data is shared, and a heartbeat (see later) is maintained between the hosts. If the primary host goes down, the backup host takes over.

### Load sharing

All hosts are active at all times and work as peers. Hosts that share data share heartbeats. Data is synchronized between all hosts.

### Heartbeat

Heartbeats are a series of signals emitted at regular intervals over a multicast channel by most servers in the VOCAL network. The Provisioning server, for example, does not heartbeat. Heartbeats are used by the Network Manager to provide Simple Network Management Protocol (SNMP) reporting on the state of each heartbeating server in the network. SNMP and the Heartbeating server are discussed thoroughly in Chapter 17.

## An Illustrated Look at a Redundant System

Figure 3-1 shows a redundant system.

The following sections describe the advantages of setting up each of these components in a redundant scheme. Each of these components was defined in Chapter 1.

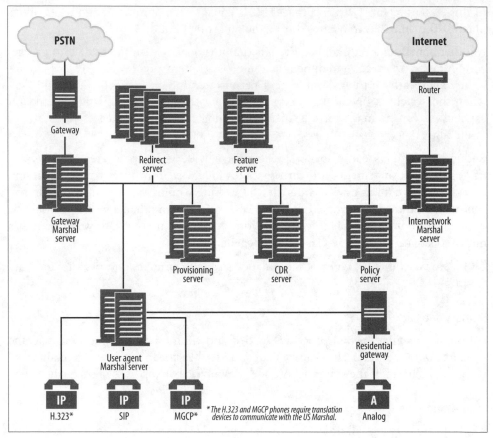

*Figure 3-1. A redundant system*

### Redirect server

For scalability, multiple Redirect servers can exist in a VOCAL system. Each Redirect server contains the same information, and registration information is shared between the Redirect servers.

Multiple Redirect servers are arranged in load-sharing redundancy schemes. If one Redirect server goes down, when it comes back up, it will resynchronize its data with the other Redirect servers before it resumes accepting SIP messages from the proxy servers.

### Provisioning server

For redundancy, a maximum of two Provisioning servers can exist in a VOCAL system. Provisioning servers use a primary/backup redundancy scheme.

### Call Detail Record (CDR) server

For redundancy, a maximum of two CDR servers can exist in a VOCAL system. Marshal servers will send billing data to both primary and secondary CDR servers. These servers are arranged in a dual redundancy scheme whereby they are not synchronized.

### Policy server

Policy servers are matched with Internetwork Marshal servers for scalability. Different Policy servers on the same network do not share information among themselves.

### Marshal server

Like the Policy servers, Redundant Marshal servers are also added for scalability. Marshal servers do not share information with other servers. When the endpoints support primary and backup proxy server schemes, Marshal servers can be grouped together for added redundancy.

### Feature server

For scalability and reliability, multiple Feature servers can exist in a VOCAL system. Feature servers do not share information with one another; each Feature server receives its data from the Provisioning server.

## Distributed Hosting Guidelines

Table 3-1 lists three different VOCAL system setups and the supported capacity based on number of calls per second and number of busy-hour call attempts. The scaled system identified assumes this hardware configuration for each server: a 700-MHz Pentium III PC with 512 MB of RAM.

*Table 3-1. Scaled VOCAL systems*

| Server types | 6-host system | 14-host system | 26-host system |
| --- | --- | --- | --- |
| Redirect servers | 1 | 2 | 5 |
| Feature servers | 1 | 2 | 5 |
| Marshal servers | 2 | 4 | 10 |
| Call Detail Record servers | 1/2 | 2 | 2 |
| Provisioning servers | 1 | 2 | 2 |
| Policy servers | 1/2 | 2 | 2 |
| Total number of hosts | 6 | 14 | 26 |
| Capacity in calls per second | 35 | 70 | 175 |
| Capacity in busy-hour call attempts | 125,000 | 250,000 | 630,000 |

## Stateless and Stateful Servers

Another point that must be made about building redundant systems is that you must know which servers are stateful and which are stateless. *Statefulness* means that past events affect current actions. With signaling, statefulness can be expressed as a quality of being able to "remember" messages after they have been forwarded to another server in the system. This is useful when a server is required to act one way if an expected reply is returned or another way if no reply is returned within a configured time frame. In VoIP systems, statefulness is further partitioned into *call statefulness*, when the server keeps track of all messages from the time that the call is initiated until it is torn down, and *transaction statefulness*, which is explained next.

In SIP, a server is transaction stateful if it keeps track of message flows for the length of a transaction. When a stateful server receives and forwards a request, it will remember the request until another server returns a terminating message. If its timer expires before receiving a terminating message, a stateful server will often generate its own terminating message and send it back along the path from where the request originated. For more information about SIP message transactions, see Chapter 7.

*Stateless* servers don't remember anything. Every request is treated like a brand-new transaction even if it repeated a transmission from another server.

Stateless servers are easier to scale than stateful servers. They are normally set up in primary and backup redundancy schemes. An issue with redundant Redirect servers is that if one goes down, it needs to resynchronize its data with the backup server before it can resume processing requests. This resynchronization was a major challenge for our developers. See Chapter 12 for more about how this was achieved.

One way that redundant Marshal servers have been effectively deployed is for isolating calls going to and coming from specific regions. If an organization were running a phone system that linked China to the United States, it would likely assign one group of Marshals to its customers in China and another group of Marshals to its customers in the U.S. On the other hand, it may be useful to keep the Redirect servers in close proximity to one another to optimize synchronicity regardless of the location of the Marshals.

# Configuring a PSTN Gateway

A wide variety of gateways is offered by a wide variety of manufacturers including Lucent, Nortel Networks, Sonus, and Cisco Systems. If you don't have access to a gateway, you can skip this section and go straight to the later section "Installing VOCAL onto a Multihost System."

# Setting Up a Cisco 5300 Gateway

In our lab, we use a Cisco 5300 gateway for its capacity: it can connect to four T1 lines. Other Cisco IOS* gateways, such as the 26xx, 36xx, 5350, 5400, and 7200, have similar instructions. For official hardware and software configuration instructions, we suggest that you contact the manufacturer of whatever gateway you choose to install into your system. What we provide next is a description of how we configured the software on a Cisco 5300 gateway to match our requirements. At press time, Cisco recommended using IOS Version 12.2(2)XB.

We expect you to use the manufacturer's guides to help you set up your gateway and to make it available on your network. In this section, we will show you how we specifically configured a Cisco 5300 to work with VOCAL. Most of that work involved configuring dial peers. Dial peers make more sense with some background information about dial plans.

## Dial plans

A dial plan is a tool used by phone systems to make routing decisions. In its simplest form, a dial plan tells the routing software where to send calls that are not addressed to local subscribers. In VOCAL, dial plans consist of two parts, the key and the contact information.

The *key* is a variable that can match patterns of dialed numbers such as 7- or 10-digit dialing for North America or variable-length phone numbers in Europe. Dial plans can also signal the routing software that all numbers that start with 9 are intended for the PSTN. The keys are sorted in the order that the routing system searches for a matching variable.

The *contact* provides the routing information for each key. If, for example, the matching key started with a 9 for off-network dialing, the contact would provide routing information that would enable that call to go to the PSTN. In VOCAL, that contact would be the IP address of the Gateway Marshal server.

In VoIP, there can also be an IP dial plan that enables URL dialing, such as *username@domain.com*. For more information about dial plans, see Chapter 5.

## Defining dial peers

A dial-peer is a configuration entity that encapsulates a digit pattern, a physical interface, and a protocol for reaching a destination. When the digit pattern is dialed over any other interface, the 5300 connects it to the destination interface, using the selected protocol.

---

\* IOS is no longer officially an abbreviation. It once stood for Internetworking Operating System.

For our purposes, there are only two types of destinations. The first serves traditional telephone interfaces. Connections of this type use the PRI interface to connect
to the PSTN using a packetized protocol rather than circuit switching. To configure
an interface of this type, follow these steps:

```
sip5300>enable
Password:
sip5300#configure terminal
Enter configuration commands, one per line.  End with CNTL/Z.
sip5300(config)#dial-peer voice 1069 pots
sip5300(config-dial-peer)#
```

The number after voice is just an arbitrary number that you assign to the dial-peer to
let you refer to it later. POTS is an acronym that stands for *plain old telephone system*, which is another way of saying *analog circuit-switched communication*. While
POTS might appear as a quaint term for digital networks, it has persisted in IOS.

Once in config-dial-peer mode, you need to set several parameters. The first is the
destination pattern. This is the pattern of digits that, when the user enters it, causes
the system to connect to your dial-peer. In other words, it is the dial-peer's phone
number. In its simplest form, the destination pattern is a fixed sequence of digits. In
a slightly more complicated form, it can contain wildcard characters. The following
configures a dial-peer to accept any call beginning with 9 and ending with seven
more digits:

```
sip5300(config-dial-peer)#destination-pattern 9.......
```

That pattern might be useful for dialing 9 to reach the PSTN, then dialing an arbitrary PSTN number. To make the call go to the PSTN, designate that it should leave
through port 0:D (the PRI port):

```
sip5300(config-dial-peer)#port 0:D
```

In this scenario, you would also want to enable digit stripping: that way, the 5300
will strip the first, fixed digit from the dialed number and pass the rest along to the
PSTN, emulating the need to dial 9 for outside numbers.

```
sip5300(config-dial-peer)#digit-strip
```

The second type of destination is a voice over IP peer. These require more options,
since the transport type, the IP number of the recipient, and the VoIP protocol must
all be specified. All our outgoing connections use SIP over UDP and follow this
pattern:

```
sip5300(config)#dial-peer voice 42 voip
destination-pattern 42
sip5300(config-dial-peer)#destination-pattern 42
sip5300(config-dial-peer)#session transport udp
sip5300(config-dial-peer)#session target ipv4:192.168.5.170
sip5300(config-dial-peer)#session protocol sipv2
```

In the example, the number 42 can be substituted for your dial-peer's name and destination pattern. These values are independent of each other.

The `ipv4:` before the recipient's IP number is required.

Unlike POTS dial-peers, dial-peers of type VoIP do not have ports associated with them; a VoIP call implicitly travels through some network interface.

There are two more options commonly worth turning off:

```
sip5300(config-dial-peer)#no vad
sip5300(config-dial-peer)#codec g711ulaw
```

Voice activity detection (`vad`) is a technique that can reduce bandwidth usage but could also degrade the signal. To avoid poor sound quality, it may be wise to turn it off. The G711 μ-law codec is an example of one codec that we used in our trial system.

### Changing the configuration of dial-peers

The syntax for changing a dial-peer nearly matches the syntax for creating a new one:

```
sip5300#configure terminal
Enter configuration commands, one per line.  End with CNTL/Z.
sip5300(config)#dial-peer voice 1069 pots
sip5300(config-dial-peer)#
sip5300(config-dial-peer)#session transport udp
```

The difference to note is that a trailing `voip` is not required (in fact, not allowed) in the line `dial-peer voice 1069 pots`, as `dial-peer voice 1069` provides enough information to identify it uniquely. Once in config-dial-peer mode, you can change any setting you like, and this change will take effect immediately.

### Viewing information about dial-peers

Once you have dial-peers configured, you should be able to dial into the 5300, either from the PSTN via PRI or from an IP network via a SIP phone. If you have doubts about where a call is going, try the *show dialplan* command; it prints a digit pattern and shows the dial-peer to which that pattern would connect:

```
sip5300x2#show dialplan number 63831069
```

Two other useful commands are *show dial-peer voice summary* and *show dial-peer voice number*, where *number* is the arbitrary string of digits that you used to name the peer. The first prints only the most noticeable points about all configured dial-peers—their destination pattern and port and, in the case of VoIP peers, their *session target* IP—while the second gives an exhaustive list of a single dial-peer's configuration parameters.

For detailed information about all the dial-peers, you can also enter *show dial-peer voice*.

```
sip5300x2#show dial-peer voice summary
dial-peer hunt 0

PASS
  TAG TYPE   ADMIN OPER PREFIX   DEST-PATTERN   PREF SESS-TARGET    THRU PORT
  5451 voip  up    up            5451              0  ipv4:192.168.5 syst
  5450 voip  up    up            5450              0  ipv4:192.168.5 syst
  9 pots     up    up            9......           0                syst 1:1
```

## Software Configuration

Before you can point some numbers from the PSTN to your test system, you need to know the following:

- The IP address and port number of your Gateway Marshal server.
- If you have more than one Gateway Marshal server, you need to know the additional IP address and port numbers and the order/method in which you want the incoming calls to be directed toward them.
- If you have already created four-digit extensions for users from 5400 to 5499 and want to keep them, you will want to have a matching set of phone numbers assigned to you.

To dial out to a T1 line, dial 9 plus any valid PSTN number; for example, 9-5551212, 9-1-408-555-1212, or 9-411.

You must configure the dial plans in provisioning to account for the 9 at the beginning of the number. If you have a reason for not wanting to dial 9 at the beginning, you can set up the dial plans in provisioning to insert the 9 for you or to accept calls to the outside without the 9. See Chapter 5 for more information.

The following demonstrates:

- Accessing the configuration of a Cisco AS5300
- The configuration of our internal trial system when we were at Vovida

Some comments have been added for nonobvious configuration entries. These entries were purposely added to test the gateway from our internal trial site:

```
/* The Cisco AS5300 is located behind a firewall and uses a private
 * (nonroutable) address. In the configuration shown, we have two T1
 * lines (one PRI and one CAS), but only the PRI is being used. The T1
 * is configured at the Central Office to serve DIDs 408.555.1200-1299.
 * The called numbers on the PRI are presented to us as 11-digit
 * numbers of the form 140855512xx.
 * For calls we make, the called numbers do NOT need a 9.
 */
[user-lnx] telnet server-sip5300
Trying 192.168.175.208...
Connected to server-sip5300-0.domain.
Escape character is '^]'.
```

```
User Access Verification

Password:
server-sip5300-0>enable
Password:
server-sip5300-0#show run
server-sip5300-0#show running-config
Building configuration...

/* Several of the entries in the configuration are simply there by
 * default. I have annotated only those which are in my opinion both
 * important to us and nonobvious.
 */
Current configuration : 5002 bytes
!
! No configuration change since last restart
!
version 12.1
no service single-slot-reload-enable
service timestamps debug uptime
service timestamps log uptime
no service password-encryption
!
hostname server-sip5300-0
!
no logging buffered
no logging buffered
logging rate-limit console 10 except errors
no logging console /* This was added to allow the viewing of logging message
                    * Telnet sessions instead of from the trial console
                    * port.
                    */
enable secret 5 5$5$5$5$.abc123abc123abc123ab.
enable password
!
!
!
resource-pool disable
!
clock timezone GMT -8
clock calendar-valid
ip subnet-zero
no ip finger
ip domain-list domain
ip domain-name domain
ip name-server 192.168.2.133
!
ip dhcp-server 192.168.174.41
mgcp modem passthrough voaal2 mode nse
no mgcp timer receive-rtcp
isdn switch-type primary-5ess /* type of PRI lines provided by CO */
isdn voice-call-failure 0
call rsvp-sync
!
```

```
!
!
!
!
fax interface-type vfc
mta receive maximum-recipients 0
!
!
!
controller T1 0 /* configuration of PRI line, must match that from CO */
 framing esf
 clock source line primary
 linecode b8zs
 pri-group timeslots 1-24
 description T1 PRI: 408.555.9233: DIDs 408.555.12[00-99]
!
controller T1 1 /* not really used but shows how to set up a CAS T1 */
 shutdown /* because we are not using it */
 framing esf
 clock source line secondary 1
 linecode b8zs
 ds0-group 1 timeslots 1-24 type e&m-fgb
 cas-custom 1
 description T1 CAS: 408.555.6380-6399
!
controller T1 2
 shutdown /* because we are not using it */
 framing esf
 linecode b8zs
 pri-group timeslots 1-24
!
controller T1 3
 shutdown /* because we are not using it */
 framing esf
 linecode b8zs
 pri-group timeslots 1-24
!
/* This rule is needed for outgoing calls (internal SIP phone to PSTN)
 * to change the calling number we give to the CO
 * for ourselves from the 4-digit extension we use internally to a 10-
 * digit number used for caller ID services.
 * For example, it changes IDs of the form 12xx to 40855512xx.
 * It also changes the type of number from 'unknown' to 'national'.
 */
translation-rule 408555
 Rule 1 51% 40855512 unknown national
!
/* This rule is needed for outgoing calls to change the called number
 * from international to national if the called number begins with a 1.
 */
translation-rule 1
 Rule 1 1% 1 international national
!
/* This rule is needed for outgoing calls to change the called number
```

```
 * from unknown to international if the called number begins with 011.
 */
translation-rule 20
 Rule 1 011% 011 unknown international
!
!
!
interface Ethernet0 /* 10MB ethernet connection is not used */
 description Lab System Connection
 no ip address
 no ip mroute-cache
 shutdown
!
interface Serial0:23 /* D-channel for the PRI on controller T1 0 */
 no ip address
 ip mroute-cache
 dialer-group 1
 isdn switch-type primary-5ess
 isdn incoming-voice modem
 isdn disconnect-cause 1
 no cdp enable
!
interface Serial2:23 /* D-channel of PRI on controller T1 2, not used */
 no 1p address
 ip mroute-cache
 dialer-group 1
 isdn switch-type primary-5ess
 isdn incoming-voice modem
 isdn guard-timer 3000
 isdn T203 10000
 no cdp enable
!
interface Serial3:23 /* D-channel of PRI on controller T1 3, not used */
 no ip address
 ip mroute-cache
 dialer-group 1
 isdn switch-type primary-5ess
 isdn incoming-voice modem
 isdn guard-timer 3000
 isdn T203 10000
 no cdp enable
!
interface FastEthernet0 /* 100MB ethernet interface */
 description Internal Trial System Ethernet Connection
 ip address 192.168.175.208 255.255.254.0
 no ip mroute-cache
 duplex auto
 speed auto
!
ip default-gateway 192.168.174.1
ip nat translation timeout never
ip nat translation tcp-timeout never
ip nat translation udp-timeout never
ip nat translation finrst-timeout never
```

```
ip nat translation syn-timeout never
ip nat translation dns-timeout never
ip nat translation icmp-timeout never
ip classless
ip route 0.0.0.0 0.0.0.0 192.168.174.1 /* Needed for routing RTP packets,
                                        * also needed to turn on IP
                                        * routing, although it is not
                                        * specifically shown in the
                                        * configuration file.
                                        */
no ip http server
!
no logging trap
dialer-list 1 protocol ip permit
dialer-list 1 protocol ipx permit
!
!
voice-port 0:D
!
voice-port 1:1
 timeouts interdigit 4
!
voice-port 2:D
!
voice-port 3:D
!
/* dial peer for our SIP phones with extensions 12xx, primary */
dial-peer voice 12 voip
 application session.t.old /* Need to use session.t.old. instead of
                           * session in order for PRI to SIP mapping to
                           * work correctly.
                           */
 destination-pattern 12..$
 progress_ind setup enable 3 /* Specifies that ringback should be
       * generated by the gateway when no
       * progress indicator is specified by the
       * caller (CO).
                             */
 session protocol sipv2
 session target sip-server /* primary PSTN gateway marshal */
 codec g711ulaw
 no vad
!
/* dial peer for our SIP phones with extensions 12xx, backup */
dial-peer voice 121 voip
 preference 1 /* Preference is from 0-9 with 0 being the highest and the
               * default, set to 1 to give this dial-peer lower
               * precedence than dial-peer 12 (preference 0).
               */
 application session.t.old
 destination-pattern 12..$
 progress_ind setup enable 3
 session protocol sipv2
 session target ipv4:192.168.116.180:5062 /* backup PSTN gateway marshal */
```

```
 codec g711ulaw
 no vad
!
/* This dial peer exists to handle calls in which the caller provides no
 * caller ID information. In this case, whatever is specified as the
 * 'destination pattern for the first dial peer for the port (in this
 * case, port 0:D) is used as the caller's ID.
 */
dial-peer voice 99 pots
 preference 1
 application session.t.old
 destination-pattern CallerId-Blocked
 translate-outgoing calling 408555
 no digit-strip
 direct-inward-dial
 port 0:D
!
/* dial-peer for international calls */
dial-peer voice 11 pots
 application session.t.old
 destination-pattern 011.
 translate-outgoing called 20
 direct-inward-dial
 port 0:D
!
/* dial-peer for long distance (national) calls */
dial-peer voice 10 pots
 application session.t.old
 destination-pattern 1..........$
 translate-outgoing called 1
 no digit-strip
 direct-inward-dial
 port 0:D
!
/* dial-peer for local calls */
dial-peer voice 7 pots
 application session.t.old
 destination-pattern .......$
 translate-outgoing calling 408555
 no digit-strip
 direct-inward-dial
 port 0:D
!
/* dial-peer for calling 311,411,511,611,711,811 */
dial-peer voice 311 pots
 preference 1
 application session.t.old
 destination-pattern [3-8]11$
 translate-outgoing calling 408555
 no digit-strip
 direct-inward-dial
 port 0:D
!
/* dial-peer for calling local operator */
```

```
dial-peer voice 1 pots
 application session.t.old
 destination-pattern 0$
 translate-outgoing calling 408555
 direct-inward-dial
 port 0:D
!
!
sip-ua
retry invite 3
retry response 3
retry bye 3
retry cancel 3
timers trying 1000
timers invite-wait-100 1000
sip-server ipv4:192.168.116.110
!
!
line con 0
 logging synchronous
 transport input none
line aux 0
line vty 0 4
 password junk /* may want to change this */
 login
!
scheduler interval 1000
end
```

# Installing VOCAL onto a Multihost System

As Table 3-1 showed, VOCAL is scalable and, in a network with 26 hosts or more, can support many thousands of users. Our recommended method for deploying a multihost system is to get VOCAL running on one host as shown in Chapter 2. Then, using your preferred file transfer routine, copy the binaries to the other hosts, edit a configuration file, reprovision some of the servers, and then restart VOCAL. Here is an example set of instructions that uses *scp* to copy the binaries from host to host. If you prefer to use *copy* or *ftp*, it's up to you.

Assume that there is a network with four hosts named *Host1*, *Host2*, *Host3*, and *Host4*. In the instructions, we have used the abbreviations for the different server types as they appear in the configuration file.

## VOCAL Configuration File

Table 3-2 provides a list of abbreviations found in the */usr/local/vocal/etc/vocal.conf* file and their definitions.

*Table 3-2. /usr/local/vocal/etc/vocal.conf abbreviations and definitions*

| Abbreviation | Definition |
|---|---|
| snmptrapd | Simple Network Management Protocol trap daemon (see Chapter 17) |
| netMgnt | Network Management Station (see Chapter 17) |
| vmserver | Voicemail server (see Chapter 14) |
| hbs | Heartbeat server (see Chapter 17) |
| fsvm 5110 | Voice Mail Feature server (see Chapter 14) |
| uavm 5170 | Voice Mail User Agents (see Chapter 14) |
| uavm 5171 | |
| uavm 5172 | |
| uavm 5173 | |
| uavm 5174 | |
| siph323csgw 5155 5150 | SIP H.323 Call Signaling gateway (see Chapter 16) |
| ps | Provisioning server (see Chapter 19) |
| ms 5060 | User Agent Marshal server (see Chapter 11) |
| ms 5065 | Gateway Marshal server (see Chapter 11) |
| ms 5080 | Conference Bridge Marshal server (see Chapter 11) |
| fs 5080 | CPL Feature servers (see Chapter 13) |
| fs 5105 | |
| fs 5100 | |
| fs 5085 | |
| fs 5095 | |
| fs 5090 | |
| js 5160 5161 | Java Telephony Application Programming Interface (JTAPI) Feature server (see Chapter 5) |
| rs 5070 | Redirect server (see Chapter 12) |

## Deploying VOCAL onto Multihosts

You must decide how the hosts are to be distributed over your network. In this example, the *ps*, *hbs*, *netMgnt*, and *snmptrapd* will run on *Host1*; the *ms* 5060 will run on *Host2*; the *ms* 5065 will run on *Host3*; and the *rs* will run on *Host4*.

To deploy VOCAL onto a four-host network:

1. Install an *allinone* system onto *Host1* and make test calls from *ua1000* to ua1001, as shown in Chapter 2, to verify that the system is working.

2. On *Host1*, type:

   ```
   cd /usr/local/
   ```

   then run:

   ```
   tar cvzf vocal.gz vocal
   ```

3. Send the *tar* file to the other hosts. (In this example, *ssh* is enabled between all hosts. You can use whatever method suits you.)

```
scp vocal.gz root@Host2:/usr/local
scp vocal.gz root@Host3:/usr/local
scp vocal.gz root@Host4:/usr/local
```

4. Untar *vocal.gz* into each host under */usr/local*:

```
tar -xvzf vocal.gz
```

5. On each host, edit */usr/local/vocal/etc/vocal.conf* as follows:

   a. On *Host1*, comment out (#) every application except *pserver*, *hbs*, *netMgnt*, *snmptrapd*, and *snmpd*.

   b. On *Host2*, comment out all except *ms 5060*.

   c. On *Host3*, comment out all except *ms 5065*.

   d. On *Host4*, comment out all except *rs 5070*.

6. Using the Provisioning GUI, remove the Marshal and Redirect servers. Reprovision these servers with their new IP addresses. See the later clarification with GUI examples.

7. From */usr/local/vocal/bin on all hosts*, run:

```
./vocalstart restart
```

8. From */usr/local/vocal/bin*, edit *ua1000.cfg* and *ua10001.cfg* as follows:

   Find the following text and edit it to match the IP address of the host running your UA Marshal (*ms 5060*):

   Under Proxy server:

   ```
   Proxy_Server        string      192.168.10.10
   ```

   Under Registration:

   ```
   Register_From       string      192.168.10.10
   Register_To         string      192.168.10.10
   ```

9. Restart *ua1000* and *ua1001* from *Host1* and verify that you can make calls between them. Follow the instructions from Chapter 2.

If you want to deploy a redundant system, look at the suggested server distributions in Table 3-1 and copy the binaries to the desired number of machines, edit the *vocal.conf* files, and reprovision the servers appropriately.

## Reprovisioning Servers

Here is a clarification of Step 6 from the preceding installation instructions. You can perform these steps from any machine that is connected to the network, even if it is not a machine that is running VOCAL, providing that its browser supports your version of the Java Runtime Environment (must be 1.3.1_01 or later).

## Using the GUI

Navigating through the Provisioning server's GUI is thoroughly explained in Chapter 5. The screen has four different icons, which are described in Table 3-3.

*Table 3-3. Configure Servers screen: icons*

| Icon | Description |
|---|---|
| | *Folder:* Appears in the directory tree when the item can be expanded into lower levels. |
| | *Document:* Appears in the directory tree when the item cannot be expanded. |
| | *Expand=Off:* Appears beside contracted folders. Click this icon to expand the folder. |
| | *Expand=On:* Appears beside expanded folders. Click this icon to contract the folder. |

The Expand=Off and Expand=On icons are toggles: use these to see more or less content of each folder.

## Steps

To reprovision the UA Marshal, Gateway Marshal, and Redirect servers:

1. Launch Netscape and go to *http://<hostname>/vocal/provisioning.html*. The Provisioning Login screen appears.

2. Log in as a Technician, typing **vovida/vovida** for Login ID/Password. The Configure Servers screen appears.

3. Delete the User Agent Marshal, as shown in Figure 3-2, as well as the Gateway Marshal and Redirect servers. Each delete request brings up a confirming dialog box. In each case, click OK to delete these servers.

   You may also need to delete the Heartbeat server because, at the time of this writing, a bug in the Provisioning server prevented adding a new Redirect server if the Heartbeat server had the same IP address regardless of the difference in port numbers. You don't need the Heartbeat server for these tests. Deleting it is a satisfactory workaround to enable you to complete these instructions. Check with Bugzilla on *Vovida.org* to see if this bug has been fixed.

   The Redirect server (RS) sometimes appears to go through the motions of being deleted without disappearing from the screen. If this happens, delete it again; an error message appears; click OK, and the RS will disappear from the screen.

4. Add new servers by selecting a group and clicking New, as shown in Figure 3-3, and by filling in the fields and clicking OK, as shown in Figures 3-4 through 3-6. Use the IP addresses and port numbers that correspond to your system.

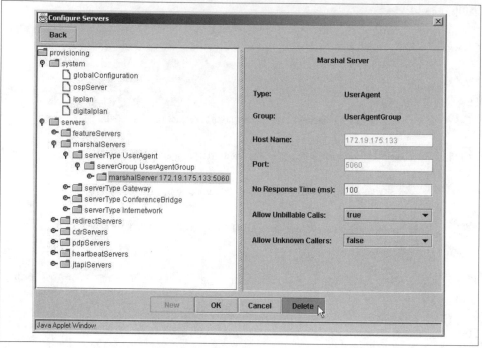

*Figure 3-2. Deleting the User Agent Marshal server*

In Figure 3-4, the hostname is the IP address of the host running *ms* 5060.

In Figure 3-5, the hostname is the IP address of the host running *ms* 5065. The PSTN gateway hostname is the gateway's IP address.

In Figure 3-6, the host is the IP address of the host that is running the Redirect server. In our environment, we set the sync port to 22002.

That completes the routines for reprovisioning these servers.

## Provisioning Users

Follow the instructions from Chapter 2 to add users to your system. The data entry fields are described in Chapter 4.

From the information provided in this chapter, you should have been able to set up a basic internal trial system. There is a lot more you can do; for example, you could set up some features, a voice mail system, or a billing system. Chapters 5 and 6 describe the different screens used for provisioning the network servers. Features are discussed in Chapter 13, voicemail in Chapter 14, billing systems and Quality of Service (QoS) in Chapter 18.

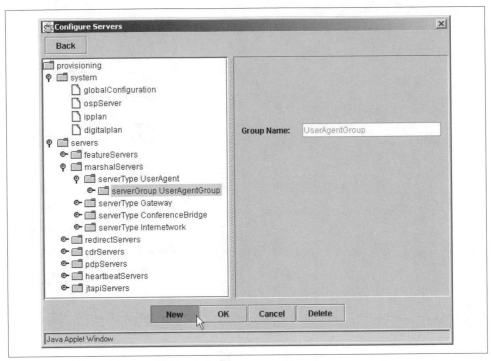

*Figure 3-3. Adding a User Agent Marshal server*

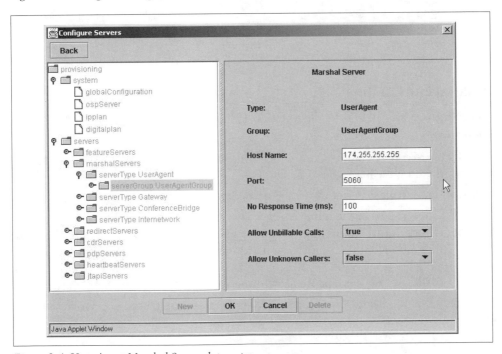

*Figure 3-4. User Agent Marshal Server data entry*

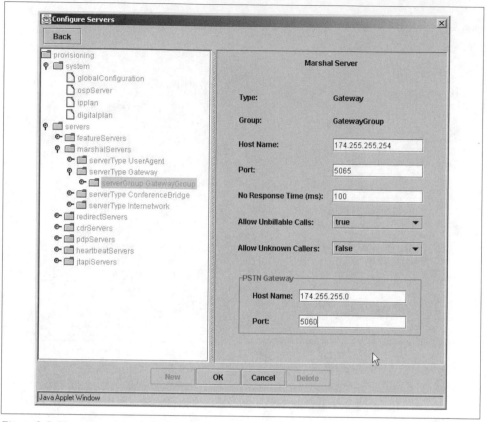

*Figure 3-5. User Agent Marshal server data entry*

## Making Calls to the PSTN

In order to be able to make calls to the PSTN from your system, you need to do the following:

1. Request PSTN numbers from either your IT people or someone who is capable of providing them for you.

2. Configure your system to use the numbers that are assigned to you.

   a. Once you have the numbers, you should use the last four digits as the usernames of the phones in provisioning and in the *ua.cfg* file, the IP phones' *tftp* files, and anywhere else that a username is required. For example, if you are assigned 555-6700 through 555-6709, your usernames should be 6700@<*your_domain_name*> through 6709@<*your_domain_name*>.

   b. Fill in the fields in the Gateway Marshal provisioning screen as follows:

      ```
      Gateway IP: <the IP address of your gateway>
      Gateway Port: 5060
      ```

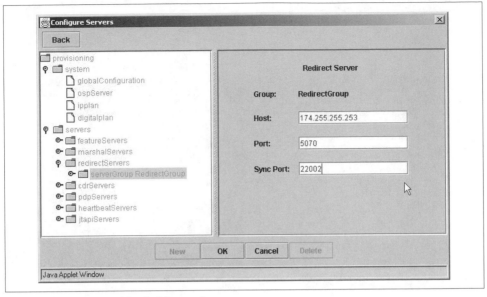

*Figure 3-6. User Agent Marshal Server data entry*

   c. Enter a digital dial plan entry for each Gateway Marshal server as follows:

```
Key:      sip:^9.{7}@
Contact: sip:$USER@<ip address of your Gateway marshal>;user=phone
```

   d. Restart the VOCAL Marshal servers by typing:

**/usr/vocal/local/bin/vocalstart restart**

   e. Launch *ua1000*, as explained earlier, and dial the local phone number; for example: a 9 5 5 5 1 2 1 2.

Unless you are using headphones and a microphone with a sound card or Quicknet card, you will not be able to speak with the person who answers the phone.

If you are using a sound device, remember to add the *-s* option when you launch the UA:

```
./ua -s -f ua1000.cfg
```

Here is a clarification of Steps 2b and 2c from above. To reprovision the Gateway Marshal server for calling out to the PSTN:

1. Launch Netscape and go to *http://<hostname>/vocal/provisioning.html*. The Provisioning Login screen appears.

2. Log in as a Technician, typing **vovida/vovida** for Login ID/Password. The Configure Servers screen appears.

3. Select *digitalplan* and click Add, as shown in Figure 3-7.

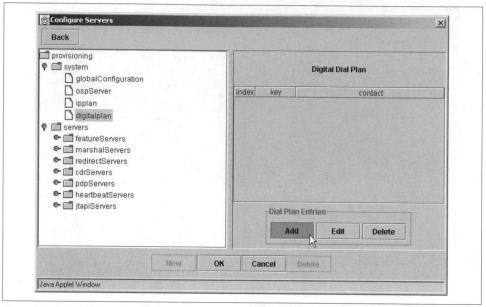

*Figure 3-7. Digital Dial Plan screen*

4. Type in the Key and Contact, as shown in Figure 3-8 (watch your syntax), and click OK.

*Figure 3-8. Digital Dial Plan data entry*

5. The Digital Dial Plan screen displaying your entry appears.

6. Click OK to accept the plan.

7. Select the Gateway Marshal server and change the gateway's hostname to match the IP address of your gateway.

8. Click OK.

9. Restart the Marshal servers. From *usr/local/vocal/bin*, type:

```
./vocalstart restart ms
```

You should be able to make calls from either *ua1000*, *ua1001*, or another provisioned IP phone device over the PSTN to a traditional phone set.

# Working with VOCAL

The difference between the system that you just tested and a full-blown trial system is simply scale. You need to provision the servers over a network that suits your needs (review Table 3-1), provision your users (see Chapters 2 and 4), configure their phone sets, and, if you want, run the Network Management utilities (see Chapter 17 and the *readme* files). If you have more than one gateway, you need to provision a separate Gateway Marshal server for each one and add each of these Marshal servers as separate contacts for each digital dial pattern that is intended for connecting to the PSTN (see Chapter 5).

This is pretty exciting stuff. In the summer of 2000, when we first started documenting our installation instructions, we compiled and deployed VOCAL onto a network of hosts and SIP/PSTN gateways. We then provisioned two IP phone sets, made them call each other, and then called home to our spouses.

Calling home from a phone system that we had just built was enormously satisfying. If you think about most people who work in industrial environments such as large-scale bakeries or automobile factories, many individuals receive rewards for a job well done, but they rarely, if ever, receive the satisfaction of participating in the entire assembly process from the component stage up to the completed end product. If you can imagine what it would be like to drive home in a new car that you had made at work that day, you can imagine how we felt the first time we got VOCAL working to the point where it could make real phone calls. The phrase, "Hi, Honey, I'm calling you from my brand-new phone system," has, for us, taken on a permanently new meaning.

# CHAPTER 4

# Provisioning Users

*Provisioning* is a term used to describe the processes and procedures used for entering data into a phone system. The *make allinone* script that you read about, and perhaps ran, in Chapter 2 provisions system configuration data to the VOCAL servers. We are currently working on an automated process to add users to the system; at this time, however, they have to be entered by hand, one-by-one.

VOCAL was built with ISPs (Internet Service Providers) and ASPs (Application Service Providers) in mind as the installing customers. These organizations have on-site staff trained in managing user accounts and IP networks. Normally, personnel with different skill sets handle each of these responsibilities; therefore, the provisioning screens are organized into two major divisions, one for the administrators and the other for the technicians. This chapter covers the administrators' screens. Chapters 5 and 6 cover the technicians' screens.

While administrators can enter usernames and data into the system, individual users may also manage certain aspects of the accounts, such as blocking calls to specific phone numbers and telling the system where to send forwarded calls. These users could be individual subscribers or employees of subscriber organizations. This chapter provides a tour of the administrator screens and includes a description of the screen that end users see when they look up their account configuration through their web browsers.

The focus of this chapter is on the structure of the GUI environment, with only a few procedural instructions. We encourage you to play around with the different fields. If your aim is to add users to the system without assigning any attributes or permissions, all you need to do is add new users and fill in the Name field with a local number, such as 1000 or 1001. Everything else is optional.

## Quick Step for Provisioning Users

This section provides instructions about adding users without going into great detail. The detail is provided by the remainder of the chapter.

## Prerequisites

Before you can run the Provisioning GUI, VOCAL must be compiled and installed, and, at least, the Provisioning server must be running. You must also have the Java Runtime Environment (JRE) Version 1.3.1_01 or later installed with the NPX_PLUGIN_PATH environment variable set and the Apache server running. See Chapter 2 for more information about all of these.

To run the Provisioning server, you can either run *vocalstart* to start VOCAL, as shown in Chapter 2, or, for testing purposes, you can run the Provisioning server by itself.

> The *make allinone* script ends with VOCAL running. So, if you just finished your installation and you haven't stopped VOCAL, it's running.

If you want to run the Provisioning server by itself, from */usr/local/vocal/bin*, type:

```
pserver -vLOG_DEBUG_STACK ps-.log
```

The log file is found in */usr/local/vocal/log*.

You must also set the environment variable in your ~/.bashrc or ~/.cshrc file to:

```
NPX_PLUGIN_PATH=<JRE path>/plugin/i386/ns4
```

For JRE 1.3.1_01, the JRE path was expressed as:

```
/user/java/jre1.3.1_01/
```

We cannot predict how this path will be changed by future versions of the JRE.

## Accessing Provisioning

After compiling and installing VOCAL, as shown in Chapters 2 and 3, do the following to provision users:

1. In the browser's location or address toolbar, type:

   ```
   http://<local host>/vocal/provisioning.html
   ```

   <local host> is the name of the host where the Provisioning server is running.

   The provisioning page comes up with a login screen, as shown in Figure 4-1.

2. Log in. Select Administrator, type **vovida** as the Login ID, and **vovida** again as the Password. The User Configuration screen, as shown in Figure 4-2, appears.

3. Right-click anywhere over the user listings on the User Configuration screen. A small menu, as shown in Figure 4-4, appears.

4. Select an option. See Table 4-4 for descriptions.

The rest of this chapter explains the screens that are involved with provisioning users and the fields found within those screens.

# Logging into the Provisioning System

Once you have brought up the login screen, follow the instructions in this section to access the provisioning system.

## The Login Screen

The VOCAL system provides password-protected access to the Provisioning server. The provisioning login screen is shown in Figure 4-1.

*Figure 4-1. Provisioning login screen*

Table 4-1 explains the screen items and fields.

*Table 4-1. Login screen item and field description*

| Item | Description |
|------|-------------|
| Access level | *Administrator* |
| | As an Administrator, you can add, edit, or delete user entries. In addition, you can set up Feature servers for users. |
| | *Technician* |
| | As a Technician, you can edit the fields that control the servers as well as add or delete servers. |
| Login ID | The default is **vovida**. |
| Password | The default is **vovida**. |
| | See the next section, "Password Maintenance," for information about setting your own user ID and password. |

# Password Maintenance

There is a separate user interface for changing passwords and adding or removing accounts for administrators and technicians. (For information about editing the passwords for end users, see "Adding New Users" later in this chapter.)

At the time this is being written, password maintenance is not set up as an applet, but you can run it in standalone mode. The executable is included in the *psClient.jar*. The main class is:

```
vocal.pw.AdminAcctManager.
```

The syntax of the command is as follows:

```
java -classpath /path/to/psClient.jar:path/to/xerces.jar
vocal.pw.AdminAcctManager pServer_host pServer_port
```

On our test system, we typed, from */usr/java/jre1.3.1_01/bin*:

```
java -classpath /usr/local/vocal/bin/psClient.jar:/usr/local/vocal/bin/xerces.jar vocal.pw.
AdminAcctManager 192.168.10.10 6005
```

 There is a space rather than a colon between *pServer_host* and *pServer_port*. If you insert a colon, the JRE will return an error.

The Manage Administrative Accounts screen appears with a list of all login IDs.

## Adding new IDs

To add a new Administrator or Technician account:

1. Right-click anywhere over the user listings. A menu appears.
2. Select Add. A dialog box appears.
3. Fill in the ID and password fields. Select one or both access levels.
4. Click OK to submit the new ID.

## Editing existing IDs

To edit the accounts of existing administrators or technicians:

1. Click to select an ID.
2. Right-click anywhere over the user listings. A menu appears.
3. From this menu, select Edit. A dialog box appears.
4. Edit the data as required.
5. Click OK to submit your changes.

## Login Procedure

To log in as an administrator:

1. Select Administrator. You will select Technician later when you work with dialing plans and server configurations in Chapter 4.

2. Enter a valid login ID and password. The default for both is **vovida**.

# User Configuration Screen

This section describes the buttons, option boxes, and data fields found on the User Configuration screen, as shown in Figure 4-2.

Figures 4-2 and 4-3 show the User Configuration screen, as it appears when you log in. Figure 4-2 shows what the screen looks like when you log in for the first time. Users 1000 and 1001 were provisioned by the *make allinone* script.

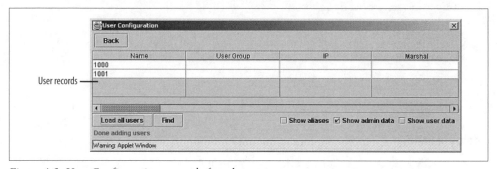

*Figure 4-2. User Configuration screen before data entry*

 Ignore the message, "Warning: Applet Window," that appears at the bottom of the GUI screen. It is a default Java message that was not removed by the developers.

Figure 4-3 shows what the screen looks like after some users have been added. For more information about adding users, see "Adding New Users," later.

## Buttons

Table 4-2 describes the buttons used on the User Configuration screen.

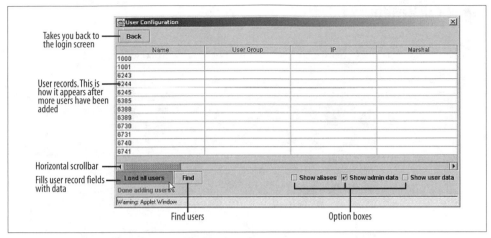

Figure 4-3. *User Configuration screen after data entry*

Table 4-2. *User Configuration screen: buttons*

| Button | Description |
|---|---|
| Back | Lets you return to the login screen. |
| Load all users | Lets you load information for all users. Clicking this button fills the user record fields with data. |
| | This function restricts user input until it has completed running, rendering your screen inaccessible for several minutes or hours, depending on the number of subscribers. |
| Find | Lets you find users. |

## Option Boxes

The option boxes filter the fields displayed on the User Configuration screen. If none of the boxes is selected, only the Name, User Group, IP, and Marshal fields appear. If all of the boxes are selected, all of the fields appear on the User Configuration screen.

Table 4-3 describes the option boxes.

Table 4-3. *User Configuration screen: option boxes*

| Option | Description |
|---|---|
| Show aliases | Displays the users that have aliases. Aliases are displayed in italics. For more information, see "Editing Users: Show Aliases," later. This must be selected in conjunction with one or both of the other option boxes. |
| Show admin data | Displays information and fields configured by the administrator. |
| Show user data | Lets you find users. For more information, see "Editing user feature: Edit User screen," later. |

## Right-Click Menu

The User Configuration screen has a hidden menu that you can bring up by right-clicking the screen over the space reserved for the user records. Figure 4-4 shows you this menu.

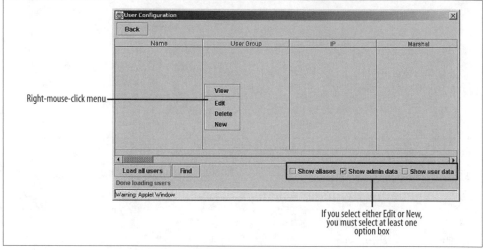

*Figure 4-4. User Configuration screen right-click menu*

Table 4-4 shows the options available from the right-click menu.

*Table 4-4. User Configuration screen: right-click menu options*

| Option | Description |
| --- | --- |
| View | Lets you view data field information in tabular format for one or more users. |
| Edit | Lets you edit information for the user. Selecting Edit opens the Edit User screen. |
| Delete | Lets you delete the user. |
| New | Lets you add a new user. Selecting New opens a blank Edit User screen. |

If you select either Edit or New, you must select at least one option box as well. The right-click menu is dependent upon these options.

# Adding, Viewing, Editing, and Deleting Users

This section provides information about using the GUI elements to perform tasks.

## Adding New Users

The following instructions cover all that you need to do to load users into your system. The Name field is the only mandatory field. Every other field described in this chapter is optional, depending upon the rules that govern your workplace.

To add new users without assigning attributes or permissions:

1. Select the Show Admin Data option box.
2. Right-click anywhere over the user listings. A menu appears.
3. Select New. The Edit User screen appears.
4. Enter the user's name in the Name field.

 You are allowed to enter and modify the Name field only when you add a new user entry. After adding the new user, you cannot modify the name. To correct a mistyped name, you can delete the user and then reenter him or her as a new user.

5. Enter or select the other fields as required.
6. Select the Add button to save the new user entry.

Figure 4-5 illustrates the Edit User screen that appears when the Show Admin Data option box is checked. A section of the screen is labeled "Features that can be enabled for the user by the administrator." The administrator simply enables or disables the availability of these features. The specific configuration, such as listing screened numbers, is done on a different screen. See "Editing Users: User Controlled," later, for more information. One exception, Return Calls, does not require any further configuration from the user.

The fields in this screen are explained next.

### Name

This field specifies the name of the user in alphanumeric characters. A unique name must be specified for each user.

### Group

This field helps you classify your users. Enter any text.

### Marshal

This field includes the following options:

*Group*
This field allows you to select a User Agent Marshal server group from the pull-down menu. The list of Marshal server groups in the pull-down menu corresponds to the Marshal server groups provisioned under *servers/marshalServer/serverType UserAgent*.

Server groups are a legacy attribute from an early design of VOCAL. They really serve no purpose in Version 1.3.0, and there is no plan to add any significance to them in the near future.

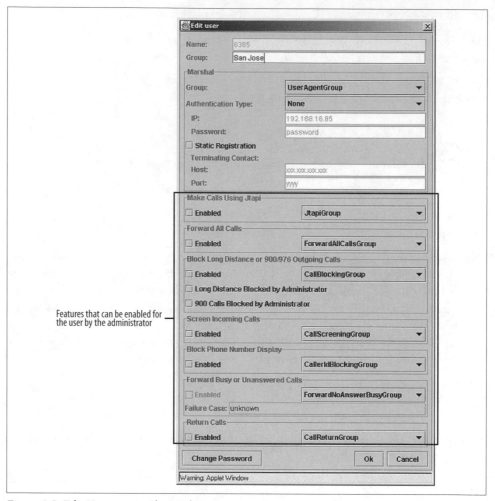

Figure 4-5. Edit User screen: Show Admin Data

*Authentication Type*

This field allows you to select the type of authentication to use:

*None*

Users are not authenticated by the User Agent Marshal server.

*Access List*

Users are authenticated by comparing the user's IP address with the address stored in the IP field.

*Digest*

Users are authenticated by comparing their password to data stored in the Password field.

*IP*
> Specifies the IP address of the user. This IP address is used to authenticate the user during registration. Used only for Access List authentication.

*Password*
> Specifies the user's password. Used only for Digest authentication.

## Static Registration

With static registration, the user agent or gateway can be located via a fixed host-name or IP address. The user agent does not have to register with the Marshal server. Static registration is primarily used for devices that do not support the SIP REGIS-TER message. The option box enables static registration.

*Terminating Contact: Host*
> Specifies the IP address of the User Agent or gateway.

*Terminating Contact: Port*
> Specifies the port number of the User Agent or gateway. You can specify the two previous fields only if the Static Registration option box is checked.

## Make Calls Using JTAPI

This Feature server is no longer being supported in VOCAL. See Chapter 6 for more information about the JTAPI server.

## Forward All Calls

This field includes:

*Option box*
> Enables the Forward All Calls feature for the user.

*Pull-down menu*
> Lets you select a Feature server group from the pull-down menu. The list of Feature server groups in the pull-down menu corresponds to the Feature server groups provisioned under *servers/featureServer/serverType ForwardAllCalls*.

## Block Long Distance or 900/976 Outgoing Calls

This field includes:

*Option box*
> Enables the Call Blocking feature, which allows:
> - Long Distance Blocked by Administrator
> - 900 Calls Blocked by Administrator

*Pull-down menu*
> Lets you select a Feature server group from the pull-down menu. The list of Feature server groups in the pull-down menu corresponds to the Feature server groups provisioned under *servers/featureServer/serverType CallBlocking*.

### Screen Incoming Calls

This field includes:

*Option box*
> Enables the Call Screening feature, which allows the user to screen specific phone numbers or names. Screened calling parties hear a busy signal when they attempt to call.

*Pull-down menu*
> Lets you select a Feature server group from the pull-down menu. The list of Feature server groups in the pull-down menu corresponds to the Feature server groups provisioned under *servers/ featureServer/serverType CallScreening*.

### Block Phone Number Display

This field includes:

*Option box*
> Enables the Caller ID Blocking feature, which allows the user to prevent her caller ID from being displayed on the called party's call display device.

*Pull-down menu*
> Lets you select a Feature server group from the pull-down menu. The list of Feature server groups in the pull-down menu corresponds to the Feature server groups provisioned under *servers/featureServer/serverType CallerIDBlocking*.

### Forward Busy or Unanswered Calls

This field includes:

*Option box*
> Lets the user forward calls to a configured number after a configured number of rings. This feature is also known as *sequential forking*; see Chapter 11 for more information.

*Pull-down menu*
> Lets you select a Feature server group from the pull-down menu. The list of Feature server groups in the pull-down menu corresponds to the Feature server groups provisioned under *servers/featureServer/serverType ForwardNoAnswer*.

*Failure case*
> Users can specify a number to which calls are forwarded when they have turned on Call Forward No Answer or Call Forward Busy.
>
> The administrator can specify a failure case number or address. Incoming calls will be forwarded to this failure case number or address if the system receives a failure message instead of busy messages or no-answer signals.

### Return Calls

This field includes:

*Option box*
> Check the Enabled option box to allow the user to return calls by pressing *69. This feature is enabled by a Call Processing Language script implemented in the Feature server.
>
> Some SIP/IP phones provide lists of missed calls that enable the user to return calls to specific numbers that are not necessarily the most recently missed calls.

*Pull-down menu*
> The pull-down menu allows you to select a Feature server group from the pull-down menu. The list of Feature server groups in the pull-down menu corresponds to the Feature server groups provisioned under *servers/featureServer/serverType CallReturn*.

### Change Password

This field allows you to change the password for the user. This password is for the end user to gain access to his feature screen, as shown in Figure 4-8. For more information, see "Editing Users: User Controlled" later in this chapter.

## Viewing Users: Individually

This section describes how to view records for individual users. If you have thousands of users loaded into your system, you will find that it is faster to load the data for individual users, or small groups of users, as required, rather than loading the data for all users every time you log in as Administrator.

Viewing individual records requires using the right-click menu, as you will see in the following instructions.

### Viewing user

To view data fields for users:

1. To select a user, click a table row. To select consecutive multiple users, hold down the Shift key while clicking the table rows. To select nonconsecutive multiple users, hold down the Ctrl key while clicking the table rows.

2. Right-click anywhere over the user listings. A menu appears.

3. Select View. Data fields with information will appear in the table. To sort the table by data type, check one or more of the option boxes:

*Show Aliases*
    Must be combined with Show Admin Data and/or Show User Data.

*Show Admin Data*
    See Table 4-6.

*Show User Data*
    See Table 4-7.

### Viewing a single user

Figure 4-6 illustrates selecting the data for a single user.

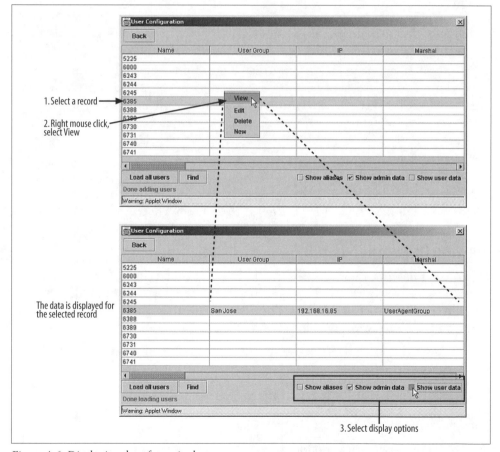

*Figure 4-6. Displaying data for a single user*

## Load all users

To view information for all users, select the Load all users button. If you are working with a few test users, this function completes itself in a few seconds. If you have a live system with thousands of users, this function may take several minutes or hours to complete, depending on the population size and network speed.

# Viewing Users: Data Field Descriptions

Different data fields appear in the User Configuration screen, depending on the option boxes selected.

Some of these fields permit either the user or administrator to set up call blocking for long distance, 1-900 numbers, or caller ID. The administrator permits or restricts user access to call blocking. The users, if permitted, set up the specific details about what phone numbers are to be blocked from use in their accounts.

### Default data fields

When none of the option boxes is checked, the User Configuration screen displays the default data fields (Table 4-5).

*Table 4-5. Default data fields*

| Field | Description |
| --- | --- |
| Name | Specifies the unique name of the user. |
| User Group | The User Group names are not used by the system but allow the administrator to logically organize the users. |
| IP | Optional: specifies the IP address of the user. If Access List is selected for Authentication Type, the IP address is used to validate the user. |
|  | If None or Digest is selected for Authentication Type, the IP address is not used. |
| Marshal | Specifies the name of the Marshal server group. |

### Admin data field

When the Show admin data option box is checked, the data fields in Table 4-6 appear in addition to the default data fields.

*Table 4-6. Admin data fields*

| Field | Description |
| --- | --- |
| Forward All Calls Enabled | Indicates whether all calls can be forwarded for the user: |
|  | *Deselected* indicates that this feature is disabled for the user. |
|  | *Selected* indicates that this feature is enabled for the user. |
| Forward All Group | Indicates the name of the ForwardAll Feature server group. |

*Table 4-6. Admin data fields (continued)*

| Field | Description |
|---|---|
| Forward Busy/No Ans. Enabled | Indicates whether the calls can be forwarded when the phone is busy or not answered: *Deselected* indicates that this feature is disabled for the user. *Selected* indicates that this feature is enabled for the user. |
| Forward Busy/No Ans. Group | Indicates the name of the ForwardBusyNoAnswer Feature server group. |
| Failure Case | Indicates the number or address to forward calls to if there is a problem with contacting the called party. |
| Call Screen Enabled | Indicates whether calls can be screened before reaching the user: *Deselected* indicates that this feature is disabled for the user. *Selected* indicates that this feature is enabled for the user. |
| Call Screen Group | Indicates the name of the Call Screen Feature server group. |
| Call Block Enabled | Indicates whether the user can be prevented from calling certain numbers: *Deselected* indicates that this feature is disabled for the user. *Selected* indicates that this feature is enabled for the user. |
| Call Block Group | Indicates the name of the Call Block Feature server group. |
| Long Distance Admin Block | Indicates whether the long-distance-blocking feature is enabled by the administrator: *Deselected* indicates that the user can control this feature if call blocking is also enabled. *Selected* indicates that the administrator controls long-distance call blocking. |
| 900 # Admin Block | Indicates whether the 900-number-blocking feature is enabled by the administrator: *Deselected* indicates that the user can control this feature if call blocking is also enabled. *Selected* indicates that the administrator controls 900-number call blocking. |
| JTAPI Enabled | This is not supported in VOCAL 1.3.0. |
| JTAPI Group | This is not supported in VOCAL 1.3.0. See Chapter 6 for more information about the JTAPI Feature server. |
| Call Return Enabled | Indicates whether the user can use *69 to return calls: *Deselected* indicates that this feature is disabled for the user. *Selected* indicates that this feature is enabled for the user. |
| Call Return Group | Indicates the name of the CallReturn Feature server group. |
| Caller ID Block Enabled | Indicates whether the user can be blocked from dialing specific phone numbers: *Deselected* indicates that this feature is disabled for the user. *Selected* indicates that this feature is enabled for the user. |
| Caller ID Group | Indicates the name of the CallerID Feature server group. |
| Authentication Type | Indicates the authentication type used to authenticate the user: None. Access List: the IP address is required. Digest: a password is required. |
| Password | Indicates the user's password when digest authentication is used. |
| Static Reg Enabled | Indicates whether static registration is enabled. |
| Terminating Host | Indicates the IP address of the terminating host when static registration is used. |
| Terminating Port | Indicates the port number on the termination host when static registration is used. |

## User data field

When the Show user data option box is checked, the data fields in Table 4-7 appear in addition to the default data fields.

*Table 4-7. User data fields*

| Field | Description |
|---|---|
| Forward All Set | Indicates whether calls can be forwarded by the user: <br><br> *Off* indicates that call forwarding is off. <br><br> *On* indicates that call forwarding is on and all calls are forwarded to a number specified by the user. |
| Forward All To | Indicates the address or number where all calls are forwarded when Forward All feature is set to on or enabled. |
| Forward Busy Set | Indicates whether Forward Busy feature is set by the user: <br><br> *Deselected* indicates that calling parties receive a busy signal if the phone is busy. <br><br> *Selected* indicates that Call Forwarding Busy is on and all calls are forwarded to a number specified by the user when the user is busy. |
| Forward Busy To | Indicates the address or number where all calls are forwarded when Forward Busy Set is enabled. |
| Forward No Ans. Set | Indicates whether Forward No Answer feature is set by the user: <br><br> *Deselected* indicates that, if the phone is not answered, it will ring until the calling party hangs up. <br><br> *Selected* indicates that Call Forwarding No Answer is on and all calls are forwarded to a number specified by the user when the user does not answer. |
| Forward No Ans. To | Indicates the address or number where calls are forwarded when the user does not answer the call and Forward No Ans. Set is enabled. |
| Long Distance User Block | Indicates whether Long Distance User Block feature is set by the user: <br><br> *Deselected* indicates that long-distance calling is allowed. <br><br> *Selected* indicates that long-distance block is on and that the user cannot make long-distance calls. |
| 900 # User Block | Indicates whether 900 Number Block feature is set by the user: <br><br> *Deselected* indicates that 900 numbers are not blocked and user can dial 900 numbers. <br><br> *Selected* indicates that 900 numbers are blocked and users cannot dial 900 numbers. |
| JTAPI Set | Not supported by VOCAL Version 1.3.0. See Chapter 6 for more information. |
| Caller ID Blocking Set | Indicates whether the Caller ID Blocking feature is set by the user: <br><br> *Deselected* indicates that the Caller ID is sent during calls. <br><br> *Selected* indicates that the Caller ID is not sent during calls. |

# Viewing Users: All Users

This section explains how to use the Load all users button and the option boxes to view user data.

For situations in which you need to compare the data between users, you can click the Load all users button. This button activates a program that reads a flat file on the Provisioning server that contains all user data and displays it in the GUI.

To load and view the data:

1. Click Load all users. The user records are filled with data.
2. Select one or any combination of the following options:
   a. Show Aliases (must be combined with either or both of the other checkboxes)
   b. Show Admin Data
   c. Show User Data
3. Use the horizontal scrollbar to view the data.

### Finding users

You can highlight any user by clicking her record. If you have thousands of users, the Find User utility will make your search easier.

The Find button activates a program that automatically searches the Name column for the first match of your criteria as you type it in. For example, if you type a 6, the first name that starts with 6 will be highlighted. If you type 63, the first name that starts with 63 will be highlighted. You can continue typing in the username until the desired name appears on the screen.

To find users:

1. Click Find. The Find User dialog box appears.
2. Type the first characters of the name. The first instance of each character will be automatically selected in the list.
3. When you are finished searching, click Done.

Figure 4-7 shows the use of the Find User utility.

## Deleting Users

To delete a user or multiple users:

1. Select one or more users.
   - To select a user, click a row in the table.
   - To select multiple users, hold down the Shift key while clicking the rows in the table.
2. Right-click anywhere over the user listings. A menu appears.
3. Select Delete.

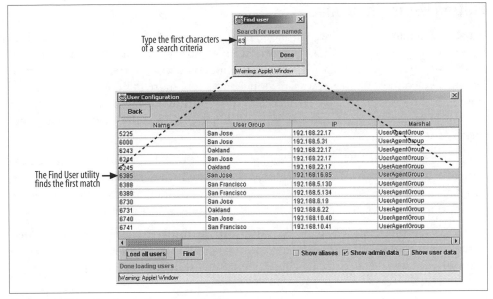

*Figure 4-7. User Configuration screen: finding users*

## Editing Users: Administrator Controlled

This section describes how to edit users.

To edit a user:

1. Select the Show Admin Data option box.
2. Select a user.
3. Right-click anywhere over the user listings. A menu appears.
4. Select Edit. The Edit User screen appears.
5. Edit the fields and option boxes as required.
6. Click OK to submit the entry.

To edit multiple users:

1. Select the Show Admin Data option box.
2. Press and hold Shift, and click multiple table rows.
3. Right-click anywhere over the user listings. A menu appears.
4. Select Edit. The Edit User screen appears.
5. Edit the fields and option boxes as required.
6. Click OK to submit the entry.

When you edit multiple users, only one Edit User screen will appear. All changes made to multiple users will be applied after you click OK.

## Editing Users: Show Aliases

An alias is an alternate address or phone number by which a user can be reached. In the early designs of VOCAL, subscribers were identified by a single name, a *master user*, which was normally the phone number but could also be any string of digits or letters. The concept of aliases was introduced to allow more than one name to correspond to the same user. This enabled a variety of different methods for calling the same user including SIP URL dialing, traditional analog phone dialing, and other forms of addressing.

With aliases, calls involving either the master user or one of its aliases are treated the same. In terms of maintenance, if aliases are deleted from the Provisioning server, this activity does not affect the master user. However, if a master user is deleted, all corresponding aliases are deleted from the Provisioning server as well.

The alias names associated with each user can be displayed using the Show Alias option box. If you select Show Aliases, you must also select Show Admin Data, or Show User Data as well. Selecting Show Aliases by itself returns an error.

## Editing Users: User Controlled

The VOCAL system provides a web page for users to maintain some of their features. These features are called *user-controlled features* and they include:

- JTAPI
- Forward all calls
- Call blocking
- Call screening
- Caller ID blocking
- Forward unanswered
- Forward busy

Features that are not enabled by the administrator appear grayed out.

To edit a user:

1. Select the Show User Data option box.
2. Select a user.
3. Right-click anywhere over the user listings. A menu appears.
4. Select Edit. The Edit User screen appears.
5. Edit the fields and option boxes as required.
6. Click OK to submit the entry.

To edit multiple users:

 When you edit multiple users, only one Edit User screen will appear. All selected users will have the same field and option box settings.

1. Select the Show Admin Data option box.
2. Press and hold Shift, and click multiple table rows.
3. Right-click anywhere over the user listings. A menu appears.
4. Select Edit. The Edit User screen appears.
5. Edit the fields and option boxes as required.
6. Click OK to submit the entry for all selected users.

### Editing user feature: Edit User screen

Figure 4-8 illustrates the Edit User screen that appears when the Show User Data option box is checked and the Edit option is selected from the right-click menu. This screen displays features that can be enabled by the user and is provided for the administrator to view the user's settings. If required, the administrator can modify the user's setting.

Users can access this screen (for their user ID only) by going to *http://<localhost>/vocal/user_configuration.html* and entering their user ID (Name field) and password. Naturally, the web server administrator could set up a hyperlink from a web page that points to this address or make it accessible through a redirected alias URL.

The fields in this screen are explained here.

**Aliases.** This field displays aliases associated with this user. To add aliases:

1. Right-click over the Aliases area. A menu appears.
2. Select Add. A dialog box appears.
3. Type the alias name for the user.
4. Click OK to submit the alias.

To remove aliases:

1. Right-click over the Aliases area. A menu appears.
2. Select Remove. A dialog box appears.
3. Type the alias name to be removed.
4. Click OK to remove the alias.

**Make Calls Using JTAPI.** JTAPI is not supported in VOCAL 1.3.0. See Chapter 6 for more information.

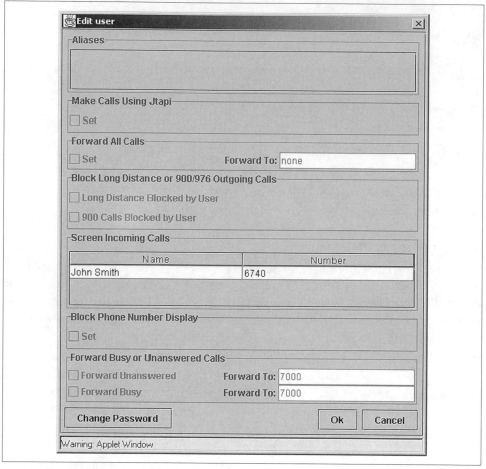

Figure 4-8. Edit User screen: Show User Data

**Block Long Distance or 900/976 Outgoing Calls.** The user can turn on call blocking for:

- Long distance numbers
- 900 numbers

If the Long Distance Blocked by User and 900 Calls Blocked by User option boxes are grayed out, this means that the administrator has not enabled these features for the user.

**Screen Incoming Calls.** The user can screen a call by name and number. To add numbers for screening:

1. Right-click near the Name and Number box. A menu appears.
2. Select Add. A dialog box appears.

3. Enter the name and number to screen and click OK. The format is the user ID; for example, 7000.

If you enter 6 in the Number field, all numbers beginning with 6 will be screened. Entering a specific phone number, such as 555-1000, will screen only that phone number.

To remove a screened number:

1. Right-click near the Name and Number box. A menu appears.
2. Select Remove. A dialog box appears.
3. Enter the name and number to be removed and click OK.

If the call-screening table is grayed out, the administrator has not enabled this feature for the user.

**Block Phone Number Display.** The user can set caller ID blocking to prevent the caller's number or address from being delivered and displayed to the called party.

If the Caller ID Blocking option box is grayed out, the administrator has not enabled this feature for the user.

**Forward Busy or Unanswered Calls.** The user can set Forward All Calls to redirect all incoming calls to a specific number. To turn on this feature for Forward Unanswered:

1. Check the Set option box.
2. Enter the user ID in the Forward To field; for example, 7000.

If the Set option box and the Forward To field are grayed out, the administrator has not enabled the Forward All Calls feature for the user.

The user can set Forward No Answer Busy to forward all incoming calls to another number if:

- The user is busy—Forward Busy
- The user does not answer the call—Forward No Answer

To turn on this feature for Forward Busy:

1. Check the Set option box.
2. Enter the number in the Forward To text box; for example, 7000.

This completes the tour of the GUI screens used for provisioning users. To gain a better understanding about using these fields, you need to look deeper into the system design. A large portion of this information is offered in the latter half of the book.

# CHAPTER 5

# Configuring System Parameters and Dial Plans

This chapter covers the technician's GUI screens. As with Chapter 4, this is a tour of the screens showing the structure and meaning of the different form fields. Unlike the users, the servers are provisioned automatically by the *make allinone* script, and the GUI screen lets you edit the provisioned data if necessary. Chapter 3 provides instructions about deploying VOCAL onto a multihost system with redundant servers.

The technician's screens are divided into *system* and *server* settings. The system settings provide access to some general system parameters and the dial plans. This chapter offers mostly description and only a few procedural instructions. We encourage you to play around with using the different fields.

The dial plans are initially empty and require some knowledge of regular expressions to make them work. We have provided some examples in this chapter with a general discussion about how dial plans work with SIP messages in VOCAL.

The *servers* folder also provides a screen for configuring the Open Settlement Protocol (OSP) server. At the time this is being written, few in the SIP community were seriously using OSP for anything except laboratory testing, although recently we have seen some renewed interest in the mailing lists. The user community has yet to grow to the stage at which OSP is widely used for brokering costs between managed IP networks. The concepts behind OSP are discussed in Chapter 18.

## Login Procedure

From the provisioning login screen (*http://<domainname>/vocal/provisioning.html*), follow these steps to access the server provisioning screens:

1. Select Technician.
2. Enter a login ID and password. The default for both is **vovida**. The Configure Servers screen, as shown in Figure 5-1, appears.

See Chapter 4 for an illustration of the login screen and for information about maintaining login IDs and passwords.

# Configuring Servers

The Configure Servers screen is divided into two frames, a directory tree frame and a data entry frame. The middle frame border is adjustable: click the frame border and drag left or right to expand the view of either frame.

Figure 5-1 shows the Configure Servers screen, highlighting its frames.

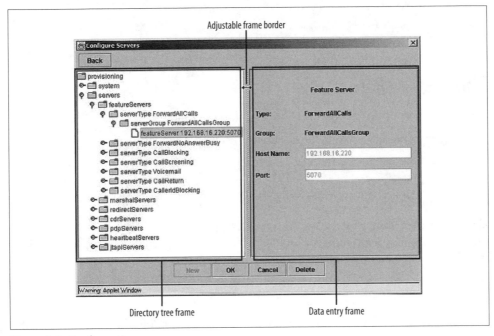

*Figure 5-1. Configure Servers screen*

There are several types of elements on this screen, including icons, buttons, and data entry fields.

Table 5-1 describes the icons.

*Table 5-1. Configure servers screen: icons*

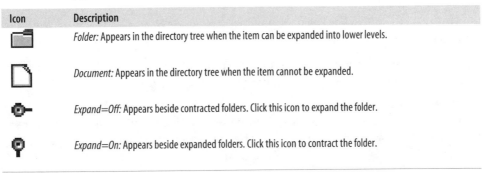

| Icon | Description |
| --- | --- |
| | *Folder:* Appears in the directory tree when the item can be expanded into lower levels. |
| | *Document:* Appears in the directory tree when the item cannot be expanded. |
| | *Expand=Off:* Appears beside contracted folders. Click this icon to expand the folder. |
| | *Expand=On:* Appears beside expanded folders. Click this icon to contract the folder. |

Table 5-2 describes the buttons.

*Table 5-2. Configure Servers screen: buttons*

| Button | Description |
| --- | --- |
| Back | Returns to the login screen |
| New | When activated, brings up a data entry screen for a new entity |
| OK | Submits data entry |
| Cancel | Exits data entry screen without submitting data |
| Delete | When activated, deletes the displayed entity from the system |

# The System Folder

This section provides information about editing the following:

- System configuration data
- OSP server
- IP dial plan
- Digit dial plan

The deployment script, described in Chapter 2, provisions the system and servers for operation when the system is installed. This section provides information about using the GUI to edit the system parameters.

In the left frame of the Configuring Servers screen are two subfolders under Provisioning:

- The system folder provides access to the System Configuration data, the OSP server, the IP dial plan, and the digital dial plan.
- The servers folder provides access to the servers.

This section provides information about the contents of the system folder.

 Throughout the tables that describe the data fields, there is some mention of default and recommended values. This type of information is not available for every field.

## System Configuration Data

The system configuration data parameters control how the VOCAL system works with registration messages and heartbeat signals.

*Registration* is the method used by SIP-based systems to keep the Redirect server informed about the location of on-network user agents. Registrations are temporary: their duration is set by an expiry timer. User agents must reregister after each expiry interval to keep their registration up-to-date. See Chapter 7 for more information.

The System Configuration Data screen includes a field for setting the registration expiry timer. Refer to Figure 5-2 and Table 5-3 for more information.

*Heartbeats* are a series of signals emitted at regular intervals, over the multicast channel, by every server on the network. Heartbeats are used by the Network Manager to provide SNMP (Simple Network Management Protocol, RFC 1157) reporting on the state of each server on the network. See Chapter 17 for more information.

The System Configuration Data screen includes fields for setting up the heartbeat broadcast, heartbeat intervals, and maximum number of missed heartbeats allowed by the system before a server is considered as being out of service.

Follow these steps to edit the system configuration data parameters:

1. Select *globalConfiguration*. The data entry fields appear in the right frame.
2. Edit the fields. These fields are discussed in Table 5-3.
3. Click OK. Your changes are submitted to the system.

Figure 5-2 shows the procedure for editing the parameters applied to a screen capture.

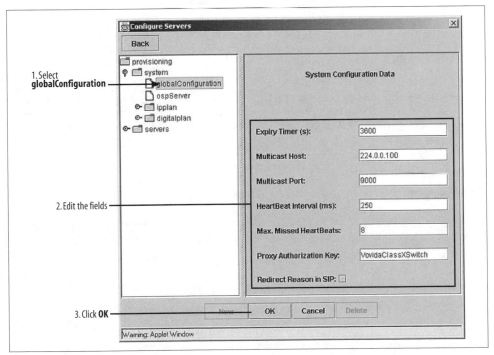

*Figure 5-2. System Configuration data: data entry fields*

Table 5-3 describes the data entry fields.

*Table 5-3. System Configuration data: data entry fields*

| Field | Description |
|---|---|
| Expiry Timer (s) | The time, measured in seconds, that user agents remain registered with the Redirect server before they must send another SIP REGISTER message. The Redirect server compares this value against a requested expiry time sent within the SIP REGISTER message. The shorter expiry time is returned to the User Agent.<br><br>*Default value:* 3600 |
| Mulitcast Host | The muliticast IP address used with the multicast port to send heartbeat broadcasts.<br><br>*Default value:* 224.0.0.100 |
| Multicast Port | The UDP port used by applications to send heartbeat broadcasts. The multicast host and port are concatenated to form a complete multicast address.<br><br>*Default value:* 9000<br><br>*Default multicast address:* 224.0.0.100:9000 |
| HeartBeat Interval (ms) | The transmission rate, in milliseconds, for heartbeats on all applications.<br><br>*Default value:* 250 ms |
| Max. Missed HeartBeats | The maximum number of heartbeats an application can miss before its status becomes Inactive.<br><br>*Default value:* 8 |
| Proxy Authorization Key | Any word or phrase used to identify the system. This phrase is added by the Marshal to all SIP messages as they enter the system and removed by the Marshal as they exit the system. Spaces are not permitted in this field.<br><br>*Example:* VocalSystem |
| Redirect Reason in SIP | If enabled, a CC-Redirect header is included in SIP messages sent through the system. This header tells the called party where the call has been redirected to and why. It may also include a limit on the number of redirections permitted through the network. This field should be disabled if the host network contains devices that cannot process the CC-Redirect header.<br><br>*Default:* disabled |

# OSP Server

These parameters enable the OSP server to communicate with a third-party settlement provider. Before you can set up this server, you will need to identify your settlement provider and the data required to connect to its server. Check the World Wide Web for lists of settlement providers. One source is *http://dir.yahoo.com/Business_and_ Economy/Business_to_Business/Communications_and_Networking/Telecommunications/ Billing_and_Customer_Service.*

OSP is one of the protocols used to enable internetwork calls. A *third-party settlement provider* is an organization that enables ISPs to receive compensation for off-network VoIP calls routed to their networks.

Follow these steps to edit the OSP server parameters:

> This appears simple minded, but it is important to show that no other steps are required to edit these fields.

1. Select *ospServer*. The data entry fields appear in the right frame.
2. Edit the fields, as discussed in Table 5-4.
3. Click OK. Your changes are submitted to the system.

Figure 5-3 shows the OSP server data entry fields.

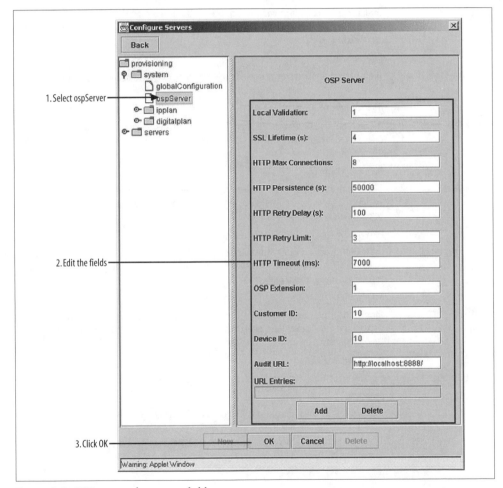

*Figure 5-3. OSP server: data entry fields*

Table 5-4 describes these fields.

*Table 5-4. OSP server: data entry fields*

| Field | Description |
|---|---|
| Local Validation | Determines how the OSP client validates tokens. |
| | *Options:* 0, 1 |
| | 0 = the OSP client authorizes token validations through a protocol exchange, where verification is done by the OSP server. |
| | 1 = the OSP client authorizes token validations locally, by verifying digital signatures. |
| | *Default:* 0 |
| | In Version 1.0 of VOCAL, the software ignores this field and the OSP client validates tokens locally. |
| SSL Lifetime (s) | The lifetime, measured in seconds, of a single Secured Socket Layer (SSL) session key. When this time limit expires, the OSP client negotiates a new session key, without interrupting any communication exchanges in progress. |
| | *Recommended:* 40 |
| HTTP Max Connections | The maximum number of simultaneous connections to be used for communication to the OSP server. |
| | *Recommended connections:* 5 to 8 |
| HTTP Persistence (s) | The time, measured in seconds, that an HTTP connection is maintained after the completion of a communication exchange. Entering a longer duration will help avoid constant teardown and establishment of connections. |
| | *Recommended:* 50,000 |
| HTTP Retry Delay (s) | The time, measured in seconds, between connection retry attempts to the OSP server. After exhausting all service points for the OSP service provider, the OSP client will apply the retry delay before resuming connection attempts. |
| | *Recommended:* 2 |
| HTTP Retry Limit | The maximum number of retry attempts for connecting to the OSP server. If no connection can be made, the OSP client will cease connection attempts and return appropriate error conditions. This number excludes the initial connection attempt. |
| | *Recommended:* 2 |
| HTTP Timeout (ms) | The maximum time, measured in milliseconds, to wait for a response from the OSP server. If no response is received before this time expires, the current connection is released and the OSP client attempts to contact the next configured service point. |
| | *Recommended:* 3000 |
| OSP Extension | Indicates whether the customer ID and device ID of the OSP service provider are known. |
| | *Options:* 0,1 |
| | 0 = You do not know the customer ID and device ID of the OSP service provider. If you type 0, these fields are not required. |
| | 1 = You know the customer ID and device ID of the OSP service provider. If you type 1, the Customer ID and Device ID fields are required. |
| | *Default:* 0 |
| Customer ID | A character string assigned by the OSP settlement provider as a unique customer identification code. Some providers may or may not require this field. |
| Device ID | A character string assigned by the OSP settlement provider as a unique device identification code. Some providers may or may not require this field. |

*Table 5-4. OSP server: data entry fields (continued)*

| Field | Description |
|---|---|
| Audit URL | The URL used for OSP audits. In Version 1.3.0 of VOCAL, the audit function is not implemented in the OSP client. However, this field requires an address such as *http://<localhost>:8888*. This field cannot remain null. |
| URL Entries | A list of character strings for the OSP client to use for sending requests. Each service point takes the form of a URL. The service points can be one of the following formats:<br><br>The domain name expressed as an octet, e.g., *http://255.255.255.255:443/<osp-server>*<br>The domain name expressed as an alias, e.g., *httpd://www.<hostname>.com/<service>/osp*<br>The domain name expressed as a local host, e.g., *httpd://<local_host>/<osp-server>/iis.dll*<br><br>Adding a new URL entry:<br>1. Click Add. A blank space appears in the URL Entries field.<br>2. Type in a URL.<br>3. Click OK. The URL is submitted to the system.<br><br>Deleting a URL entry:<br>1. Select an entry from the URL Entries field. The entry's background color changes to purple.<br>2. Click Delete. The entry disappears.<br>3. Click OK. The altered table is submitted to the system. |

## Dial Plans

The Redirect server provides routing information for all messages that pass through VOCAL. If a call is being made between two subscribers on the same system, the Redirect server has mechanisms built into its code to route the call through the system. These mechanisms are explained in Chapters 7 and 12. However, if a local user calls someone who is not a subscriber to the system—for example, a residence or business that is connected to the public switched telephone system (PSTN)—the Redirect server relies on its dial plan for final call routing.

Two types of dial plans are implemented in VOCAL, the IP and digit dial plans. The organization, syntax, maintenance procedures, and GUI screens are identical for both plan types. The difference lies in the types of calls that they anticipate. The digit dial plan is set up to handle phone numbers, and the IP dial plan handles user addresses formatted as SIP addresses; for example, *sip:user@domain.com*. The Redirect server uses information provided in the SIP request message to determine where to look for routing information. This is explained later.

Dial plans are organized as a table of indexes, keys, and contacts. The index is used to set priorities within the table, the keys indicate anticipated dial patterns, and the contacts provide routing information. To support load balancing and redundancy, each key can have several contacts to provide routing alternatives. Just as the index determines the key's priority within the table, the contacts are arranged by priority within each key. For more information about these fields, see "Keys and contacts" later in this chapter.

## Regular expressions

If you are familiar with regular expressions, skip ahead to "Keys and contacts."

The keys use regular expressions, also known as *regexes*, which are made up of special characters, as shown in Table 5-5, and ordinary characters. Any character used in a regular expression that is not a special character is an ordinary character. Special characters become ordinary when they are preceded by a backslash (\).

This is the same regex used in many Unix programming environments, such as *vi* or *grep*. The syntax of *regexes* is explained more thoroughly in the following online reference: *http://www.math.utah.edu/docs/info/regex_1.html* and in the manpages for *grep* and regex.

Table 5-5 describes the most commonly used special characters in dial plan keys. These characters are used in the examples shown later under "Sample dial plans."

*Table 5-5. Regex characters most commonly used in dial plans*

| Special character | Description |
|---|---|
| [  ] | Indicates a valid range. For example: [3-5]11 means 311, 411, and 511. |
| . | Matches any character except a newline. |
| {  } | Indicates a multiplier of the previous item. For example: {10} means 10 characters; 1{4} means 1111. |
| * | Indicates that the preceding regular expression can be repeated as many times as possible. For example: 011.* means 011 followed by zero or any number of any characters. |
| + | Indicates that at least one match from the preceding regular expression is required. For example: 1[01]+2 does not match 12, but matches 102, 112, or any other expression that matches for 1[10]*2 except 12. |
| ? | Indicates that zero or one match from the preceding regular expression is required. For example: 1[01]?2 matches 12, 102, or 112 and nothing else. |
| \ | Indicates that the next character is a literal expression. For example: \*69 means dialing "* 6 9". |
| ^ | Beginning of a line. |
| $ | End of a line. |

## Keys and contacts

Looking at how the keys and contacts are constructed becomes straightforward when you look at how the elements from the keys and contacts map to the elements that make up the SIP uniform resource identifier (URI).

Two types of calling are permitted in VOCAL: the called party is either a phone or a device with an IP address. Each of these calling types provides two alternative calling methods. These methods are significant in how the Redirect server routes the call.

*Calling a phone*
Both of these alternatives are sent with the parameter *user=phone*.
- If the called address is another local phone user who is provisioned and registered on the same system, you can dial a local extension number, such as

1000. This is how you made one SIP UA call another in the testing from Chapter 2. The Redirect server looks up this number on its subscriber list and routes the call through the User Agent Marshal server that was used by the called party for its registration.

- If the called address is on the PSTN, the Redirect server looks up the routing information in its digit dial plan and then routes the call through a Gateway Marshal server on its way to the PSTN.

*Calling an IP device*

Both of these alternatives are sent with the parameter *user=ip*.

- If the called address is entered using SIP URL dialing, such as user@domain. com, the Redirect server looks it up in its IP dial plan.
- If the called address is entered as a directly dialed IP address, such as @192. 168.20.200, the call is sent directly to the other device without going through VOCAL. This is possible, but unusual. Normally this type of dialing is done for testing purposes between softphones. In practical terms, this is no different than URL dialing except that a DNS server is not required to translate the URL into an IP address.

Let's look at an example in which a calling party has dialed a phone number for a user who is attached to the PSTN, such as 9-1-408-555-1212. (In North American business phone systems, it is common to have to dial 9 to get a dial tone from the central office, bypassing the private branch exchange (PBX).)

The first line of the SIP request message will look like this:

```
INVITE sip:914085551212@192.168.26.180;user=phone SIP/2.0
```

From this example, the first line contains the identity of the message, INVITE, followed by the SIP URI, sip:914085551212@192.168.26.180;user=phone and ends with the version of SIP that the calling party supports, SIP/2.0. The *user=phone* parameter tells the Redirect server that the called party is a phone and to use the digit dial plan to find the routing information. Had this parameter been *user=ip*, the Redirect server would have gone to its IP dial plan for routing information.

In VOCAL Version 1.3.0, the IP dial plan has been disabled. See "IP dial plan" later in this chapter for more information.

Looking at the digit dial plan, the Redirect server looks for a match within the keys entered into the dial plan, and finds, among the many keys listed there:

```
^sip:9.*@
```

The key starts with a caret (^), followed by a variable (sip:9.*@ ), representing SIP URIs that are addressed to any string of characters that begins with 9 and ends with the at sign (@). The key stops reading the SIP URI at the at sign, therefore ignoring the domain part of the IP address.

As far as regular expressions are concerned, the @ sign is not a special character; it is a device used in the VOCAL system to fully specify phone numbers. The @ sign appears at the end of the user portion of the SIP URI. For example: 0@ means 0 is dialed by itself. The expression 0@ does not refer to longer phone numbers that start with 0 such as collect calls and international calls.

Going back to the example: as you can see, the user portion of the SIP URI, sip:914085551212, matches the key ^sip:9.*@ . Therefore, in literal terms, the Redirect server will take everything that comes before the at sign in the SIP URI of the incoming request message and combine it with everything that comes after the @ sign in the contact to create a new SIP URI that will help route the call to its final destination.

In this example, the first contact field that corresponds to ^sip:9.*@ could be something like:

```
sip:$USER@192.168.26.180:5065;user=phone
```

The line starts with the message identifier, sip:, followed by a token, $USER, and the remainder of a SIP URI. As this contact is intended to send calls to the PSTN, it includes the domain address for the SIP/PSTN gateway. The Redirect server replaces the token $USER with the user ID from the SIP URI in the INVITE message sip:914085551212 and sends this new, combined address, sip:914085551212@192.168.26.180:5065;user=phone, in a response message to the next hop in the system.

See Chapter 8 for full examples and explanations of SIP messages, including a fully annotated call message flow. See Chapter 12 for a full explanation of how the Redirect server executes routing.

## Sample dial plans

Here are some other examples of digit dial plans:

*Dialing locally*

Key: ^sip:.{7}@

Contact: sip:$USER@192.168.116.110:5060;user=phone

When a user dials a seven-digit phone number, the system forwards the call to the Gateway Marshal server.

*Dialing the operator*

Key: ^sip:0@

Contact: sip:$USER@192.168.116.110:5060;user=phone

When a user dials 0, the system forwards the call to the Gateway Marshal server. This is used to call an operator at an incumbent local exchange carrier.

*Dialing long distance*

Key: ^sip:1.{10}@

Contact: sip:$USER@192.168.116.110:5060;user=phone

When a user dials 1 plus a 10-digit phone number, the system forwards the call to the Gateway Marshal server.

*Dialing a long-distance operator*

Key: `^sip:00@`

Contact: `sip:$USER@192.168.116.110:5060;user=phone`

When a user dials 00, the system forwards the call to the Gateway Marshal server. This is used to call a long-distance operator.

*Dialing international long distance*

Key: `^sip:011.*`

Contact: `sip:$USER@192.168.116.110:5060;user=phone`

When a user dials 011 followed by any number of any digits, the system forwards the call to the Gateway Marshal server.

*Dialing 9 to get outside of the local system and onto the PSTN*

Key: `^sip:9.*@`

Contact: `sip:$USER@192.168.116.110:5060;user=phone`

The entire number, including the 9, is sent to the gateway via the Gateway Marshal server, which strips the 9 off before forwarding the call to the PSTN. The dial plan shows that any characters of any character length can follow the 9. See Chapter 3 for more information about gateway configurations.

This type of dial plan raises the question, "How does the system know that the user has dialed a complete number if there is no specified character length?" In VOCAL, the User Agent uses dial patterns and timers to determine when the user has finished dialing. See Chapter 3 and Appendix A for more information.

*Dialing long distance for call return*

Key: `^sip:.{10}@`

Contact: `sip:1$USER@192.168.116.110:5060;user=phone`

When a user dials a 10-character phone number (for example, area code + local number) the system adds a 1 to the beginning of the string and then forwards the call to the Gateway Marshal server. This pattern has been added to enable the Call Return feature.

*Dialing long distance for call return: numbers only*

Key: `^sip:[0-9]{10}@`

Contact: `sip:1$USER@192.168.116.110:5060;user=phone`

This is the same as the previous example, except that this key requires all dialed characters to be numbers.

*Dialing 311 through 811*

Key: `^sip:[3-8]11@`

Contact: `sip:$USER@192.168.116.110:5060;user=phone`

When a user dials 311, 411, 511, 611, 711, or 811, his call is forwarded to the Gateway Marshal server.

*Dialing voice mail*

Key: `^sip:7000@`

Contact: `sip:9999@192.168.116.220:5078;user=phone`

When a user dials 7000, the system forwards the call to the voicemail user agents. This is used by the Call Forwarding Feature servers. The actual phone number for the Voice Mail User Agents does not have to be 7000. In this example, that number is 9999. For more information about the voice mail system, see Chapter 14.

*Call return*

Key: `^sip:\*69`

Contact: `sip:$USER@192.168.116.220:5074;user=phone`

When a user dials *69, the system forwards the call to the Call Return Feature server (at 192.168.116.220:5074), which is enabled by a call processing language script. This is only one way to implement Call Return. Another popular method is by making a list of callers available on the user agent and enabling the user to select one of many calls to return through a feature button.

*Four-digit dialing for local exchange dialing*

Key: `^sip:[0-9]{4}@`

Contact: `sip:$USER@192.168.10.10:5060;user=phone`

This rule matches any four-digit call and if the called number was a local SIP UA, the call would be routed to a UA Marshal server.

*Five-digit dialing for local exchange dialing between sites within the same enterprise*

Key: `^sip:5[0-9]{4}@`

Contact: `sip:$USER@192.168.20.10:5060;user=phone`

This rule matches any five-digit call that starts with 5, as an example. This call would be routed to the site that corresponds to 5 perhaps through a UA Marshal server or maybe through an Internetwork Marshal server depending on how the system has been configured.

## Procedures

The following procedures are identical for both the IP and digital dial plans. Figure 5-4 shows the procedure for editing a digital dial plan table entry. The procedure and the dialog box are the same for IP dial plans.

To add a new key to the IP plan table:

1. Select *digitalplan*. The data entry fields appear in the right frame.
2. From the Dial Plan Entries group, click Add. A dialog box appears.

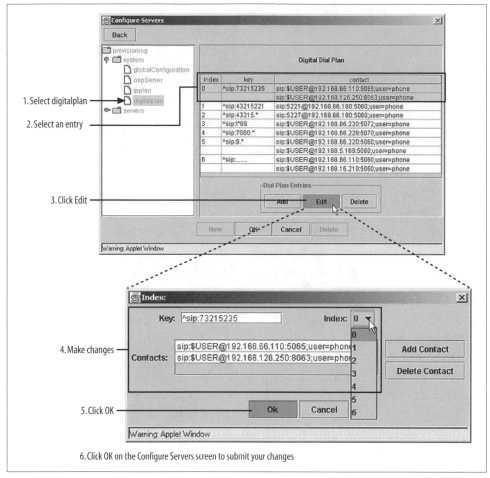

*Figure 5-4. Editing digital dial plans*

3. Enter a key and a contact.

    a. To add additional contacts, click Add Contact.

    b. To remove a contact, click Delete Contact.

4. Select an Index position.

    The default is 0. If this is unchanged, your new key is given top priority. If you choose a position that is already being used, your new key will assume this position and push all other keys, with the same position or lower, down one position.

5. Click OK. The dialog box disappears.

6. Click OK on the Configure servers screen to submit your new entry.

To edit existing keys:

1. Select either *ipplan* or *digitalplan*. The data entry fields appear in the right frame.
2. Select a table entry. You can select the Index number, the key, or any of the contacts.
3. From the Dial Plan Entries group, click Edit. A dialog box appears.
4. Make your changes.
   a. Edit the Key and Contacts by double-clicking on their fields.
   b. To add additional contacts, click Add Contact.
   c. To remove a contact, click to select it, then click Delete Contact.
   d. To move a contact's position, cut the contact information and paste it into a new line.
   e. To change the key's priority, select a different Index value.
5. Click OK. The dialog box disappears.
6. Click OK on the Configure servers screen to submit your new entry.

To delete keys:

1. Select either *ipplan* or *digitalplan*. The data entry fields appear in the right frame.
2. Select a table entry. You can select the Index number, the Key, or any of the Contacts.
3. From the Dial Plan Entries group, click Delete. A dialog box with the following message appears:

   ```
   Do you want to delete <key> at index <index_value>.
   ```

4. Click Yes. The key is removed from the table.
5. Click OK on the Configure servers screen to submit your new entry.

Table 5-6 describes the table fields.

*Table 5-6. dial plans: table fields*

| Field | Description |
|---|---|
| Index | The search order for the dial plan records. Given a request URI, the Redirect server searches the dial plan for the first key that matches it. The search order determines priority within the plan. |
| | In the add and edit dialog boxes, you can assign and change the index numbering of the keys. In the example shown in Figure 5-4, the index for this key can be changed to any of the existing index numbers. Choosing a different number will not overwrite any of the existing keys; it will simply shift their position within the index. |
| Key | A series of characters representing a regular expression. When the user field of the request URI does not match a user configured in your system, this string is compared against the Request URI field of the SIP INVITE message. |
| Contact | The contact list is the list of contacts in SIP format that will be traversed when an INVITE message comes in containing a Request URI field that matches the key. The format should look something like this:<br><br>`sip:$USER@198.176.54.32:5060;user=phone`<br><br>$USER is replaced with the user portion of the Request URI. |

Table 5-7 describes the Dial Plan Entries group buttons.

*Table 5-7. Dial Plan Entries group: buttons*

| Button | Description |
| --- | --- |
| Add | Brings up a dialog box for adding new entries. |
| Edit | Brings up a dialog box for editing existing entries. |
| | Select a table entry before clicking Edit. |
| Delete | Brings up a confirm prompt with options to delete (Yes) or cancel (No) the request. |
| | Select a table entry before clicking Delete. |

Table 5-8 describes the buttons found in the dialog box.

*Table 5-8. Dial Plan Entries dialog box: buttons*

| Button | Description |
| --- | --- |
| Add Contact | Adds a blank field to the contacts. |
| Delete Contact | Removes a contact field from the table entry. |
| | Select a contact before clicking Delete. |
| OK | Submits the entry to the table and closes the dialog box. |
| | You must click OK from the Configure servers screen to submit the table to the system. |
| Cancel | Closes the dialog box. |

## Digit dial plan

The digit plan contains prefixes and phone numbers for any special handling for phone numbers not related to a specific user. The Redirect server checks the digit plan to provide routing information to the other servers in the system.

Figure 5-5 shows the digit dial plan table.

## IP dial plan

The IP plan contains the SIP URLs of subscribers; it is used when the dialed number looks more like an email address than a traditional phone number. The Redirect server checks the IP plan to provide routing information to the other servers in the system.

 The IP plan is under development and is not in use with Version 1.3.0 of VOCAL.

Figure 5-6 shows the IP Dial Plan screen.

In the IP dial plan examples, the dots in the domain names need to be escaped with a backslash, as in *vovida\.org*, or else they take on a different meaning.

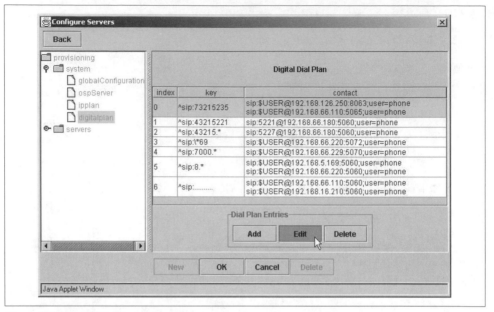

*Figure 5-5. Digit plan data entry screen*

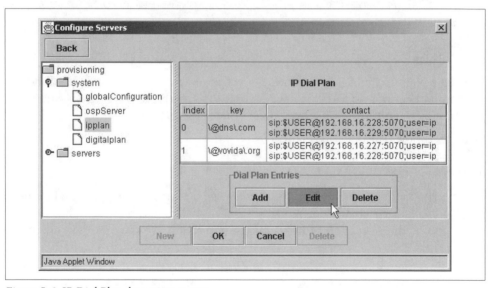

*Figure 5-6. IP Dial Plan data entry screen*

At the time of this writing, there is a bug in the Redirect server that prevents it from distinguishing between different domain names in the SIP URI. We're looking for help from the community to fix this bug.

This concludes our tour of the *system* folder on the Configuring servers screen. In Chapter 6, we look at the contents of the *servers* folder.

# Provisioning Servers

Chapter 5 covered the *systems* folder within the technician's GUI screens. This chapter covers the screens found within the *servers* folder.

## The Servers Folder

This section provides information about:

- Adding server groups
- Adding servers
- Editing servers
- Deleting servers

Provisioning is the series of tasks required for setting up the servers to communicate with one another, and with off-network entities. The *make allinone* script performs these tasks during system installation.

New entities can be added, edited, and deleted through the GUI. The procedure is the same for each server type.

In the left frame of the Configure Servers screen are two subfolders under *Provisioning*:

- The *system* folder provides access to the system configuration data, the OSP server, the IP dial plan, and the digital dial plan. For more information, see Chapter 5.
- The *servers* folder provides access to the servers.

This section provides information about the contents of the *servers* folder.

The servers are organized into the following levels of hierarchy:

*Process*
  A general description of the server's function, such as the Feature servers or the CDR server.

*Type*

> A specific description of the server's function such as the Forward All Calls Feature server or User Agent Marshal server. Only the Feature and Marshal servers are divided by type.

*Group*

> A method for assigning users to servers. Each group can contain multiple servers. Providing users with primary and backup servers within assigned groups enhances system redundancy and reliability.

Figure 6-1 is an enhanced view of the directory tree, emphasizing the servers and their layers of hierarchy.

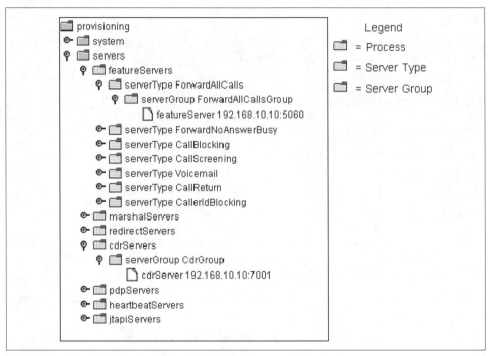

*Figure 6-1. Enhanced directory tree showing the directory hierarchy*

## Adding New Server Groups

The VOCAL system provides one default server group per server type. New server groups can be added to the system. However, these are the legacy of an old design specification, and the usefulness of server groups has not persisted in VOCAL.

We recommend that the administrator use these default groups without adding any more to the system. We also do not recommend editing or deleting the default groups because it is difficult to incorporate those changes into the configuration files.

# Adding New Servers

New servers can be added to server groups as the system scales up to process more users. Servers within the same group are distinguished from one another by their port numbers.

All new servers must be manually added to the */usr/local/vocal/etc/vocal.conf* file on the host. Otherwise, the servers will not come up when the system is rebooted. They also need to be started before the other servers can start using them.

To add a new server:

1. From any of the server folders, select a server group.
2. Click New. The data entry fields appear in the right frame.
3. Fill in the fields.
4. Click OK.

The server is added to the group.

# Editing Servers

After a new server has been added, its provisioning data can be edited by selecting from the directory tree and changing its fields.

Feature servers should not be edited or deleted after users have been added to the system. Otherwise, all users that are assigned this feature may have to be regenerated to accept the changes. More information follows.

### Regenerating users

*Regenerating users* means logging into the user provisioning as an administrator (as described in Chapter 4) and editing the server descriptions for all affected users. Saving these changes requires the GUI to write new XML and CPL scripts for all the selected users to the Provisioning server. If many thousands of users are affected, saving these changes will take an undesirable length of time. This is the result of a non-database system in which users can find out what features they have, but a feature cannot find out what users are using it. The new provisioning system, as described in Chapter 19, solves many of these problems, but this new system is not implemented in VOCAL 1.3.0.

### How to edit servers

To edit a server:

1. From a server group, select a server. The data entry fields appear in the right frame.
2. Make your changes.
3. Click OK. The changes are submitted to the system.

## Deleting Servers

Servers can be deleted from the system at any time, but once they are in service with user agents, deleting servers is not recommended.

To delete a server:

1. Select a server. The data entry fields appear in the right frame.
2. Click Delete. The following prompt appears:

        Are you sure you want to delete this CDR server?

3. Click Yes. The server is deleted.

# Call Detail Record Servers

The Call Detail Record (CDR) server collects information from the Marshal servers indicating when calls start and their duration. If a third-party billing system has been installed into the system, the CDR server can send billing records to it in the Remote Authentication Dial-In User Service (RADIUS) format. The CDR server is discussed more thoroughly in Chapter 18.

Table 6-1 shows the provisioning tasks that can be performed with the CDR server.

*Table 6-1. CDR server tasks*

| Task | Comments |
| --- | --- |
| Adding a CDR server group | CDR server groups are the first sublevel below *cdrServers* on the directory tree. In most systems, there is one CDR server group containing one or two servers. |
| Adding a CDR server | If a second CDR server is required, add it to the existing group. Most systems have two CDR servers mirrored for reliability. |
| Editing a CDR server | Select a server, edit the fields, and then click OK. Tables 6-7, 6-8, and 6-9 describe the fields. |
| Adding additional CDR servers | Systems that can process up to 50 calls per second do not require more than two CDR servers. Additional servers can be added as the system grows in size or if greater reliability is desired. Follow the instructions in "Adding New Servers." |
| Deleting CDR servers | Select a server and click Delete. |

Figure 6-2 shows the CDR server's data entry fields.

Tables 6-2 through 6-4 describe the fields.

*Table 6-2. CDR server data entry fields: CDR group*

| Field | Description |
| --- | --- |
| Host Name | The IP address of the CDR server |
| Port | The port number of the CDR server |

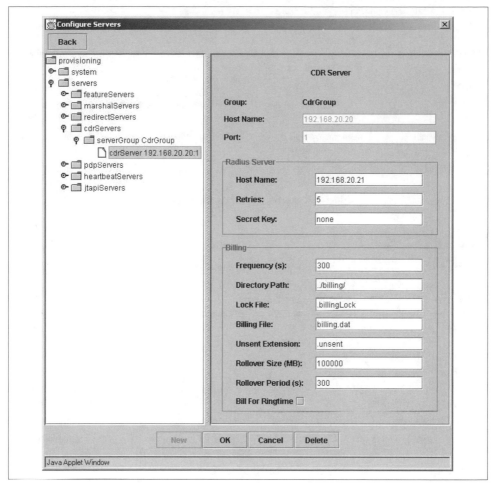

*Figure 6-2. CDR server data entry screen*

*Table 6-3. CDR server data entry fields: RADIUS server*

| Field | Description |
| --- | --- |
| Host Name | The IP address of the RADIUS server. |
| Retries | The number of times that the CDR server will attempt to connect to the RADIUS server before ceasing attempts and returning error messages. *Default:* 5 |
| Secret Key | A text string used for MD5 Digest security. |

*Table 6-4. CDR server data entry fields: billing*

| Field | Description |
| --- | --- |
| Frequency (s) | The frequency, measured in seconds, with which the CDR server sends records to the billing system. *Default:* 86,400 seconds (1 day) |
| Directory Path | Location of stored billing files on the CDR server. |

*Table 6-4. CDR server data entry fields: billing (continued)*

| Field | Description |
|---|---|
| Lock File | If there are two CDR servers, this file informs each server if the other is writing to the directory path. This prevents file allocation errors. |
| Billing File | CDRs are written to this file. |
| Unsent Extension | A file extension appended to billing files that are to be sent to the billing system. *Default:* unsent |
| Rollover Size (MB) | Maximum permitted size, measured in megabytes, of a billing file before it is automatically rolled over. *Default:* 5 |
| Rollover Period (s) | Maximum permitted age, measured in seconds, of a billing file before it is autoRange. *Default:* 86,400 seconds (1 day) |
| Bill for Ringtime | If selected, the billing starts when the phone starts ringing. |

# Redirect Server

The Redirect server stores contact and feature data for all registered subscribers and a dialing plan to enable routing for off-network calls.

Table 6-5 shows the provisioning tasks that can be performed with the Redirect server.

*Table 6-5. Redirect server tasks*

| Task | Comments |
|---|---|
| Adding a Redirect server group | Redirect server groups are the first sublevel below *redirectServers* on the directory tree. |
| Editing a Redirect server | Select a server, edit the fields, and then click OK. |
| Adding additional Redirect servers | Additional servers can be added as the system grows in size, or if greater reliability is desired. Follow the instructions in "Adding New Servers." |
| Deleting Redirect servers | Select a server and click Delete. |

Figure 6-3 shows the Redirect server's data entry fields.

Table 6-6 describes the fields.

*Table 6-6. Redirect server data entry fields*

| Field | Description |
|---|---|
| Host | The IP address of the Redirect server |
| Port | The port number used by the Redirect server |
| Sync Port | A UDP port used by the Redirect server to synchronize its data with the other Redirect servers on the system |

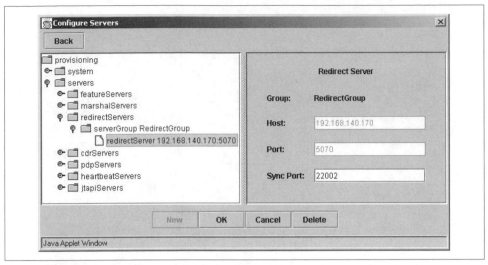

*Figure 6-3. Redirect server data entry screen*

# User Agent Marshal Server

The User Agent Marshal server is the primary point of contact for all User Agents, such as SIP phones. The User Agent Marshal server is a SIP proxy server that authenticates UAs and forwards messages on their behalf. See Chapters 1 and 7 for more information.

Table 6-7 shows the provisioning tasks that can be performed with the User Agent Marshal server.

*Table 6-7. User Agent Marshal server tasks*

| Task | Comments |
| --- | --- |
| Adding a User Agent Marshal server group | User Agent Marshal server groups are the second sublevel below *marshalServers* and *serverType User Agent* on the directory tree. |
| Adding a User Agent Marshal server | Depending on the subscriber base size, a system may include several Marshal servers in the same server group or dispersed over several groups. |
| Editing a User Agent Marshal server | Select a server, edit the fields, and then click OK. Table 6-8 describes the fields. |
| Adding additional User Agent Marshal servers | Systems that can process up to 50 calls per second do not require more than two User Agent Marshal servers. Additional servers can be added as the system grows in size or if greater reliability is desired. Follow the earlier instructions in "Adding New Servers." |
| Deleting User Agent Marshal servers | Select a server and click Delete. |

Figure 6-4 shows the User Agent Marshal server's data entry fields.

Table 6-8 describes the fields.

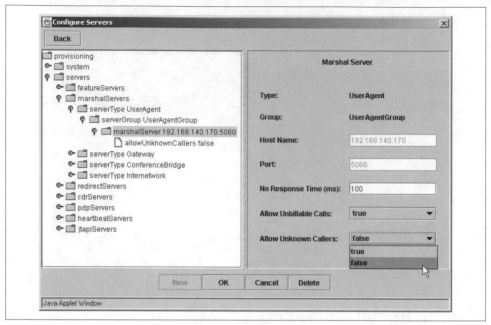

*Figure 6-4. User Agent Marshal server data entry screen*

*Table 6-8. User Agent Marshal server data entry fields*

| Field | Description |
|---|---|
| Type | A notice describing this server's Marshal server type. |
| Group | A notice describing this server's Marshal server group. |
| Host Name | The hostname or IP address of this server. |
| Port | The SIP port number of this server. |
| No Response Time (ms) | This timer keeps track of the connection to the CDR server. If the connection cannot be established before the time value in this cell expires, the call is considered unbillable. |
| Allow Unbillable Calls | If the call is considered unbillable by the rules implemented in the billing system and this field is set to true, the call proceeds.<br><br>If the call is considered unbillable and this field is set to false, the Marshal server rejects the call with a 402 Payment Required message. |
| Allow Unknown Callers | Not implemented. Set to false. |

# Gateway Marshal Servers

The Gateway Marshal servers connect the VOCAL system to the PSTN gateways.

Table 6-9 shows the provisioning tasks that can be performed with the Gateway Marshal server.

*Table 6-9. Gateway Marshal server tasks*

| Task | Comments |
| --- | --- |
| Adding a Gateway Marshal server group | Gateway Marshal server groups are the second sublevel below *marshalServers* and *serverType Gateway* on the directory tree.<br><br>Gateway Marshals that are dedicated to specific PSTN area codes must be separated by area code into different groups. |
| Adding a Gateway Marshal server | The number of Gateway Marshals may depend on how you have configured your system to best serve customers in different regions. |
| Editing a Gateway Marshal server | Select a server, edit the fields, and then click OK. Table 6-10 describes the fields. |
| Adding additional Gateway Marshal servers | Additional servers can be added as the system grows in size or if greater reliability is desired. Follow the earlier instructions in "Adding New Servers." |
| Deleting Gateway Marshal servers | Select a server and click Delete. |

Figure 6-5 shows the Gateway Marshal server's data entry fields.

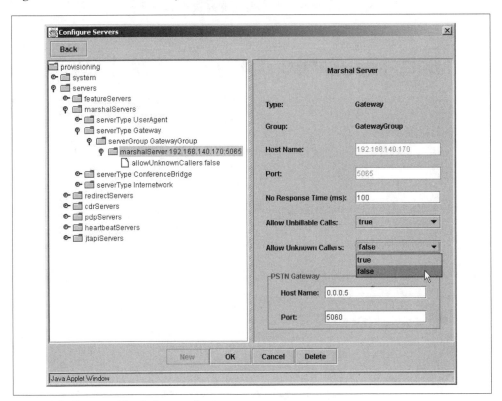

*Figure 6-5. Gateway Marshal server data entry screen*

Table 6-10 describes the fields.

Table 6-10. *Gateway Marshal server data entry fields*

| Field | Description |
|---|---|
| Type | A notice describing this server's Marshal server type. |
| Group | A notice describing this server's Marshal server group. |
| Host Name | The hostname or IP address of this server. |
| Port | The SIP port number of this server. |
| No Response Time (ms) | This timer keeps track of the connection to the Call Detail Record server. If the connection cannot be established before the time value in this cell expires, the call is considered unbillable. |
| Allow Unbillable Calls | If the call is considered unbillable by the rules implemented in the billing system and this field is set to true, the call proceeds. |
| | If the call is considered unbillable and this field is set to false, the Marshal server rejects the call with a 402 Payment Required message. |
| *PSTN Gateway* | |
| Host Name | IP address of the PSTN-to-SIP gateway device that communicates with the marshal. |
| Port | The SIP port of the PSTN-to-SIP gateway device. *Default:* 5060 |
| Allow Unknown Callers | Not implemented. Set to false. |

# Conference Bridge Marshal Server

The Conference Bridge Marshal server connects to gateways or routers that lead to third-party conferencing systems. The VOCAL system supports both *meet me* and *ad hoc* conference calls. We tested VOCAL's conferencing functionality with a conference bridge provided by Voyant Technologies (*http://www.voyanttech.com*). We are not aware of any open source conference bridges available at this time. Maybe someone in the community will write one.

For meet me conference calls, any Marshal type can be used.

Table 6-11 shows the provisioning tasks that can be performed with the Conference Bridge Marshal servers.

Table 6-11. *Conference Bridge Marshal server tasks*

| Task | Comments |
|---|---|
| Adding a Conference Bridge Marshal server group | Conference Bridge Marshal server groups are the second sublevel below *marshalServers* and *serverType ConferenceBridge* on the directory tree. Marshal servers are called randomly within the same group; therefore, adding more servers to the same group will enhance performance as well as reliability. |
| Adding a Conference Bridge Marshal server | Conference Bridge Marshal servers are the third sublevel below *marshalServers*, *serverType Conference Bridge*, and *marshalServerconferenceBridge<IP address:port>* on the directory tree. |
| Editing a Conference Bridge Marshal server | Select a server, edit the fields, and then click OK. Table 6-12 describes the fields. |

*Table 6-11. Conference Bridge Marshal server tasks (continued)*

| Task | Comments |
|------|----------|
| Adding additional Conference Bridge Marshal servers | Additional servers can be added as the system grows in size or if greater reliability is desired. Follow the earlier instructions in "Adding New Servers." |
| Deleting Conference Bridge Marshal servers | Select a server and click Delete. |

Figure 6-6 shows the Conference Bridge Marshal server's data entry fields.

*Figure 6-6. Conference Bridge Marshal server data entry screen*

Table 6-12 describes the fields.

*Table 6-12. Conference Bridge Marshal server data entry fields*

| Field | Description |
|-------|-------------|
| Type | A notice describing this server's Marshal server type. |
| Group | A notice describing this server's Marshal server group. |
| Host Name | The hostname or IP address of this server. |

| Field | Description |
|---|---|
| Port | The SIP port number of this server. |
| No Response Time (ms) | This timer keeps track of the connection to the Call Detail Record server. If the connection cannot be established before the time value in this cell expires, the call is considered unbillable. |
| Allow Unbillable Calls | If the call is considered unbillable by the rules implemented in the billing system and this field is set to true, the call proceeds. |
| | If the call is considered unbillable and this field is set to false, the Marshal server rejects the call with a 402 Payment Required message. |
| Allow Unknown Callers | Not implemented. Set to false. |
| *Gateway* | |
| Host Name | IP address of the PSTN-to-SIP gateway device that is connected to the Time Division Multiplexing (TDM) conference bridge. This can also be the IP address of a SIP conference bridge. |
| Port | The SIP port of the PSTN-to-SIP gateway device. *Default:* 5060 |
| *Conference* | |
| Bridge Number | A well-known phone number used by user agents to make *ad hoc* conference calls. This number can be any length. Dashes are not required. |
| Access Numbers | A list of numbers that matches the access numbers for the conference bridge. The Conference Bridge Marshal server maps the access numbers to the bridge numbers. |

Table 6-13 describes the buttons.

*Table 6-13. Conference Bridge Marshal server buttons*

| Button | Description |
|---|---|
| Add | Adds bridge numbers to the access list |
| Delete | Deletes selected bridge numbers from the access list |

# Internetwork Marshal Server

The Internetwork Marshal servers connect the VOCAL system to other VoIP systems over IP networks.

Table 6-14 shows the provisioning tasks that can be performed with the Internetwork Marshal servers.

*Table 6-14. Internetwork Marshal server tasks*

| Task | Comments |
|---|---|
| Adding an Internetwork Marshal server group | Internetwork Marshal server groups are the second sublevel below *marshalServers* and *serverType interNetwork* on the directory tree. |
| Adding an Internetwork Marshal server | Internetwork Marshal servers are the third sublevel below *marshalServers*, *serverType Internetwork*, and *Internetwork <IP address:port>* on the directory tree. |
| Editing an Internetwork Marshal server | Select a server, edit the fields, and then click OK. Table 6-15 describes the fields. |

*Table 6-14. Internetwork Marshal server tasks (continued)*

| Task | Comments |
|---|---|
| Adding additional Internetwork Marshal servers | Additional servers can be added as the system grows in size or if greater reliability is desired. Follow the earlier instructions in "Adding New Servers." |
| Deleting Internetwork Marshal servers | Select a server and click Delete. |

Figure 6-7 shows the Internetwork Marshal server's data entry fields.

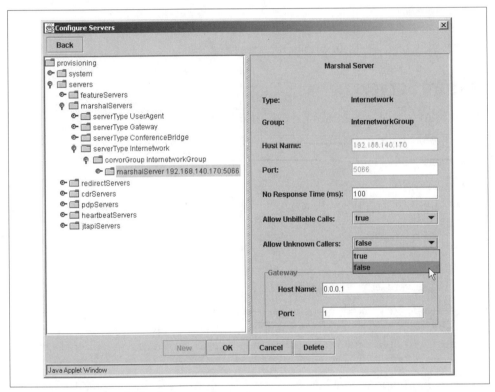

*Figure 6-7. Internetwork Marshal server data entry screen*

Table 6-15 describes the fields.

*Table 6-15. Internetwork Marshal server data entry fields*

| Field | Description |
|---|---|
| Type | A notice describing this server's Marshal server type. |
| Group | A notice describing this server's Marshal server group. |
| Host Name | The hostname or IP address of this server. |
| Port | The SIP port number of this server. |

*Table 6-15. Internetwork Marshal server data entry fields (continued)*

| Field | Description |
|---|---|
| No Response Time (ms) | This timer keeps track of the connection to the Call Detail Record server. If the connection cannot be established before the time value in this cell expires, the call is considered unbillable. |
| Allow Unbillable Calls | If the call is considered unbillable by the rules implemented in the billing system and this field is set to true, the call proceeds. |
| | If the call is considered unbillable and this field is set to false, the Marshal server rejects the call with a 402 Payment Required message. |
| Allow Unknown Callers | Not implemented. Set to false. |
| *Gateway* | |
| Host Name | IP address of the other VoIP system. |
| Port | The SIP port of the other VoIP system. *Default:* 5060 |

# Feature Servers

The VOCAL system supports the following features:

- Call Blocking
- Caller ID Blocking
- Call Forward All Calls
- Call Forward No Answer Busy
- Call Return
- Call Screening
- Voice Mail
- JTAPI server (see its later section)

These features are created through Call Processing Language (CPL) scripts, and each of these features has its server. On smaller systems, the features may all reside on the same host, but unlike the Provisioning and Redirect servers, there is no master Feature server.

For more information about how to assign features to users, see Chapter 4.

For more information about how features are used by subscribers, see Chapter 13.

Table 6-16 shows the provisioning tasks that can be performed with the Feature servers.

*Table 6-16. Feature server tasks*

| Task | Comments |
|---|---|
| Adding a Feature server group | Feature server groups are the second sublevel below *featureServers* and *serverType <feature name>* on the directory tree. |
| Adding a Feature server | Every Feature type must be separated into its own group. |
| Editing a Feature server | Select a server, edit the fields, and then click OK. Table 6-17 describes the fields. |

*Table 6-16. Feature server tasks (continued)*

| Task | Comments |
|------|----------|
| Adding Additional Feature servers | Additional servers can be added as the system grows in size or if greater reliability is desired. Follow the earlier instructions in "Adding New Servers." |
| Deleting Feature servers | Select a server and click Delete. |
| | Deleting Feature servers in live systems is not recommended unless absolutely necessary. If you delete a Feature server that has users assigned to it, you will have to regenerate all of those users through the provisioning system. |

Table 6-17 describes the fields found on the Feature server data entry screen.

*Table 6-17. Feature server data entry fields*

| Field | Description |
|-------|-------------|
| Type | A notice describing this server's Feature server type |
| Group | A notice describing this server's Feature server group |
| Host Name | The hostname or IP address of this server |
| Port | The SIP port number of this server |

# Voice Mail Feature Servers

The Voice Mail Feature server provides the VOCAL system with the capability to forward calls to a voice mail system. This server is required only if you intend to use the system provided with VOCAL.

Table 6-18 shows the provisioning tasks that can be performed with the Feature servers.

*Table 6-18. Feature server tasks*

| Task | Comments |
|------|----------|
| Adding a Voice Mail Feature server group | Voice Mail Feature server groups are the second sublevel below *featureServers* and *serverType voiceMail* on the directory tree. |
| Adding a Voice Mail Feature server | Voice Mail Feature servers are the third sublevel below *featureServers* and *serverType voiceMail* on the directory tree. |
| Editing a Voice Mail Feature server | Select a server, edit the fields, and then click OK. Table 6-19 describes the fields. |
| Adding additional Voice Mail Feature servers | Additional servers can be added as the system grows in size or if greater reliability is desired. Follow the earlier instructions in "Adding New Servers." |
| Deleting Voice Mail Feature servers | Select a server and click Delete. |
| | Deleting Feature servers in live systems is not recommended unless absolutely necessary. If you delete a Voice Mail Feature server that has users assigned to it, you will have to regenerate all of those users. |

Figure 6-8 shows the Voice Mail Feature server's data entry fields.

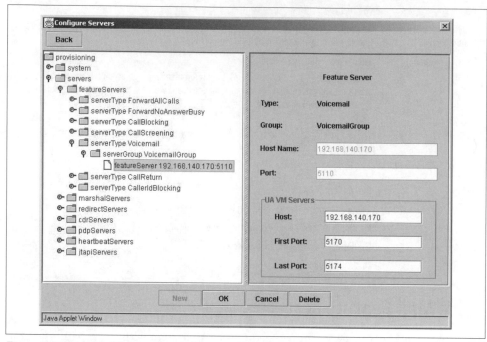

*Figure 6-8. Voice Mail Feature server data entry screen*

Table 6-19 describes the fields.

*Table 6-19. Voice Mail Feature server data entry fields*

| Field | Description |
| --- | --- |
| Type | A notice describing this server's Feature server type. |
| Group | A notice describing this server's Feature server group. |
| Host Name | The hostname or IP address of this server. |
| Port | The SIP port number of this server. |
| *UA VM Servers* | |
| Host | The hostname or IP address of the User Agent Voice Mail server. |
| First Port | The first available UDP port for receiving voice mail messages. |
| Last Port | The last available UDP port for receiving messages. |
| | The First Port and Last Port fields define the number of available ports for voice mail. If there are 10 ports, the system will accept 10 voice mail users at any one time and return a busy signal for all other callers. |

# JTAPI Servers

Java Telephony Application Programming Interface (JTAPI) is a Sun Microsystems specification for providing computer telephony intelligence (CTI). CTI applications are designed for call centers to provide functions such as controlled call redirection and automated dialing. While Sun provides the specification, there is no implementation library.

The JTAPI specification describes five packages:

- Core
- Call Control
- Phone
- Media
- Call Center

Early releases of the VOCAL system included an implementation of the Core package that supported basic third-party call control capability and a sample application, the Vocalpad, that utilized the implementation. This meant that a user could control a user agent (UA) by running Vocalpad on her PC and instructing the UA to call the calling party. However, this functionality is no longer supported, mostly because the JTAPI server is dependent upon the TRANSFER message, which has been deprecated within the SIP standard.

If you are interested in working with third-party call control (3PCC), consider looking at the definition of the back-to-back user agent offered at the end of Chapter 10.

# Heartbeat Server

The majority of VOCAL servers transmit regular pulses called *heartbeats*, which allow the other hosts on the network to determine whether any of the servers are down. The Heartbeat server manages the heartbeat flow. The following servers simply send heartbeats:

- Call Detail Record server
- Conference Bridge Marshal server
- Feature server
- Gateway Marshal server
- Internet Marshal server
- Policy server
- User Agent Marshal server
- Voice Mail User Agents

The following servers perform more complex tasks:

- Heartbeat server listens for heartbeats from all other components except for the Voice Mail User Agents.
- Provisioning server sends heartbeats and listens for heartbeats from the other Provisioning server (if a redundant server exists).
- Redirect server sends heartbeats and listens for heartbeats from the other Redirect server (if a redundant server exists), as well as from all Feature servers and Marshal servers.
- Voice Mail Feature server sends heartbeats and listens for heartbeats from only the Voice Mail User Agents.

The following servers neither send nor listen for heartbeats:

- Network Manager server
- Voice Mail server

The Heartbeat server is used by the Network Manager to update the GUI table of servers and server states. If your system does not include a Network Manager, the Heartbeat server is not required. For more information about network management, see Chapter 17.

Table 6-20 shows the provisioning tasks that can be performed with the Heartbeat servers.

*Table 6-20. Heartbeat server tasks*

| Task | Comments |
| --- | --- |
| Adding a Heartbeat server group | Heartbeat server groups are the first sublevel below *heartbeatServers* on the directory tree. There is normally only one Heartbeat server on a system. |
| Adding a Heartbeat server | Heartbeat servers are the second sublevel below *heartbeatServers* on the directory tree. There is normally only one Heartbeat server on a system. |
| Editing a Heartbeat server | Select a server, edit the fields, and then click OK. Table 6-21 describes the fields. |
| Adding additional Heartbeat servers | There is no need for additional Heartbeat servers. |
| Deleting Heartbeat servers | Select a server and click Delete. |

Table 6-21 describes the fields found on the Heartbeat server data entry screen.

*Table 6-21. Heartbeat server data entry fields*

| Field | Description |
| --- | --- |
| Host | The hostname or IP address of this server |
| Port | The port number for this server |

# Policy Servers

The Policy server is the key component used to achieve *Quality of Service* (QoS). Service providers typically will ensure QoS only if authorizations and payments are guaranteed by either a third party or through a direct peer-to-peer agreement. The Policy server administers admission control for QoS requests and provides the Internetwork Marshal (policy client) with the information necessary to enforce the admitted QoS requests. The Policy server outsources the *authorization, authentication, and accounting* (AAA) requests to a third-party clearinghouse, which then acts as a trusted broker for a large number of network providers. Quality of Service is discussed in detail in Chapter 18.

Table 6-22 shows the provisioning tasks that can be performed with the Policy servers.

*Table 6-22. Policy server tasks*

| Task | Comments |
| --- | --- |
| Adding a Policy server group | Policy server groups are the first sublevel below *pdpServers* on the directory tree. There is normally only one Policy server on a system. |
| Adding a Policy server | Policy servers are the second sublevel below *pdpServers* on the directory tree. There is normally only one Policy server on a system. |
| Editing a Policy server | Select a server, edit the fields, and then click OK. Table 6-23 describes the fields. |
| Deleting Policy servers | Select a server and click Delete. |

Figure 6-9 shows the Policy server's data entry fields.

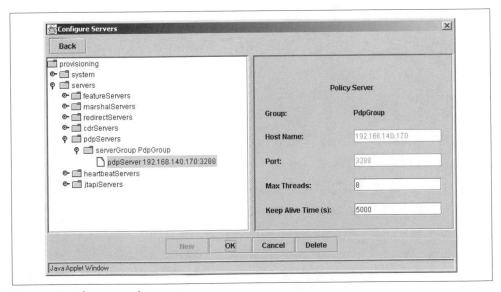

*Figure 6-9. Policy server data entry screen*

Table 6-23 describes the fields.

*Table 6-23. Policy server data entry fields*

| Field | Description |
| --- | --- |
| Host Name | The hostname or IP address of this server. |
| Port | The port number for this server. |
| Max Threads | The maximum number of permitted connections to the Policy server. *Recommended: 5–8.* |
| Keep Alive Time (s) | The time, measured in seconds, to maintain the TCP/IP connection to the Policy server. |

This completes the tour of the GUI screens used for provisioning servers.

# CHAPTER 7

# Session Initiation Protocol and Related Protocols

This chapter provides a technological overview of the Session Initiation Protocol (SIP) that focuses on its ability to enable VoIP. If you take the time to read the SIP standard, Request For Comments (RFC) 3261, you will notice that SIP can do much more than set up basic calls. Rather than writing an exhaustive reinterpretation of the normative statements found within the standard, it seems more useful to explain SIP in its simplest form. After reading this chapter, you should be able to read through a SIP call flow, interpret a log of the message text, and gain some new insight into some of the activity that transpires within VOCAL when users pick up their IP phones and call one another. If you are already familiar with SIP, you may want to skip ahead to Chapter 8 where we discuss the data structures found within the Vovida SIP stack.

While most of the facts covered in this chapter will remain static for a long time, there is always a potential for change. The Internet Engineering Task Force (IETF)'s web site, *http://www.ietf.org*, is your best source for keeping up with the latest RFCs and drafts. Refer to RFC 2026 for an explanation of the Internet standards process.

## What Is SIP?

SIP is a rendezvous protocol for finding users and setting up and modifying multimedia sessions. VoIP is only one of many possible applications for SIP. In our initial discussion of SIP message flows, we attempt to describe SIP generically. Later in this chapter, our SIP examples become directly related to calling between endpoints on a VoIP system.

Since March 1999, when SIP was accepted by the IETF as a proposed standard, SIP has gained a wide following. There is a SIP Working Group that you can access on the Web at *http://www.ietf.org/html.charters/sip-charter.html*.

The strengths of SIP lie in its simplicity and basic assumptions:

*Component reuse*

> In many ways, SIP can be considered a child of the Simple Mail Transfer Protocol (SMTP, RFC 2821) and HyperText Transfer Protocol (HTTP, RFC 2616). Like SMTP and HTTP, SIP uses Multipurpose Internet Mail Extensions (MIME) to carry extra information. SIP also uses universal resource identifiers (URIs) for addressing in the same way as HTTP.

*Scalability*

> Users can be anywhere on the Internet and invited to many different sessions at once.

*Interoperability*

> As SIP is an open standard, the development community is able to build a wide variety of implementations that can communicate with other SIP-based products.

Since RFC 3261 was published, there has been tremendous activity by many vendors to develop SIP implementations. Interoperability between implementations has been tested at many SIP Interoperability Test Events, where developers test how their code interacts with that of others. For a listing of past and future events, see *http://www.cs.columbia.edu/sip/sipit*.

The SIP specifications were first published in 1999 as RFC 2543. In 2002, a second version of these specifications was released as RFC 3261.

## What SIP Does

SIP permits interaction between devices through signaling messages. These messages can fulfill many purposes, including:

- Registering a user with a system
- Inviting users to join an interactive session
- Negotiating the terms and conditions of a session
- Establishing a media stream between two or more endpoints
- Terminating sessions

As with many other signaling protocols, there are a few types of requests and many types of responses. These message types are listed and explained in the standard and, in lesser detail, later in this chapter. The following description mentions only the registration, invitation, and termination messages.

There's a protocol based on SIP called SIMPLE that is used for Instant Messaging and Presence. Many companies, including MSN and AOL, have claimed that they are going to support this in the future.

### Registering a user with a system

A SIP-based system can be as simple as two devices sharing the same Ethernet cable or a distributed system of servers and endpoints spanning several geographical locations. As systems become larger, the methods used to invite parties to new sessions become more complex. In addition, most people are mobile and may require the ability to reply to session invitations from a variety of locations and devices.

Registration links users to their service. When a SIP-based telephone or endpoint, called a User Agent (UA) in SIP, comes online, the first thing it normally does is send out a REGISTER message looking for a Registration server, which keeps track of the location of all subscribed users. The Registration server uses the information carried in the REGISTER message to bind the user's ID with an IP address where it can be contacted. Registrations are not permanent; they normally expire within many minutes, and are continuously renewed as long as the endpoint remains in contact with its service.

### Inviting users to join an interactive session

A registered SIP-based endpoint is ready to use its service to create sessions. In SIP terms, creating a session means sending an INVITE to one or more devices to create a connection that enables a media path to transmit voice, video, and/or data between users. INVITE's many forms of addressing include standard E.164 phone numbers (the precise way of describing phone numbers used on the public switched telephone network (PSTN)), direct-dialed IP addresses, and SIP URIs. When a user wants to invite another user to a session, he enters the other user's address into his device and waits for a response.

While the user is waiting, the device creates an INVITE message and typically sends it to a VoIP softswitch where it is received by either a Proxy or a Redirect server, depending on the switch's architecture. The server then uses a location service to determine how to route the invitation and whether it should go to another user who is registered to the same system, an IP address on another network, or a phone number on the PSTN.

### Negotiating the terms and conditions of a session

When a SIP-based system locates an invited device, it must inform it about the type of session that is being offered. SIP carries this information as an attachment and, like an email service, SIP is concerned only with delivery of the message and not with the contents of the attachment. The most commonly used attachment in SIP messages is a Session Description Protocol (SDP, RFC 2327) body. This is not an exclusive relationship: the body text can include data that is supported by a number of different protocol stacks.

The different types of SIP messages and their ability to work with attachments are explained more clearly throughout this chapter.

When a called party receives an INVITE, it has the option to accept or reject it. Sometimes it will reject the message because it is incapable of working with the proposed session media. For example, a phone may not be able to accept a session of streaming video. Other times, the reason for rejecting the invitation may be subtle differences in the session coding, which determines, among other things, the audible quality of the signal and its compression. Some coding-decoding standards (codecs) are free for use by everyone; others are proprietary and not supported by all IP devices.

Some organizations and individuals define codecs as being short for compression-decompression. In practical purposes, this is the same as coding-decoding.

Rejecting a request, for whatever reason, is perfectly within the rules of the protocol. The SDP in the initial INVITE offers a list of multiple codecs that it's willing to use. The far end picks an acceptable codec from the list and returns this as an answer in the 200 message. It is also possible for the initial INVITE to have no SDP and the 200 to contain an offer with the answer returned in the ACK.

### Establishing a media stream between two or more endpoints

Hopefully, the invitation is accepted by the far end, and the inviting party will see or hear an indication that the invited party has been located. This may be a ringing tone or a graphic indication. In many cases, this indication is being sent by the device at the far end over a media stream that is established separately from the SIP message flow. This is known as *activating early media*.

In voice calls, the media stream uses a protocol called Real-time Transport Protocol (RTP, RFC 1889). This protocol works in two separate, one-way streams between the endpoints. In some sessions, both streams are set up at the same time after the called party picks up the phone. If early media is activated, the first stream is set up when the far end starts ringing, enabling the caller to hear the ringing and the called party saying, "Hello." If early media is not activated, the near end voice service may provide a ringing tone for the caller. The media stream is then set up for both directions at virtually the same time as when the called party answers.

While the media stream is started and terminated by the SIP messages, the stream can take its own route between endpoints over the IP network. It does not flow through the same servers that are used to establish the call.

### Terminating sessions

When the users have completed their communication, the device that "hangs up" first sends a SIP BYE message to the other device to tear down the media stream and ready both endpoints for creating or receiving future sessions.

# SIP Architecture Components

The standard defines several SIP components, and there are many valid ways to implement these into a call control system. Chapter 1 includes an illustrated overview of how these components were implemented in VOCAL.

SIP User Agents are the endpoints used to make and receive calls. They can either be IP phones, software running on a PC, or a combination of an analog telephone adaptor with a basic black phone. SIP servers are contained with the call control system provide location, proxy, redirection, and registration services. This chapter provides a detailed discussion about how VOCAL combines these components into a call control system.

## SIP User Agents

RFC 3261 defines the telephony devices as User Agents (UAs), which are combinations of user agent clients (UACs) and user agent servers (UASs). The UAC is the only entity on a SIP-based network that is permitted to create an original request. The UAS is one of many server types that are capable of receiving requests and sending back responses. Normally, UAs are discussed without any distinction made between their UAC and UAS components.

SIP UAs can be implemented in hardware such as IP phone sets and gateways or in software such as softphones running on the user's computer. It is possible for two user agents to make SIP calls to each other with no other software components. When we start talking about message flows, we'll look at examples that include just two IP phones. Later in the chapter, we will look at the more complex configurations that involve other system components.

When you're writing about the actors in a call flow, if you're not careful, your text can read like a long-winded legal contract with "party of the first part" and so on. Many developers refer to the two parties in a single call as the *caller* and *callee*, which is syntactically consistent with *employer* and *employee* and other similar relationships. When we were writing our first user guides, we felt that *callee* really wasn't part of most people's spoken language. We decided that most people could better relate to *calling party* and *called party*, and we have used those terms in this book.

## SIP servers

Even though the UA contains a server component, when most developers talk about SIP servers, they are referring to server roles usually played by centralized hosts on a distributed network. Here is a description of the four types of SIP servers that are discussed in the RFC:

*Location server*
> Used by a Redirect server or a Proxy to obtain information about a called party's possible location.

*Proxy server*

Also referred to as a Proxy. Is an intermediary program that acts as both a server and a client for the purpose of making requests on behalf of other clients. Requests are serviced internally or transferred to other servers. A proxy interprets and, if necessary, rewrites a request message before forwarding it.

*Redirect server*

An entity that accepts a SIP request, maps the address into zero or more new addresses, and returns these addresses to the client. Unlike a Proxy, it cannot accept calls but can generate SIP responses that instruct the UAC to contact another SIP entity.

*Registrar server*

A server that accepts REGISTER requests. A registrar is typically colocated with a Proxy or Redirect server and may offer location services. The Registrar saves information about where a party can be found.

In VOCAL, the SIP Location, Redirect, and Registrar servers are combined together into a single server called the VOCAL Redirect server. SIP servers can provide a security function by authenticating users before permitting their messages to flow through the network. Frequently, all four server types are included in one implementation. Proxies can also provide features such as Call Forward No Answer (CFNA).

# Sample Message Flows

As mentioned earlier, SIP phones can be connected together in pairs with an Ethernet cable, or they can be nodes on a LAN or a WAN that supports SIP. The SIP-based network can also work with translators to talk to VoIP phones and networks that work with other protocols such as the Media Gateway Control Protocol (MGCP) and H.323 (the International Telecommunications Union (ITU) VoIP standard). Finally, since most people and organizations are connected to the PSTN, VoIP networks normally contain gateways to translate the protocol messages into Integrated Services Digital Network (ISDN) or Signaling System 7 (SS7) signals for transmission to the phone company's central office.

Many request and response messages are identified in the RFC. This section introduces you to some of the more common messages and shows how they are used by the network components.

## SIP Request Messages

The following are the most common request messages described in the standard and the only ones that are used in our example call flows:

*INVITE*

Indicates that the user or service is being invited to participate in a session.

*ACK*
> Confirms that the client has received a final response to an INVITE request.

*BYE*
> Indicates that the user wishes to terminate the session.

*CANCEL*
> Cancels a previous request. This can be confused with BYE; see "BYE and CAN-CEL: Similar but Not Interchangeable" later in this chapter.

*REGISTER*
> Registers the address listed in the To header field with a SIP server.

## SIP Response Messages

There are many different responses, arranged into six different types. Here is a list of the types with an example of each:

*1xx Responses: Informational Responses*
> Example: 180 Ringing

*2xx Responses: Successful Responses*
> Example: 200 OK

*3xx Responses: Redirection Responses*
> Example: 302 Moved Temporarily

*4xx Responses: Request Failure Responses*
> Example: 404 Not Found

*5xx Responses: Server Failure Responses*
> Example: 503 Service Unavailable

*6xx Responses: Global Failure Responses*
> Example: 600 Busy Everywhere

Refer to RFC 3261 for a complete list of messages.

## Call Components

As we get further into this discussion, we would like to make a distinction between the following elements of a call. The SIP standard does not formally recognize the term *call*; instead, *call* is defined as follows:

*Call*
> An informal term that refers to some communication between peers, generally set up for the purposes of a multimedia conversation

Despite the informality, we usually refer to the purpose of our work as setting up calls, and we measure the performance of VOCAL by how many calls per second it can generate. In SIP, we really should discuss setting up dialogs rather than making calls; however, in VoIP, *call* is a legacy term from the PSTN, and most people use it in conversation rather than the more proper term, *dialog*.

In SIP, calls are set up by messages that can be grouped together in transactions. The standard defines *message* and *transaction* as follows:

*Message*
> A message is the data sent between SIP elements as part of the protocol. SIP messages are either requests or responses.

*SIP Transaction*
> A SIP transaction occurs between a client and a server and comprises all messages from the first request sent from the client to the server up to a final (non-1xx) response sent from the server to the client. When the UAC receives a response, other than a 1xx response, to an INVITE, it sends an ACK. For 2xx responses, this is considered a new transaction. For all other responses, it is considered part of the initial INVITE transaction.

Figure 7-1 shows three transactions: the first and second establish the call and the third tears it down. Each transaction contains a series of messages that go back and forth between the User Agents or phones.

In the first transaction (Steps 1–3 in Figure 7-1), one phone sends an INVITE to the other, which sends back a 180 Ringing response to indicate that the phone is ringing, followed by a 200 OK response to indicate that the user has answered it. When the first phone receives the OK message, it returns an ACK to let the other know that a conversation has been enabled between the two endpoints. This pattern of INVITE-180-200 makes up the first transaction. The ACK that is sent in response to the 200 (Step 4 in Figure 7-1) is, by itself, the second transaction.

In the third transaction (Steps 6 and 7), the first phone to hang up sends a BYE request to the other side. When the second phone hangs up, it closes the call by sending a 200 OK response. As this 200 is the final response in the call, the first phone to hang up is not required to return an ACK, and, after all, by the time it receives the 200, it has hung up and torn down its communication path. In this case, the pattern BYE-200 contains only two messages but still represents a transaction.

This concept of transactions becomes more important later in this chapter when we look at the actual text sent by the messages. The next section looks at this basic call in slightly more detail.

## Basic Call Setup

When we talk about working with IP phones, we say "goes off-hook" when the phone is picked up and "goes on-hook" or "hangs up," even if it is a softphone with no physical receiver or hook to hang it on. For these next examples, if you imagine each phone as being a physical desktop SIP IP phone, rather than a software package running on a PC, the explanations will be less confusing.

For the first example, take two SIP IP phones with nothing more than an Ethernet cable connected between them. Then have one phone call the other by dialing its IP address. Figure 7-1 shows the messages that are generated between the phones.

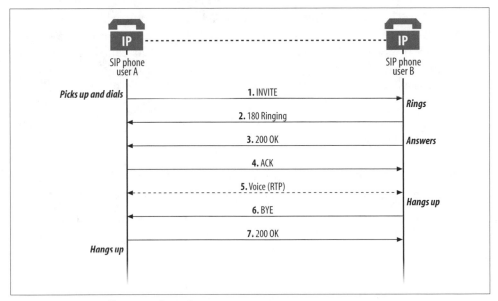

*Figure 7-1. Basic call setup and teardown*

Here is a description of the messages shown in Figure 7-1:

1. *INVITE*

    A user takes phone A off-hook and dials phone B. This causes phone A to send an INVITE to phone B. The INVITE contains an SDP offer describing where Phone A will receive RTP.

2. *180 Ringing*

    Phone B receives the INVITE, activates its ringer, and sends the response message, 180 Ringing, back to phone A. This message has SDP answer revealing which of the offered codecs was used and where Phone B will receive RTP.

3. *200 OK*

    The user at phone B answers the phone. This causes phone B to send the response message, 200 OK, back to phone A.

4. *ACK*

    Phone A receives the 180 Ringing and 200 OK responses and replies to the 200 OK message with an ACK request to complete the triple handshake.

5. *Voice (RTP)*

    As soon as the user answers phone B, a voice channel is established between the two phones. The Real-time Transport Protocol (RTP, RFC 1889) is used to establish the voice channel.

In real life, early media is set up between phone B and phone A. This ensures that, when the called party answers the phone and says "Hello," the entire word comes through clearly to the calling party. Actually, the voice channel consists of two one-way paths. The first path may be established from the ringing phone to the caller when session descriptions, attached as headers that are supported by the Session Description Protocol (SDP, RFC 2327), appear in both the INVITE and 1xx response. The second path from the caller to the called party is established immediately after the called party answers. We drew the two one-way channels as a single two-way voice channel in the illustration for clarity.

6. *BYE*

At the end of the conversation, when the user hangs up phone B, the phone sends a BYE request to phone A.

7. *200 OK*

Phone A receives the request and responds with a 200 OK message. The call is terminated and the voice channel is torn down. Afterwards the user hangs up.

While you might set up two phones this way for testing, normal SIP-based systems use a network of servers for call control, operational system support, and features.

## Proxying

The idea behind Proxy servers is to do the work on behalf of the User Agent. The Proxy can also protect the Redirect server from direct exposure to off-network entities by authenticating users when it receives INVITEs. Being a server, the Proxy cannot generate its own original requests; however, it can forward requests to other servers or reject requests if they do not fulfill a set of criteria. Proxies can also extend requests in parallel or serial in a process called *forking*, which is explained later in this chapter.

Figure 7-2 shows a Proxy interacting with a SIP phone and a Redirect server.

Here is a description of the messages shown in Figure 7-2:

 If Figure 7-2 appears confusing, turn to Figures 7-4 through 7-6 to see how it fits into the big picture.

1. *INVITE*

The user takes phone A off-hook and dials another phone. This causes phone A to send an INVITE to the network.

2. *100 Trying*

Proxy A receives the INVITE and immediately sends a 100 Trying response back to the phone. This response informs phone A that the network has received its INVITE and that another transmission of the INVITE is not required.

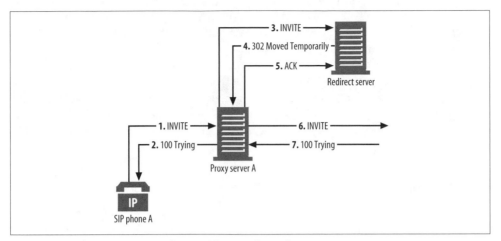

*Figure 7-2. Proxy interacting with a SIP phone and a Redirect server*

*3. INVITE*

Proxy A, having authenticated the INVITE as being from a legitimate user, forwards it to the Redirect server for routing.

There are two different methods for authenticating users, digest and access list, and we recommend using digest. Authentication is discussed in Chapter 11.

*4. 302 Moved Temporarily*

The Redirect server receives the INVITE and looks up the called party on its table of registered users. If the called party is not a registered user, the Redirect server consults a dialing plan to determine where the INVITE should be sent. The Redirect server generates a 302 Moved Temporarily message from the data provided by the INVITE, with the routing information added to one of the message headers. The 302 Moved Temporarily response is sent back to Proxy A. How this is specifically accomplished is explained later in Chapter 11.

*5. ACK*

Proxy A sends an ACK request back to the Redirect server as soon as it receives the 302 Moved Temporarily message to complete the three-way handshake.

*6. INVITE*

Proxy A takes the routing information from the 302 Moved Temporarily message, generates a new INVITE, and sends it to the next stop on the network.

*7. 100 Trying*

The next server in the network, a Proxy that is not shown in Figure 7-2, receives the INVITE and returns a 100 Trying response to indicate that the message was received and that a retransmission is unnecessary.

For the time being, we're asking you to accept that some of these processes work like magic. Some of this magic is explained later in this chapter, and more of it is covered later in this book.

The way that a SIP server deals with an incoming INVITE is partly defined in the RFC and partly defined by the choices the system architects need to make to achieve their deployed implementation. One choice that's available is whether to return a 302 Moved Temporarily message or to proxy the INVITE on to another server or endpoint. One thing that needs to be considered: "Should this server remain in the call path?" If your server is doing billing, or something stateful, you should proxy the call to stay in the call path. If you don't need to stay in the call path, the 302 message pushes intelligence out toward the endpoints, thereby removing your server from the call path.

## Redirection

The idea behind a Redirect server is to instruct the Proxies about the next step in the route to the intended recipient of the message. Redirect servers do not propagate messages; this means that the Redirect server will not forward the INVITE or other messages to another hop in the network. Redirect servers always generate a 3xx, or error, response message in reply to requests.

The Redirect server figures out routing information. In some SIP implementations, the Redirect server may query a Location server, especially those using Telephony Routing over IP (TRIP, RFC 2871). The Location server can be a non-SIP server that keeps records of least-cost routing information. See Chapter 12 for more information.

## Registration

When SIP IP phones are connected to SIP-based phone systems, the first thing they do is register with the Registrar server to initiate its use of the system for outgoing calls and to provide location information to permit incoming calls to be properly directed. In most systems, all messages sent to the Registrar server must go through an authenticating server, which could be a proxy server or the Redirect server itself. The registration matches a provisioned user with an IP address where she can be reached to enable incoming messages to reach her.

Here is a description of the messages shown in Figure 7-3:

1. *REGISTER*

    Immediately after phone A is powered on and connected to the network, it sends a REGISTER to Proxy A.

2. *100 Trying*

    Proxy A receives the REGISTER and immediately sends a 100 Trying response back to phone A. This response informs phone A that its REGISTER has been received by the network and that a retransmission is unnecessary.

3. *REGISTER*

    In the simplest authentication schemes, Proxy A forwards the REGISTER to the Registration server. In other schemes, such as digest authentication, the Proxy

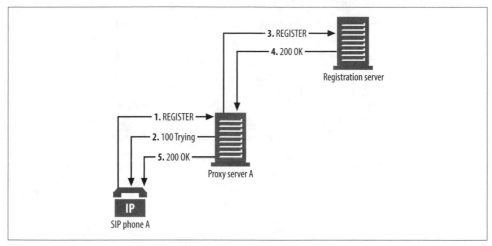

*Figure 7-3. Registering an IP phone*

trades messages with the SIP phone to request authentication data. The Registration server saves this data. See Chapter 11 for more information about authentication and security.

*4. 200 OK*

The Registration server sends a 200 OK response to Proxy A. In the case of registrations, there is no ACK.

*5. 200 OK*

Proxy A forwards the 200 OK message back to phone A, which is now registered with the system.

## Putting It All Together

If you take two SIP phones and connect them to a distributed network, the basic call setup and teardown involves a lot more messages than those shown earlier in Figure 7-1.

Figure 7-4 shows an INVITE going through a SIP-based network and arriving at the called party's phone. In this call, Proxy server A is the *ingress* proxy because it is receiving the INVITE from outside of the system and forwarding it to the Redirect server. Proxy server B is the *egress* proxy because it is forwarding the INVITE to an endpoint that is outside of the system. It is possible for the ingress and egress server to be running on the same host, as illustrated in Figure 7-7.

Here is a description of the messages shown in Figure 7-4:

*1. INVITE*

Phone A sends an INVITE to the network.

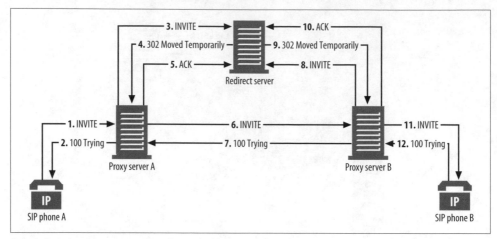

*Figure 7-4. Sending an INVITE through a basic SIP-based system*

*2. 100 Trying*

Proxy A receives the INVITE and immediately sends a 100 Trying response back to the phone.

*3. INVITE*

Proxy A, having authenticated the INVITE as being from a legitimate user, forwards it to the Redirect server for routing.

*4. 302 Moved Temporarily*

The Redirect server generates a 302 Moved Temporarily message from the data provided by the INVITE and the routing information provided by the Location server.

*5. ACK*

Proxy A sends an ACK request back to the Redirect server as soon as it receives the 302 Moved Temporarily message to complete the triple handshake.

*6. INVITE*

Proxy A takes the routing information from the 302 Moved Temporarily message, generates a new INVITE, and sends it to Proxy B.

*7. 100 Trying*

Proxy B receives the INVITE and returns a 100 Trying response.

*8. INVITE*

Proxy B forwards the INVITE to the Redirect server for routing.

*9. 302 Moved Temporarily*

The Redirect server generates a 302 Moved Temporarily message and sends it to Proxy B.

*10. ACK*

Proxy B sends an ACK request back to the Redirect server as soon as it receives the 302 Moved Temporarily message to complete the triple handshake.

**11. INVITE**

Proxy B takes the routing information from the 302 Moved Temporarily message, generates a new INVITE, and sends it to phone B.

**12. 100 Trying**

Phone B receives the INVITE and returns a 100 Trying response.

Figure 7-5 shows a continuation of the message flow as phone B starts ringing and is answered.

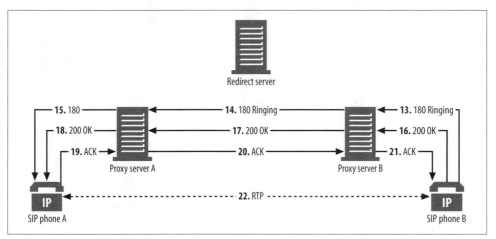

*Figure 7-5. Basic call setup through a SIP-based phone system*

Here is a description of the messages shown in Figure 7-5:

**13–15. 180 Ringing**

Phone B starts ringing and sends a 180 Ringing message back through the network to phone A. The user listening on phone A can hear ringing tones.

**16–18. 200 OK**

The user at phone B picks up the phone. Phone B generates a 200 OK message and sends it through the network to phone A.

**19–21. ACK**

Phone A returns an ACK message through the network to complete the triple handshake.

**22. RTP Voice Channel**

The voice channel is set up, and the calling and called parties can talk to each other.

Figure 7-6 shows how the call is torn down.

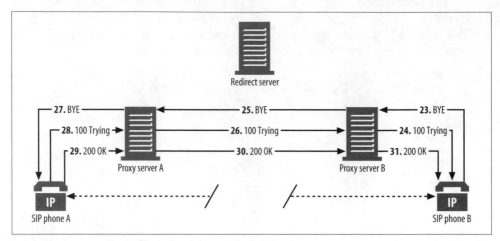

*Figure 7-6. Basic call teardown through a SIP-based phone system*

Here is a description of the messages shown in Figure 7-6:

*23–28 BYE and 100 Trying*
  The user on phone B hangs up. Phone B generates a BYE message and sends it through the network to phone A. As it passes through each proxy server, a 100 Trying responses is immediately sent back to stop retransmission of the BYE.

*29–31 200 OK*
  Phone A hangs up and sends a 200 OK message to phone B. The call and the voice channel are torn down.

The Redirect server is not involved in the call path after sending the 302 message.

# Message Headers

In the previous section, the message flows began with an INVITE message. Earlier in this chapter, we looked at how the INVITE message contains SIP headers and an SDP attachment. In this section, we will look at these components more closely.

## The INVITE

The following is an INVITE that has been generated by a User Agent and sent to a Proxy:

```
----------------------------------------------------------------
INVITE sip:6713@192.168.26.180:6060;user=phone SIP/2.0
Via: SIP/2.0/UDP 192.168.22.36:6060
From: UserAgent<sip:6710@192.168.22.36:6060;user=phone>
To: 6713<sip:6713@192.168.26.180:6060;user=phone>
Call-ID: 96561418925909@192.168.22.36
CSeq: 1 INVITE
Subject: VovidaINVITE
```

```
Contact: <sip:6710@192.168.22.36:6060;user=phone>
Content-Type: application/sdp
Content-Length: 168

v=0
o=- 238540244 238540244 IN IP4 192.168.22.36
s=VOVIDA Session
c=IN IP4 192.168.22.36
t=3174844751 0
m=audio 23456 RTP/AVP 0
a=rtpmap:0 PCMU/8000
a=ptime:20
------------------------------------------------------------------
```

# Breakdown of the Headers

The INVITE is divided into two sections. The top section contains the SIP headers, and the bottom section contains the body of the SIP message, in this case, an SDP session. The headers are separated from the body by a blank line. SDP session description information is discussed later in this chapter.

### INVITE

The first line identifies the message type, the address of the called party, and the version of SIP that the caller supports. In the example, it is:

```
INVITE sip:6713@192.168.26.180:6060;user=phone SIP/2.0
```

The message is an INVITE. The called party's address is expressed as a SIP URI, and the caller supports Version 2.0 of SIP.

The SIP URI contains the following elements:

- An identifier, sip:, which indicates that this is a SIP URI and not, for example, an email address.
- The IP address of the call's destination, which can be broken down as:
  — The user portion of the address 6713, which is what the user dialed.
  — The @ sign, which separates the user portion from the domain portion of the address, just as it does in an email address.
  — The domain portion of the IP address, which is actually the address of the VoIP service provider. In this example, it is the IP address and port number of the assigned User Agent Marshal server.
- A semicolon indicating the start of the URI parameters, which can include, among other things, a user parameter.
- A user parameter, user=phone, which indicates that the call is a phone number and not someone's SIP IP address that happens to resemble a phone number.

## Via

This field provides a history of the message's path through the network or networks. This helps prevent looping and ensures that replies take the same route back to the originator. In the example, it is:

```
Via: SIP/2.0/UDP 192.168.22.36:6060
```

The message is being transported over UDP, and the IP address and port of the sender is 192.168.22.36:6060.

Via headers are used to determine the routing for SIP responses. It is not unusual to have more than one Via in the same message. If two, three, or more Vias are listed, their sequence is significant. As messages are sent through proxy servers, each server adds a Via line to indicate that the message has been through that proxy.

The Marshal servers also add a user parameter to their Via line, called *branch*, indicating where the message is coming from with respect to the system. This helps avoid the appearance of looping messages if the ingress and egress Marshal server are the same process, as illustrated in Figure 7-7. The branches are:

*Branch=1*
> The message has come from outside of the system, has been proxied by the ingress Marshal server, and is on its way to the Redirect server.

*Branch=2*
> The message has come from the ingress Marshal server, after being processed by the Redirect server, and is en route to another server within the system.

*Branch=3*
> The message has come from inside the system, and has been proxied by the egress Marshal server and is on its way to the Redirect server for final routing.

*Branch=4*
> The message has been proxied by the egress Marshal server, after receiving its final routing, and is on its way to an endpoint, such as a gateway.

Branching appears in the later section, "Sample SIP Call Message Flow," and is discussed more thoroughly with more illustrations in Chapter 11.

## From

This field is required in all request and response messages. It provides the identity of the request's initiator. In the example, it is:

```
From: UserAgent<sip:6710@192.168.22.36:6060;user=phone>
```

The initiator is identified as being named UserAgent, and its SIP URI is given.

## To

This provides the identity of the intended recipient of the request. In the example, it is:

```
To: 6713<sip:6713@192.168.26.180:6060;user=phone>
```

The name of the intended recipient is 6713, and its SIP URI is given.

The From and To messages indicate the sender and the recipient of the transaction but not necessarily the flow of each message within the transaction. For example, in a call from *UserA* to *UserB*, the INVITE sent by *UserA* to the Marshal contains From: UserA and To: UserB. The 100 message sent back from the Marshal also contains the same From and To even though this message is going in the opposite direction. This has caused some confusion among those who are learning to code SIP messages.

## Call-ID

This field provides a globally unique identifier to distinguish specific invitations or multiple registrations of the same user.

Some Unix users are accustomed to creating unique numbers by combining the process ID with the IP address of their machine. Using processID@IPaddress is not recommended: it is unique if the IP address is public, but if we are behind a Network Address Translation (NAT) firewall, it is not. For example, many IP phones use 192. 168.0.1 as their internal IP address. It is better to use 32-bit cryptographically random numbers. In the example, it is:

```
Call-ID: 96561418925909@192.168.22.36
```

The name of the caller ID is given as ID@IPaddress.

## CSeq, or command sequence

This field is required in request messages and in response messages: it provides the request method with a unique decimal sequence number. If new requests are sent by the user with the same Call-ID, but different methods or content, the CSeq value has to increase by 1 to let the other network entities know that they are receiving a new message. Otherwise, the entities will think they are receiving a retransmission. Retransmissions of the same request keep the original CSeq number.

When either an ACK or a CANCEL message is sent in response to an INVITE, the CSeq value must be the same as the CSeq field within the INVITE. When a BYE request cancels an INVITE, it must have a CSeq value that is at least 1 higher than the CSeq value of the previous message. In the example, it is:

```
CSeq: 1 INVITE
```

This is the first instance of an INVITE from this caller.

## Subject

This field is optional. It provides a summary of the nature of the call. In theory, the caller can provide a meaningful subject for the call, but in practice, we have not seen it used this way. The Vovida SIP stack fills it with a default value. In the example, it is:

```
Subject: VovidaINVITE
```

We used this field in our lab to identify an INVITE sent by one of our UAs.

### Contact

This field provides the caller's contact information in the form of a SIP URI, which lets the far end know where to reach the caller. In the example, it is:

```
Contact: <sip:6710@192.168.22.36:6060;user=phone>
```

The caller's name in our lab was 6710 and can be reached at port 6060 on the machine that has the IP address 192.168.22.36.

### Content-Type

This field provides the media type of the message body. In the example, it is:

```
Content-Type: application/sdp
```

In this case, the message body is in SDP format.

### Content-Length

This field provides the size of the message body measured in decimal number of octets. In the example, it is:

```
Content-Length: 168
```

The length of the body is 168 octets.

## Other SIP Message Headers

Here are a few other commonly used message headers, plus some obscure headers that inspired comments from our developers. There are many more headers listed in the RFC.

### Proxy-Authorization

This header is important for REGISTERs providing authentication data during registration after receiving a 407 Unauthorized message from the proxy. This header contains the credentials that support permitting the message to flow into the network.

### Expires

This is important for REGISTERs providing the date and time of expiry for the registration.

### Organization

This header is optional: it is a write-only field with no known useful purpose.

## Priority

This header is optional; however, there are high hopes for a useful purpose, like enabling 911 calling. There are also fears that telemarketers and other malcontents could abuse it.

## Proxy-Require

This header is optional: it may be used to indicate features that must be supported by the Proxy, such as Quality of Service (QoS).

## Record-Route

This is added by any proxy that is on the path of a request message routing to indicate that it wishes to be included in future signaling.

## Retry-After

We have not seen this in use, but it looks as if it could have a useful purpose. For instance, it could be used within a 486 Busy Here message to suggest a callback interval.

## Route

In RFC 2543, this provides instructions about where the message is supposed to go. Each network host removes the first entry, then forwards the message to the host specified in the removed entry. In RFC 3261, the handling is more complex with the introduction of a concept called loose routing.

## Server

This is optional: it provides information about the message-handling software used by the user agent. This field could include branding for debugging code.

## Timestamp

This is optional: it provides information about when the request was sent by the client. This value can be used with the Authorization header during registration.

## Unsupported

This is optional: it provides a list of features not supported on the server. It can contain useful information about why an error may have occurred.

## Warning

This is optional: it provides a warning message in plain language that can display on a caller's phone and indicate a problem or status.

**Proxy-Authenticate**

This is important for 407 Unauthorized messages: it provides a challenge to clients that are required to register under a specific security scheme. The client normally retransmits a REGISTER with the required security information.

# SDP Messages

The Session Description Protocol (SDP, RFC 2327) is used to describe the components of the communication channel that are under negotiation when an endpoint is attempting to set up a call. These components include codecs, ports, and the streaming protocol that will enable one-way or two-way transmission. In a voice call originating from a SIP-based device, SDP headers are usually sent with the INVITE and the 200 OK message and normally describe how a data stream is going to be supported through the Real-time Transport Protocol (RTP: RFC 1889).

It is also possible to send an INVITE without SDP headers and, instead, send these headers with the 200 OK and ACK messages. This is sometimes useful when the originating device must receive data about the quality or capabilities of the party it is trying to contact before it can finish setting up the audio stream.

## The INVITE

Here is the INVITE that was described earlier. To bring more attention to the SDP portion, most of the SIP headers have been replaced with an ellipsis (...).

```
---------------------------------------------------------------
INVITE sip:6713@192.168.26.180:6060;user=phone SIP/2.0
...
Content-Type: application/sdp
Content-Length: 168

v=0
o=- 238540244 238540244 IN IP4 192.168.22.36
s=VOVIDA Session
c=IN IP4 192.168.22.36
t=3174844751 0
m=audio 23456 RTP/AVP 0
a=rtpmap:0 PCMU/8000
a=ptime:20
---------------------------------------------------------------
```

## Breakdown of the Lines

The following is a brief description of the lines as they appear in the sample message.

### v: version number

This is always set to zero and indicates the start of the SDP content.

## o: session origin and owner's name

If the user is unknown, it is the session ID and version that are typically set to the Network Time Protocol (NTP, RFC 1305) timestamp of the start time. The syntax of this message is as follows:

```
o = <username> <session ID> <version> <net type> <address type> <addr>
```

The value *<addr>* is the address of the machine that is creating the session. In the example, it is:

```
o=- 238540244 238540244 IN IP4 192.168.22.36
```

In this example, the session ID and version are the same. The username is set to - , which is a placeholder for the real name.

## s: session name

This is simply an identifier. In the example, it is:

```
s-VOVIDA Session
```

The session is called Vovida Session.

## c: connect information

This specifies the IP address of a session. There are several other RTP fields, but these are the critical ones for VoIP. In the example, it is:

```
c=IN IP4 192.168.22.36
```

The session has an IP Version 4 address, and it is given.

## t: time the session is active

This field expresses the duration of the session. The syntax is as follows:

```
t = <start time> <stop time>
```

If the stop time is unbounded, it is set to 0. Time values are NTP times in seconds. In the example, it is:

```
t=3174844751 0
```

The start time is expressed in seconds, and the stop time is unbounded.

## m: Media name and transport address

This field describes the type of media, the transport address, and the codec supported by the sender. This field sometimes contains codecs that the receiver doesn't support. When that happens, the receiver sends back a message that lists the codecs it supports in the m: field of the SDP.

The syntax is as follows:

```
m = <media> <port>/<number of ports> <transport> <fmt list>
```

In the example, it is:

```
m=audio 23456 RTP/AVP 0
```

When RTP/AVP is the transport, the `<fmt list>` list is a list of integers that specify the codec that can be used.

Table 7-1 shows some of the possible codecs and their identifiers.

*Table 7-1. Some common codec identifiers*

| Identifier | Codec |
| --- | --- |
| 0 | G.711 μLaw |
| 2 | G.726-32 |
| 3 | GSM |
| 4 | G.723 |
| 8 | G.711 aLaw |
| 9 | G722 |
| 10 | CD-quality audio (L16/44100/2) |
| 15 | G.728 |
| 18 | G.729 |

This list can be found in RFC 1890 and on the Internet Assigned Numbers Authority (IANA) web site (*http://www.iana.org/assignments/rtp-parameters*) under Protocol Number Assignment, R, and RTP Parameters.

### a: attribute lines

In addition, each codec can have additional information supplied via attribute lines. In the example, it is:

```
a=rtpmap:0 PCMU/8000
a=ptime:20
```

The first line refers to a Real-time Transport Protocol mapping (*rtpmap*) for the codec type of PCMU /8000. The PCMU indicates that it is G.711 μLaw and the 8000 indicates that the sample rate is 8KHz. The second line refers to a packet time expressed in milliseconds. The example means that each packet will contain 20 ms of voice traffic.

### Alternate attributes

Here are some alternative SDP attributes:

```
a = rtpmap: <payload type> <encode name>/<clock rate>[<encode parameter>]
```

Payload types in 96–127 are dynamically assigned using the *rtpmap* attribute, so the actual codec is specified in the a line:

```
m = audio 4000 RTP/AVP 97
a = rtpmap:97 L16/44100/2
```

This defines CD-quality audio. It is Linear 16 encoding at 44100 Hz with two channels. This could also have been expressed as:

```
m = audio 4000 RTP/AVP 10
```

Dynamically assigned numbers are preferable to predefined numbers.

# Sample SIP Call Message Flow

Figure 7-7 illustrates a simple call flow with two users, one Marshal server, and one Redirect server.

The following text is a listing of the messages from the call flow illustrated in Figure 7-7. Many of the message headers are repeated throughout the transactions.

In this scenario, the calling phone is extension 5121 and the called phone is local 5120. There is only one Marshal server, 192.168.36.180:5060, and one Redirect server, 192.168.36.200:5060. For the sake of brevity, in this setup, the Marshal server acts as both ingress and egress proxy. You will notice some messages being sent to and from the same IP address. We have also used the informal term, *call*, instead of the proper term, *dialog*, to express the setup, media negotiation, establishment, and teardown of a conversation.

1. An INVITE message is sent by a calling party (5121) to the VoIP service, intended for a called party (5120). The key fields in this message have been highlighted in bold text.

```
INVITE sip:5120@192.168.36.180 SIP/2.0
Via: SIP/2.0/UDP 192.168.6.21:5060
From: sip:5121@192.168.6.21
To: <sip:5120@192.168.36.180>
Call-ID: c2943000-e0563-2a1ce-2e323931@192.168.6.21
CSeq: 10 INVITE
Expires: 180
User-Agent: Cisco IP Phone/ Rev. 1/ SIP enabled
Accept: application/sdp
Contact: sip:5121@192.168.6.21:5060
Content-Type: application/sdp
Content-Length: 219

v=0
o=CiscoSystemsSIP-IPPhone-UserAgent 16264 18299 IN IP4 192.168.6.21
s=SIP Call
c=IN IP4 192.168.6.21
t=0 0
m=audio 25282 RTP/AVP 0 101
a=rtpmap:0 pcmu/8000
a=rtpmap:101 telephone-event/8000
a=fmtp:101 0-11
```

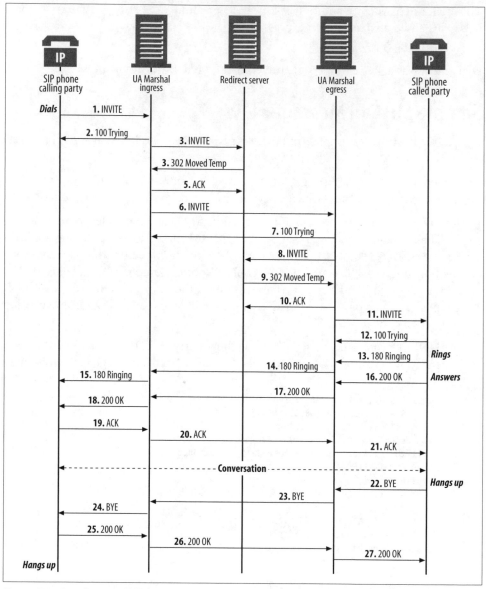

*Figure 7-7. Simple SIP call flow*

Looking at the INVITE followed by the SIP URI:

```
INVITE sip:5120@192.168.36.180
```

the phone is not aware of how to locate the called party, 5120; it knows only one address to which to send all messages, 192.168.36.180, and this could be the address of the service where it is registered or a different address that has been manually configured into the phone.

The phone adds itself to the Via:

```
Via: SIP/2.0/UDP 192.168.6.21:5060
```

because that it how it gets responses sent back to it.

Many people talk about what goes in the From and To fields:

```
From: sip:5121@192.168.6.21
To: <sip:5120@192.168.36.180>
```

They're really not used for many things. The From field contains the number of this phone and can be used for authorization. It is also used by the Redirect sever to build a calling contact list, which enables routing to any calling features implemented and configured for this user; for example, call blocking. The To field repeats who was dialed, the 5120. The URI at the top is much more important. That is what the other network elements are going to use to route the call. The To field is a historical record, and if it contains incorrect information, it makes no difference to the routing.

As for the Call-ID:

```
Call-ID: c2943000-e0563-2a1ce-2e323931@192.168.6.21
```

you can see that we've generated a large, unique, random number, and we've added the IP address of the calling party.

The command sequence number:

```
CSeq: 10 INVITE
```

could have started at any number that we wanted, and we arbitrarily chose 10. For each future message, we will increment this number. The rules state that it must increment, but do not specify the interval. We could increment by 1 or by any other number that suited our needs.

The next interesting field is the Contact field:

```
Contact: sip:5121@192.168.6.21:5060
```

which provides information about reaching the calling party. This has nothing to do with where the response to this INVITE is sent; that is determined by the first Via field. The contact is sent only for directing future transactions such as BYEs.

The Content-Type and -Length:

```
Content-Type: application/sdp
Content-Length: 219
```

say that the body is SDP and contains 219 bytes of data.

Looking at the SDP data, the first key part is:

```
c=IN IP4 192.168.6.21
```

which gives the IP address of where we want the other side to send the media.

The m= line:

```
m=audio 25282 RTP/AVP 0 101
```

says that the media is audio and to send it on port 25282, RTP. The 0 is a well-defined number that indicates the G.711 codec, μLaw PCM data. The higher numbers are dynamically assigned. The 101 has been picked by the phone randomly; on a later line in the SDP, we will find out what the 101 means.

The next three fields are attribute (a=) fields, which can be used for a variety of different types of information. In many cases, the attributes define items that were not anticipated in the SDP draft. In our example, the phones have used them to clarify the contents of the m= field.

The first thing that comes after an a= is the type of attribute, and the usual type is Real-time Transport Protocol map (*rtpmap*), which tells it how to line up the other information.

The first attribute tells us what 0 means:

```
a=rtpmap:0 pcmu/8000
```

This is not required because 0 is defined in the SDP standard. This explains an *rtpmap* that defines 0, with the definition, pulse code modulation (pcm) μlaw with a sampling of 8000 samples per second.

The 101:

```
a=rtpmap:101 telephone-event/8000
```

says that these are going to be telephone events, which is our way of sending DTMF and similar events. It also says that the time base for that is going to be 8000.

There is some additional information about the type of telephone event:

```
a=fmtp:101 0-11
```

that says we can work with telephone event types 0 through 11, which correspond to the dialing keys on the telephone, 0–9, *, and #. The abbreviation, fmtp, stands for *format-specific parameters*, and it means that characters can be conveyed in a way that the SDP doesn't understand but that is specific to the media format; in this case, it is specific to RTP.

2. The User Agent Marshal server, acting as the ingress Marshal server, replies that it has received the message, by sending a 100 Trying response. This stops the calling party from retransmitting another INVITE message.

```
SIP/2.0 100 Trying
Via: SIP/2.0/UDP 192.168.6.21:5060
From: <sip:5121@192.168.6.21:5060>
To: <sip:5120@192.168.36.180:5060>
Call-ID: c2943000-e0563-2a1ce-2e323931@192.168.6.21
CSeq: 10 INVITE
Content-Length: 0
```

The 100 message mirrors the request to which it is responding. It must have the same Call-ID and command sequence number so that the phone's SIP transceiver (explained in Chapter 8) can match it to the INVITE that was sent out and stop retransmission. The 100 Trying message does not carry SDP.

As this message is going back directly to the originator of the call, the Via is not as important as it would be if this message had to travel through several proxy servers on its way to the phone. As you will see later, if a message has to return through several proxies, it has several Via headers, each of which is stripped off as it passes through the proxies on the way to the phone.

Notice that the From and To fields are identical to those of the INVITE message even though this message is going in the opposite direction.

3. The User Agent Marshal server forwards the INVITE message to the Redirect server:

```
INVITE sip:5120@192.168.36.200:5060;user=phone SIP/2.0
Via: SIP/2.0/UDP 192.168.36.180:5060;branch=1
Via: SIP/2.0/UDP 192.168.6.21:5060
From: <sip:5121@192.168.6.21:5060>
To: <sip:5120@192.168.36.180:5060>
Call-ID: c2943000-e0563-2a1ce-2e323931@192.168.6.21
CSeq: 10 INVITE
Proxy-Authorization: Basic VOCALSystem
Expires: 180
Record-Route: <sip:5120@192.168.36.180:5060;maddr=192.168.36.180>
Contact: <sip:5121@192.168.6.21:5060>
Content-Type: application/sdp
Content-Length: 219

(The SDP has not changed.)
```

Although the SIP URI is now addressed to the Redirect server, most of the other message content has not changed since the original INVITE.

The Marshal has added itself to the Via header with a branch=1 parameter. As we explained earlier, this branch identifies this header as having come from the ingress Marshal server.

A Record-Route field:

```
Record-Route: <sip:5120@192.168.36.180:5060;maddr=192.168.36.180>
```

has been added. This field contains the Marshal server's address and is used to ensure that future messages related to this call go through this server.

The SDP (not shown) is exactly the same as it was in the original transmission of the INVITE message.

4. The Redirect server replies with a 302 Moved Temporarily message containing information about where the ingress Marshall must send the INVITE message:

```
SIP/2.0 302 Moved Temporarily
Via: SIP/2.0/UDP 192.168.36.180:5060;branch=1
Via: SIP/2.0/UDP 192.168.6.21:5060
From: <sip:5121@192.168.6.21:5060>
To: <sip:5120@192.168.36.180:5060>
Call-ID: c2943000-e0563-2a1ce-2e323931@192.168.6.21
CSeq: 10 INVITE
Contact: <sip:5120@192.168.36.180:5060>
Content-Length: 0
```

The Redirect server looks up the SIP URI in its registry and dial plan and builds a contact list that helps it figure out where this call should be sent to next. If a calling feature is implemented and configured for the user, such as call blocking, the Call Blocking Feature server would be the next hop for this message. The Redirect server also looks at any called features that might be implemented and configured for the called party, such as call forward all calls. If any of these are required, the server that provides the required called feature becomes the next hop after the calling Feature server.

Finally, the Redirect server looks at the terminating contact information, which was provided by the users when they registered with VOCAL. If the called party is a local user, the address of the User Agent Marshal server that processed its registration is added as the final stop on the contact list. If the called party is on the PSTN, the address of the Gateway Marshal server becomes the final hop. If neither the calling nor called party was a registered user on the system, that would mean that VOCAL was being used as a transit system between two other domains, and the Redirect server would simply provide routing based on its dial plans. While using VOCAL as a transit system would be unusual, compared to our original intentions, VOCAL could fulfill this role in compliance with the SIP standard.

In our example, both the calling and called parties are local subscribers, and there are no features configured for either; therefore, the contact list contains only the terminating contact and the User Agent Marshal server that processed the called party's registration. This User Agent Marshal server thereby becomes the egress Marshal server, and the Redirect server puts the URI of this Marshal server into the Contact header:

```
Contact: <sip:5120@192.168.36.180:5060>
```

The Redirect server has not added itself to the Via header because it does not forward the transaction but instead sends a response to the request. You will see later that, eventually, the Redirect server stops participating in the message flow. This message does not carry SDP.

5. The User Agent Marshal server replies with an ACK message that completes the transaction between this server and the Redirect server:

```
ACK sip:5120@192.168.36.200:5060;user=phone SIP/2.0
Via: SIP/2.0/UDP 192.168.36.180:5060;branch=1
From: <sip:5121@192.168.6.21:5060>
To: <sip:5120@192.168.36.180:5060>
Call-ID: c2943000-e0563-2a1ce-2e323931@192.168.6.21
CSeq: 10 ACK
Content-Length: 0
```

The key items in the message have to do with what is being acknowledged. The ACK informs the Redirect server that it can stop retransmission of the 302 message, in the same way that the 100 Trying message did earlier. The Call-ID and Command Sequence Number have to match those found in the 302 message.

While the Command Sequence number remains the same, the message name following the number has changed:

```
CSeq: 10 ACK
```

In previous messages, it said 10 INVITE. The appearance of ACK in this header closes the transaction. This is where the ACK differs from the 100 Trying message. Whereas the 100 Trying message is a provisional response to a request, the ACK is a request that completes a transaction after a final response, such as 302 Moved Temporarily or 200 OK.

ACK is a request because of its structure, not necessarily because of its behavior or content. According to the RFC:

> SIP requests are distinguished by having a Request-Line for a start-line. A Request-Line contains a method name, a Request-URI, and the protocol version separated by a single space (SP) character.

As for the responses:

> SIP responses are distinguished from requests by having a Status-Line as their start-line. A Status-Line consists of the protocol version followed by a numeric Status-Code and its associated textual phrase, with each element separated by a single SP character.

Look at the request line for the ACK:

```
ACK sip:5120@192.168.36.200:5060;user=phone SIP/2.0
```

and compare it to the status line of the 100 Trying message:

```
SIP/2.0 100 Trying
```

Even though the ACK is sent in response to the 302, by its structure, it is a SIP request. There is no data transmitted in the ACK; therefore, the framers of the SIP standard have decided that ACK can complete the transaction without requiring a further 200 OK response. You will see that 200 OK responses are used later in this call flow to close other transactions.

This instance of the ACK message does not carry SDP. In other cases in which ACK is being used in session negotiation between peers, it will carry SDP.

6. The ingress Marshal server forwards the INVITE to the egress Marshal server, as instructed by the Redirect server:

```
INVITE sip:5120@192.168.36.200:5060;user=phone SIP/2.0
Via: SIP/2.0/UDP 192.168.36.180:5060;branch=2
Via: SIP/2.0/UDP 192.168.6.21:5060
(The remaining SIP headers have not changed.)

(The SDP has not changed.)
```

The ingress Marshal has added itself to the Via header and there are no changes to any other field.

7. The egress Marshal server responds with a 100 Trying to stop retransmission:

```
SIP/2.0 100 Trying
(The remaining SIP headers have not changed.)
```

As with the earlier 100 Trying message, this one mirrors the previous INVITE message, except that it does not carry SDP.

8. The egress Marshal server forwards the INVITE message to the Redirect server for new routing information:

```
INVITE sip:5120@192.168.36.200:5060;user=phone SIP/2.0
Via: SIP/2.0/UDP 192.168.36.180:5060;branch=3
Via: SIP/2.0/UDP 192.168.36.180:5060;branch=2
Via: SIP/2.0/UDP 192.168.6.21:5060
(The remaining SIP headers have not changed.)

(The SDP has not changed.)
```

The egress Marshal also inserts itself in the Via header with a branch=3 parameter.

9. The Redirect server replies with a 302 Moved Temporarily message like before:

```
SIP/2.0 302 Moved Temporarily
Via: SIP/2.0/UDP 192.168.36.180:5060;branch=3
Via: SIP/2.0/UDP 192.168.36.180:5060;branch=2
Via: SIP/2.0/UDP 192.168.6.21:5060
From: <sip:5121@192.168.6.21:5060>
To: <sip:5120@192.168.36.180:5060>
Call-ID: c2943000-e0563-2a1ce-2e323931@192.168.6.21
CSeq: 10 INVITE
Contact: <sip:5120@192.168.6.20:5060>
Content-Length: 0
```

The Redirect server once again checks the user ID of both the calling and called parties against its subscriber list and builds a contact list. The Redirect server can tell by looking at the Via headers that this call is ready to be forwarded outside of the system and thereby adds the address of a destination outside of the system to the Contact header. In this case, it is the specific address of a registered subscriber, and the Redirect server does not need to look it up on a dial plan.

If the destination had been a phone on the PSTN, the Redirect server would have looked for a matching key in its digit dial plan and used the address of the associated contact to create a new SIP URI containing the address of the SIP/PSTN gateway. This mechanism is explained in Chapter 5. If the destination were an IP phone on a different system, the Redirect server would have checked its IP dial plan for routing information. However, in VOCAL 1.3.0, the digit dial plan has been disabled. See Chapter 5 for more information.

10. The Marshal server replies with an ACK message to complete the transaction:

```
ACK sip:5120@192.168.36.200:5060;user=phone SIP/2.0
...
```

11. The egress Marshal server forwards the INVITE message to the called party, who happens to be a locally provisioned user on the system:

```
INVITE sip:5120@192.168.6.20:5060 SIP/2.0
Via: SIP/2.0/UDP 192.168.36.180:5060;branch=4
Via: SIP/2.0/UDP 192.168.36.180:5060;branch=2
Via: SIP/2.0/UDP 192.168.6.21:5060
```

(The remaining SIP headers have not changed except for Record-Route.)
**Record-Route: <sip:5120@192.168.36.180:5060;maddr=192.168.36.180>,<sip:5120@192.168.36.180:5060;maddr=192.168.36.180>**

(The SDP data has not changed.)

Notice that the address in the SIP URI matches the address found in the Contact header of the previous 302 message. Notice also that the top Via header is followed by branch=4, indicating that this message is leaving the system. The Record-Route header becomes interesting when we look at the BYE message later in this flow.

12. The called party replies with a 100 Trying message to indicate to the Marshal server that it has received the message and to stop retransmissions:

    SIP/2.0 100 Trying
    (The remaining SIP headers have not changed.)

13. The called party accepts the session description as is and replies with a 180 Ringing message:

    SIP/2.0 180 Ringing
    **Via: SIP/2.0/UDP 192.168.36.180:5060;branch=4,SIP/2.0/UDP 192.168.36.180:5060;branch=2,SIP/2.0/UDP 192.168.6.21:5060**
    (The remaining SIP headers are the same as those shown in the INVITE.)

    (There is no SDP.)

This 180 is interesting because it is going to be routed across the system to the other side using the Via headers. There is no SDP attached to this message because we did not require early media in our lab. If there had been an SDP, a one-way RTP path could have been established from the called party to the calling party.

14. The egress Marshall server forwards the ringing message to the ingress Marshal server:

    SIP/2.0 180 Ringing
    **Via: SIP/2.0/UDP 192.168.36.180:5060;branch=2,SIP/2.0/UDP 192.168.6.21:5060**
    (The remaining SIP headers have not changed.)

    (There is no SDP.)

Notice that there is one less Via header than in the previous 180.

15. The ingress Marshal server forwards the ringing message to the calling party:

    SIP/2.0 180 Ringing
    **Via: SIP/2.0/UDP 192.168.6.21:5060**
    (The remaining SIP headers have not changed.)

The Redirect server is no longer in the message flow because it did not include itself in the Via headers; there is only one Via header in this message, and it contains the address of the calling party.

16. The called party answers and sends a 200 OK message to the egress Marshal server:

```
SIP/2.0 200 OK
(The remaining SIP headers are the same as those shown in the INVITE.)

v=0
o=CiscoSystemsSIP-IPPhone-UserAgent 13045 2886 IN IP4 192.168.6.20
s=SIP Call
c=IN IP4 192.168.6.20
t=0 0
m=audio 30658 RTP/AVP 0 101
a=rtpmap:0 pcmu/8000
a=rtpmap:101 telephone-event/8000
a=fmtp:101 0-11
```

Looking at this SDP, the c= line contains the address of this phone, the called party; the m= audio line has the port number. The codecs that it has selected to use are 0 and 101, which were the only two options available to it. If it had been given four options, it probably would have picked just one of them. Now we can start having media in both directions.

17. The egress Marshal server forwards the 200 OK message to the ingress Marshal server:

```
SIP/2.0 200 OK
(The remaining SIP headers have not changed.)

(The SDP has not changed.)
```

18. The ingress Marshal server forwards the 200 OK message to the calling party:

```
SIP/2.0 200 OK
(The remaining SIP headers have not changed.)

(The SDP has not changed.)
```

19. The calling party replies with an ACK message to complete the transaction:

```
ACK sip:5120@192.168.36.180:5060 SIP/2.0
Via: SIP/2.0/UDP 192.168.6.21:5060
From: <sip:5121@192.168.6.21:5060>
To: <sip:5120@192.168.36.180:5060>;tag=c29430002e0620-0
Call-ID: c2943000-e0563-2a1ce-2e323931@192.168.6.21
Route: <sip:5120@192.168.36.180:5060;maddr=192.168.36.180>, <sip:5120@192.168.6.
20:5060;maddr=192.168.36.180>
CSeq: 10 ACK
Content-Length: 0
```

This message mirrors the 200 OK to enable the far end to stop retransmission. The ACK does not carry an SDP.

20. The ingress Marshal server forwards the ACK to the egress Marshal server:

```
ACK sip:5120@192.168.36.180:5060 SIP/2.0
(The remaining SIP headers have not changed.)
```

21. The egress Marshal server forwards the ACK to the called party:

```
ACK sip:5120@192.168.6.20:5060 SIP/2.0
(The remaining SIP headers have not changed.)
```

A two-way RTP path has been established, and a conversation can take place between the two users.

22. The conversation ends and the called party hangs up first, thereby sending a BYE message to the Marshal server:

```
BYE sip:5120@192.168.36.180:5060 SIP/2.0
Via: SIP/2.0/UDP 192.168.6.20:5060
From: <sip:5120@192.168.36.180:5060>;tag=c29430002e0620-0
To: <sip:5121@192.168.6.21:5060>
Call-ID: c2943000-e0563-2a1ce-2e323931@192.168.6.21
User-Agent: Cisco IP Phone/ Rev. 1/ SIP enabled
CSeq: 11 BYE
Route: <sip:5120@192.168.36.180:5060;maddr=192.168.36.180>, <sip:5121@192.168.6.
21:5060;maddr=192.168.36.180>
Content-Length: 0
```

This tears down the RTP path from the called party to the calling party. The BYE has a Route header:

```
Route: <sip:5120@192.168.36.180:5060;maddr=192.168.36.180>, <sip:5121@192.168.6.
21:5060>
```

which goes through both Marshal servers. When the INVITE arrived, it had a Record-Route header:

```
Record-Route: <sip:5120@192.168.36.180:5060;maddr=192.168.36.180>,<sip:5120@192.
168.36.180:5060;maddr=192.168.36.180>
```

This header said, "here are the elements that we want in this call path." The called party reversed it so that it would go back through the servers in the opposite direction. The called party saves that state through the entire call, and every time it sends another message, such as a re-INVITE (another INVITE sent during the same call with a different SDP) or a BYE, it needs to insert this path in the Route header.

Here's how the Route header works: the first element to receive the message looks at the top address on the Route header, removes it, and forwards the message to that address. When there are no more Route headers, the Contact header is used to forward the message to the final destination. If there is no Contact header, the URI is used.

In the old versions of the standards, the User Agents did not have to list themselves in the Contact header, when the Route header was exhausted, the message went to its final destination. In the most recent drafts, the User Agents are required to list themselves in the Contact field.

That is a simple description; it can be more complicated in different implementations.

23. The egress Marshal server forwards the BYE to the ingress Marshal server:

```
BYE sip:5121@192.168.6.21:5060 SIP/2.0
Via: SIP/2.0/UDP 192.168.36.180:5060;branch=4
Via: SIP/2.0/UDP 192.168.36.180:5060;branch=2
Via: SIP/2.0/UDP 192.168.6.20:5060
From: <sip:5120@192.168.36.180:5060>;tag=c29430002e0620-0
To: <sip:5121@192.168.6.21:5060>
Call-ID: c2943000-e0563-2a1ce-2e323931@192.168.6.21
CSeq: 11 BYE
Route:<sip:5121@192.168.6.21:5060;maddr=192.168.36.180>
Content-Length: 0
```

Notice that the Route header has been reduced to one address.

24. The ingress Marshal server forwards the BYE to the calling party:

```
BYE sip:5121@192.168.6.21:5060 SIP/2.0
Via: SIP/2.0/UDP 192.168.36.180:5060;branch=4
Via: SIP/2.0/UDP 192.168.36.180:5060;branch=2
Via: SIP/2.0/UDP 192.168.6.20:5060
From: <sip:5120@192.168.36.180:5060>;tag=c29430002e0620-0
To: <sip:5121@192.168.6.21:5060>
Call-ID: c2943000-e0563-2a1ce-2e323931@192.168.6.21
CSeq: 11 BYE
Content-Length: 0
```

Notice that there is no longer a Route header.

25. The calling party receives the BYE, hangs up, and responds with a 200 OK, which tears down the other half of the RTP path:

```
SIP/2.0 200 OK
Via: SIP/2.0/UDP 192.168.36.180:5060;branch=4,SIP/2.0/UDP 192.168.36.180:
5060;branch=2,SIP/2.0/UDP 192.168.6.20:5060
From: <sip:5120@192.168.36.180:5060>;tag=c29430002e0620-0
To: <sip:5121@192.168.6.21:5060>
Call-ID: c2943000-e0563-2a1ce-2e323931@192.168.6.21
Server: Cisco IP Phone/ Rev. 1/ SIP enabled
CSeq: 11 BYE
Content-Length: 0
```

As before, the 200 mirrors the BYE to enable the far end to stop retransmission.

26. The ingress Marshal server forwards the 200 OK message to the egress Marshal server:

```
SIP/2.0 200 OK
(The remaining SIP headers have not changed.)
```

27. The egress Marshal server forwards the 200 OK message to the called party:

```
SIP/2.0 200 OK
(The remaining SIP headers have not changed.)
```

The call has been terminated.

In conversations about BYE messages and 200s, some people ask, "When a phone hangs up and sends a BYE,does it have to wait for the 200 before it can participate in another call?" A traditional telephone system point of view can be expressed as, The person has hung up, and, therefore, the phone must be ready for the next call right away and cannot wait for the 200.

How a hang-up appears to work with a traditional phone system has nothing to do with SIP. In order to maintain compliance with the standard, when you send a BYE, you need to wait until you receive the 200 before you can consider the call as being torn down. You can hide this delay from users by enabling them to set up a new call in parallel while the previous call completes its teardown. However, dumping the previous call, for the sake of convenience, is possible, but it is also irresponsible and noncompliant with the standard. The irresponsibility lies in how a dumped call causes unnecessary problems for other SIP-compliant entities involved in the call flow.

There are many more call flows like this one in the user documentation on *Vovida. org*. Hopefully, this walk through the messages will help you interpret the others found in the System Administration Guide.

# Forking

In more complex implementations, SIP Proxies can be programmed to fork INVITEs to more than one destination. This is useful for people who are mobile or who use multiple phones in their work. For example, someone may have a phone in a lab and another in an office, on a different floor or in a different building, and may prefer to have both phones ring simultaneously for every call. This is known as *parallel forking*. This permits her to move freely from one work environment to another without having to remember to call-forward one phone to the other every time she changes location.

In other situations, the user might prefer to use a hunting method, also known as *sequential forking*, for receiving calls on multiple phone sets. Here's how this works: All calls are first sent to a primary phone, maybe in an office. If that phone is not picked up within a configured number of rings, then the call is sent to a secondary phone, maybe a cellular phone. Eventually, if the call is not answered, it is usually sent to voice mail.

## A Forking Example

The example shown in Figures 7-8 through 7-11 illustrates a parallel search for a called party.

The first 10 messages shown in Figure 7-8 are the same as the first 10 that are discussed under Figure 7-4. The 11th and 12th message, the INVITE sent by Proxy B to both phone B and phone C, are where the forking occurs. These messages are sent simultaneously to both phones, although they may not be received simultaneously.

In Figure 7-9, both SIP phones, B and C, react to the received INVITE by ringing simultaneously.

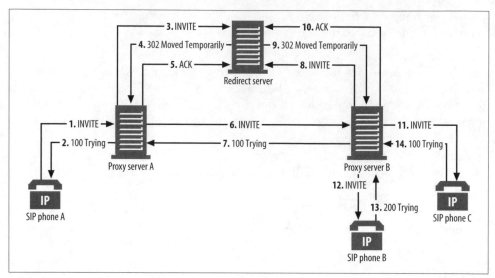

*Figure 7-8. Forking, sending the INVITE to two phones*

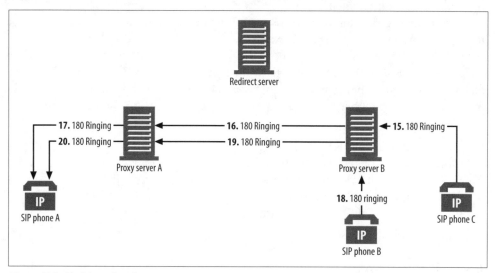

*Figure 7-9. Forking; the phones ring*

In Figure 7-10, phone C is answered and sends back a 200 OK message to Proxy B. That message is sent through the network back to phone A.

Proxy B, the one that forked the call, now has to do something intelligent; it has to stop phone B from ringing, without disrupting the connection to phone C. It sends a CANCEL message to phone B to stop the ringing, and the phone replies with a 200 OK. Unlike the 200 message that was sent by phone C, this message is intended only for Proxy B, and it is not forwarded back to phone A.

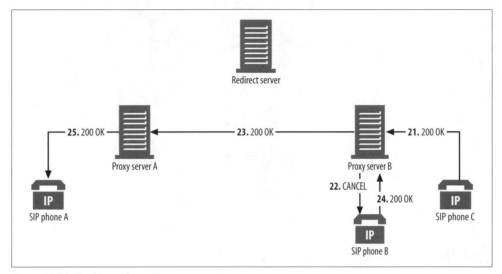

*Figure 7-10. Forking; phone C answers*

What if phone A hangs up while both phones B and C are ringing? Phone A would send a CANCEL to its original INVITE that would arrive at Proxy B. This Proxy has to keep track of all entities that it has forwarded the INVITE to, and the forking proxy has to fork the CANCEL to those same entities. In other words, the forking proxy has to keep a list of where it forwards messages, so it can send a CANCEL to the same addresses.

## Forking to a Forwarded Phone

This is a weird situation but entirely plausible. A proxy has forked an INVITE to two other proxies, but both proxies are forwarded to the same device (see Figure 7-11). The called party wants to answer the phone without forcing one of the forked INVITEs to voice mail. This looks simple on the surface, but it has accounted for more than its fair share of head scratching.

Since both INVITEs have been forwarded to the same phone, both will have the same Call-ID and Command Sequence number; however, the Vias will be different. The phone has to ignore the second INVITE, by comparing the fields and realizing that the same INVITE is coming from both proxies. To do this, it has to send a 100 to the first proxy and then send a 480 Temporarily Unavailable message.

# Weird Situations

Despite the best intentions of the developers, unusual situations occur that sometimes test the limits of the protocol and the patience of the customers.

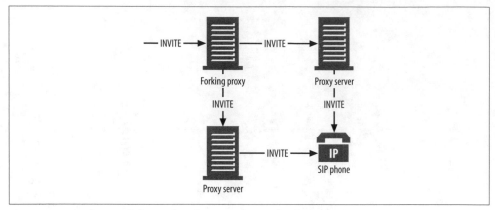

*Figure 7-11. Forking, call forwarding from two proxies to the same phone*

## Two Users Hang Up Simultaneously

Case #1, shown in Figure 7-12, is what usually happens.

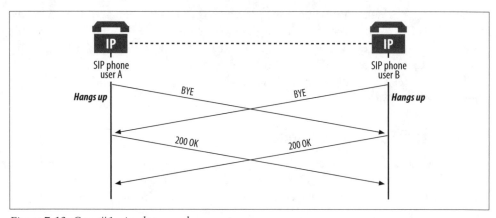

*Figure 7-12. Case #1, simultaneous hang-up*

## BYE and CANCEL: Similar but Not Interchangeable

In Case #2 (Figure 7-13), when phone A receives the BYE, it may not know about the call anymore. It could return a 404 or 200 message. It's weird because the 200 had to pass the BYE in the network. This seldom happens, but it is possible.

Hanging up a ringing call should cancel the INVITE and not send a BYE. In our testing, we have come across some buggy implementations where a phone would continue ringing long after the caller had sent a BYE. We would have to pick the phone up and hang it up to stop its ringing.

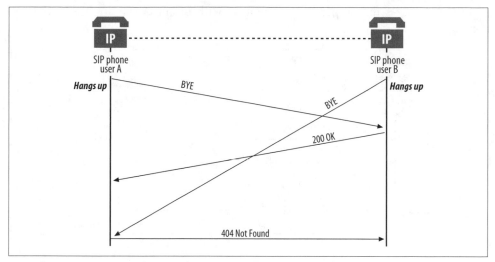

*Figure 7-13. Case #2, simultaneous hang-up*

This is not a problem with the protocol. This was a problem with how the SIP User Agent was written. The point we're making is that while the BYE and CANCEL messages might appear to have a similar purpose, they are not interchangeable. If the call is being canceled before it is established, CANCEL is the only message that should be sent.

Figure 7-14 illustrates what might happen if a BYE instead of a CANCEL is sent before the call is established. Whether phone B sends repeated 180 Ringing messages or another message would depend on how it was implemented.

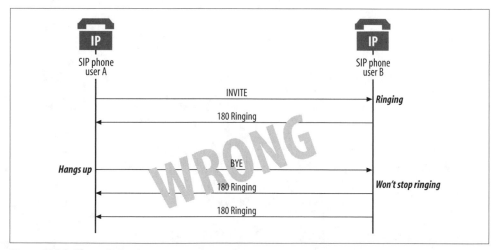

*Figure 7-14. Phone B won't stop ringing*

Figure 7-15 shows the proper implementation with phone A sending the CANCEL message.

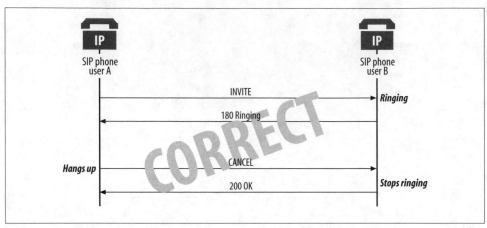

*Figure 7-15. Problem solved*

# CHAPTER 8

# Vovida SIP Stack

Now that you've had the user guide tour about how SIP works, let's go a little deeper into the architecture and talk about the classes, modules, and threads that we implemented into our SIP stack.

Throughout this chapter, when we refer to a certain protocol stack, we're referring to Vovida's implementation of the standard that governs that stack.

If you don't have a background in software development, you may find the remainder of this book's content difficult to digest. You could help yourself by brushing up on the Unified Modeling Language (*http://www.rational.com/uml/index.jsp*), object-oriented design, C++, and data structure and theory, which, unfortunately, require a few years of schooling to understand.

## Architecture

Figure 8-1 shows how the major components of the SIP stack work together and with other protocol stacks.

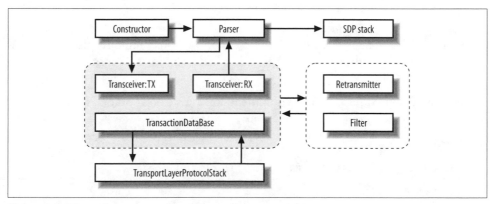

*Figure 8-1. SIP stack components*

These processes are defined as follows:

*Constructor*
Builds messages, either as brand-new requests or as reconstructions of received responses.

*Parser*
Encodes and decodes messages.

*SDP stack*
Encodes, decodes, and alters Session Description Protocol (SDP) bodies.

*SIP transceiver*
Includes five components, listed next. An intelligent structure, it matches requests with responses, and responses with acknowledgments.

> *Transceiver: TX*
> Transmits messages over a transport layer–independent interface.
>
> *Transceiver: RX*
> Receives messages over a transport layer–independent interface.
>
> *Retransmitter*
> Retransmits messages if the proper response or ACK has not been received. The transceiver retransmitter is constantly in the data flow but does not act unless there is a problem. The retransmitter queries the database for any messages that require responses.
>
> *Transaction database*
> Manages the currently active and recently completed SIP transactions. There are two databases: the first matches requests to responses; the second matches responses to acknowledgments.
>
> *Filter*
> Filters out duplicate messages from outside entities. The filter queries the transaction database for messages. If any messages appear in the database, the filter filters out duplicate transmissions.

*Transport layer protocol stack*
While User Datagram Protocol (UDP) is the most commonly used transport layer protocol within VOCAL, the system also supports Transmission Control Protocol (TCP).

Each of these definitions is expanded in the following sections of this chapter.

## Classes

There are five common types of classes used within the SIP stack:

- Messages
- Headers

---

- Body
- Network management (see Chapter 18)
- Transport

These class types are described next.

## Messages

*SipMsg* is the base class for all SIP messages. *SipCommand* and *StatusMsg* extend *SipMsg*, as shown in Figure 8-2.

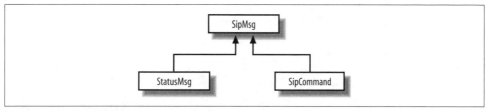

*Figure 8-2. Simplified view of the message classes*

These classes are defined as follows:

*SipMsg*
> The base class for all SIP messages

*StatusMsg*
> The class that represents all the 1xx, 2xx, 3xx, 4xx, 5xx, and 6xx SIP response messages

*SipCommand*
> The class that forms the base class for SIP requests, such as INVITE, ACK, and CANCEL

## Headers

The header classes are data containers, which keep track of data contained in the headers of the SIP messages. These containers are derived from the *SipHeader* class, as shown in Figure 8-3. Each known SIP header has a corresponding class derived from *SipHeader*.

This diagram is simplified; in the true diagram, there are 51 classes pointing to *SipHeader*. The positions of the classes that aggregate to *SipContentDataContainer* are not shown to scale. If you want to see the full diagram, you can download it from *Vovida.org* (*http://www.vovida.org*). A sampling of these classes is defined as follows:

*SipHeader*
> All header classes are derived from this class.

*SipContentDataContainer*
> Aggregates with *SipContentType*, *SipContentLength*, and *SipContentDisposition*.

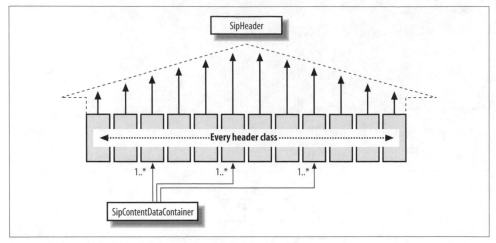

*Figure 8-3. Simplified view of the header classes*

There are also 51 header classes, including *SipTo*, *SipFrom*, *SipVia*, *SipContentType*, *SipContentLength*, and *SipContentDisposition*, which work with the corresponding header fields found within SIP messages. For example, *SipTo* works with the To header field. If a SIP message arrives with an unknown header, which means that no *SipHeader* class exists for that header, the *SipUnknownHeader* class parses the unknown header.

## Body

The body of the SIP message is essentially an attachment such as a list of SDP headers or other information that is carried by, but not supported by, SIP (see Figure 8-4).

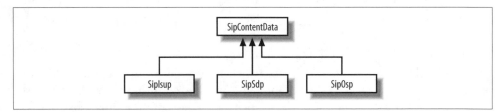

*Figure 8-4. Simplified view of the body classes*

These classes are defined as follows:

*SipContentData*

Decodes body content from incoming messages into objects and encodes body content from outgoing messages into message strings.

*SipIsup*

Used by SIP for Telephones (SIP-T) to transmit the Signaling System 7 (SS7) Integrated Services User Part (ISUP) within a SIP message across a network. *SipIsup* blindly passes proprietary PSTN information across the network.

*SipSdp*

Works with the SDP headers attached to SIP messages.

*SipOsp*

Used to transmit Open Settlement Protocol (OSP) tokens. See Chapter 19.

## Network management

Every process that works with SIP has its own SIP stack, and each SIP stack has a Simple Network Management Protocol (SNMP) agent that communicates with the Network Manager and Heartbeat server (see Figure 8-5). The *SipAgent* maintains a table of the Management Information Base (MIB) entries of the particular instance of the SIP stack. When the transceiver is instantiated, a *SipAgent* is also instantiated. The *SipAgent* class talks to the SNMP Manager via a lightweight UDP protocol called *AgentAPI*. The SNMP Manager queries the *SipAgent* regarding the statistics of the SIP stack. In theory, the SNMP Manager can also set certain attributes in the SIP stack; however, we have not implemented this functionality into our code. The class that handles all this activity is the *SipTransceiver*.

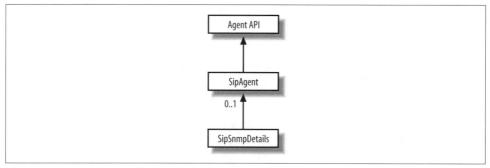

*Figure 8-5. Simplified view of the network management classes*

These classes are defined as follows:

*AgentAPI*

A lightweight UDP protocol used by the SIP agent to communicate with the SNMP Manager. For more information about SNMP, see Chapter 18.

*SipAgent*

*SipAgent* contains a table of counters. The *SipTransceiver* class updates these counters by calling the *updateCounter* method from the *SipSnmpDetails* class, which calls the *updateCounter* method from the *SipAgent* class.

*SIPSnmpDetails*

*SipSnmpDetails* maintains the up-to-date information about the stack. The constructor within the *SipAgent* passes a map and a reference, which updates the *SipAgent* about the stack. *SipSnmpDetails* has all of the related information about the stack. It is instantiated by the application, which is stored in the map.

## Transport

The stack supports sending and receiving messages over both UDP and TCP. The application indicates which transport protocol to use via the transport parameter in the SIP URI.

The stack also permits response and request message matching over UDP by maintaining the state of all requests and final responses within transactions for a maximum of 32 seconds. This matching also permits the stack to keep track of responses coming back for INVITE messages that require an ACK to complete the three-way handshake. If the timer expires before an expected response is received, and the message is being sent over UDP, the stack performs an automatic retransmission of the request. The maximum number of retransmissions is a configurable parameter with *exponential timing*.

Exponential timing works as follows: The first retransmission is sent after 500 milliseconds (ms). The delay between each additional retransmission is double the previous interval. For example, the second retransmission is sent after 1000 ms, the third after 2000 ms, and every subsequent retransmission after 4000 ms until the last retransmission, as defined in the constants within *vocal-1.3.0/sip/sipstack/ TransceiverSymbols.hxx*, has been sent.

In case of requests, such as INVITE, if the *SipTransceiver* does not receive a response after the maximum number of retransmissions has transpired, it sends a 408 Request Timeout message to the application. However, in the case of responses, the application must take care of any timeouts.

# Constructing and Deconstructing Messages

SIP is basically a communication protocol that provides the means for applications to send and receive messages. In the *Vovida.org* implementation, some of these messages are constructed from scratch, while others are constructed from the components of received messages.

When a SIP device receives an INVITE message, it uses the contents of the message to construct new messages. For example, to construct a 200 OK message, the application passes the received INVITE, thereby automatically filling in a large portion of required fields with data that is useful to the transaction. If some of these fields need to be changed, the application can make these changes before sending the message.

Building responses based on the request headers is a big time saver. ACKs are normally constructed out of SIP status responses to INVITE messages, and BYEs are built from the original INVITEs that were used to initiate the calls.

The following is a simplified look at how the SIP stack builds messages.

# Sending

Here is a well-commented example of a sending program:

```
#include "VovidaSipStack.hxx"
#include "InviteMsg.hxx"

int main( )
{
    // Here is the SIP stack for this message.  The argument is the
    // SIP port for this program to receive SIP messages on.

    Sptr<SipTransceiverFilter> sipStack = new SipTransceiverFilter(5060);

    // Construct a SIP URI for the message.
    Sptr<SipUrl> toUrl = new SipUrl(Data("sip:user@domain.com"));

    // Construct a simple SIP INVITE.
    InviteMsg msg(toUrl);

    // Here is an example of setting a field on a SIP message.
    SipSubject subject;
    subject.set("New call");
    msg.setSubject(subject);

    // Here is a more complex example.  In this example, we will set
    // the Display Name and From URL in the From line of this message.
    //Now,create and set the From URL.
    Sptr<SipUrl>fromUrl =new SipUrl(Data("sip:bob@example.com"));
    //Now,set the display name.
    from.setDisplayName("Bob Smith");
    from.setUrl(fromUrl );

    // Now, you must set the From in the message to the contents of from
    msg.setFrom(from);.

    // In this example, we will set the SDP body of this INVITE.
    // sdp will point to the SDP body in the message.  If we change
    // sdp, we change the SDP contents of the body.
    Sptr<SipSdp> sdp;
    sdp.dynamicCast( msg.getContentData(0) );

    // When you construct an INVITE by passing a URL, the INVITE
    // constructed should contain an SDP body already.  We have gotten
    // it in sdp--if it is not there, or it is not an SDP body, this
    // assert would fail, but this should never happen.
    assert (sdp != 0) ;

    // set the RTP port to 6000.
    sdp->setRtpPort(6500);

    // Now, we will send the INVITE message.
```

```
    sipStack->sendAsync(msg);
    sipStack =0; //dereferences the stack so smart pointers can garbage collect it
    return 0;
}
```

## Receiving

Here is a well-commented example of a receiving program:

```
#include "VovidaSipStack.hxx"
#include "InviteMsg.hxx"
#include <iostream>

int main()
{
    // Here is the SIP stack for this message.  The argument is the
    // SIP port for this program to send on.
    Sptr<SipTransceiverFilter> sipStack = new SipTransceiverFilter(5060);

    // Get the message--this comes in a queue of messages.
    Sptr<SipMsgQueue> queue = sipStack->receive();

    Sptr<SipMsg> msg = queue->back();

    // Convert the object to one of class SipInvite, so that we can
    // access the headers for it particularly.
    Sptr<InviteMsg> invite;
    invite.dynamicCast(msg);

    if(invite != 0)
    {
        // If invite != 0 (e.g., if invite is pointing to a message)
        // then it is a SIP INVITE.  Examine the From: field.
        SipFrom from = invite->getFrom();

        // Print the Display Name of the From:
        cout << "Display Name: " << from.getDisplayName() << endl;

        // Now, get the From: URL.
        Sptr<BaseUrl> url = from.getUrl();

        // Verify that this URL is a SIP URI.
        if(url->getType() == SIP_URL)
        {
            // Convert to a Sptr<SipUrl> so we can access SIP-specific fields.
            Sptr<SipUrl> sipUrl;
            sipUrl.dynamicCast(url);
            // This assertion should always be true.
            assert(sipUrl != 0);

            cout << "Domain name: " << sipUrl->getHost() << endl;
        }
```

```
// Now, retrieve the RTP port from the SDP body of the message.
Sptr<SipSdp> sdp;
sdp.dynamicCast(invite->getContentData(0));
if(sdp != 0)
{
    int port = sdp->getRtpPort( );
    cout << "port: " << port << endl;
}

// Now, construct a 200 OK response.  The first argument here
// is the message to respond to, the second is the response type.
StatusMsg status(*invite, 200);

// You can (and probably should) add an SDP body containing
// the information for your RTP streams here.  However, we
// will send the reply without filling out the status message.
sipStack->sendReply(status);
    }
  sipStack = 0;
  return 0;
}
```

# Examples

Here are examples of how some messages are created. In the examples, *vocal-1.3.0* is a placeholder for the directory named after the current version. For example, for VOCAL Version 1.3.0, this directory is named *vocal-1.3.0*.

### Example 1

This API is used for creating a SIP INVITE message; for more information, and a useful test program, see *vocal-1.3.0/sip/test/InviteMsgTest.cxx*.

```
InviteMsg( Sptr<BaseUrl> sendToUrl, int SIPlistenPort, int rtpPort)
```

*BaseUrl* contains the address of the intended recipient.

### Example 2

This API is used for creating a SIP 200 OK message; for more information, and a useful test program, see *vocal-1.3.0/sip/test/StatusMsgTest.cxx*.

```
StatusMsg(SipCommand& command, int statusCode)
```

SipCommand is any of the request objects.

### Example 3

This is used for creating a SIP ACK message; for more information, and a useful test program, see *vocal-1.3.0/sip/test/AckMsgTest.cxx*.

```
AckMsg(StatusMsg& statusMsg)
```

StatusMsg is the response object.

**Example 4**

This is used for creating a SIP BYE message; for more information, and a useful test program, see *vocal-1.3.0/sip/test/ByeMsgTest.cxx*.

```
ByeMsg(StatusMsg&)
```

or:

```
ByeMsg(AckMsg&)
```

ByeMsg must be created with either StatusMsg (when the calling party hangs up) or AckMsg (when the called party hangs up), because these messages include the most recent contact information from the far end.

# Parsing

SIP messages arrive from the network as character strings, which the parser transforms into objects. The Vovida SIP stack uses a method called *lazy parsing*, which, rather than constructing all header objects at the time the parser is invoked, constructs them the first time another structure accesses them. This is explained later. Each header object takes care of parsing the corresponding header line. The *SipContentDataContainer* class parses the MIME data, which normally consists of an SDP message.

SIP headers are kept in *SipMsg* objects via the container *SipRawHeaderContainer*, which is a vector of a smart pointer (Sptr, explained later). *SipRawHeader* is a container that holds two representations of a SIP header—the raw, textual representation, as well as a pointer to a *SipHeader* object. The *SipRawHeader* class can translate between the raw and parsed forms of the headers, compare the headers, and perform other housekeeping tasks.

Figure 8-6 illustrates the method for handling To headers.

This method is known as lazy parsing, because instead of parsing the entire message before processing the data, the data for any given header is parsed from the message only after it is requested by another process. This change has made a big difference in the performance of the stack.

Lazy parsing uses smart pointers, and for those of you not familiar with these clever little fellows, let's take a brief time-out to discuss them.

## Smart pointers

While *pointers* are simply addresses in memory of alternate locations, *smart pointers* include a reference counter to ensure that the memory is freed once the pointer is no longer being used. This reference counter is a solution to the common problem with pointers: when pointers are copied, there is no way to keep track of the number of pointers that are pointing to the same address, which creates a memory leak: a persistent allocation of unused memory.

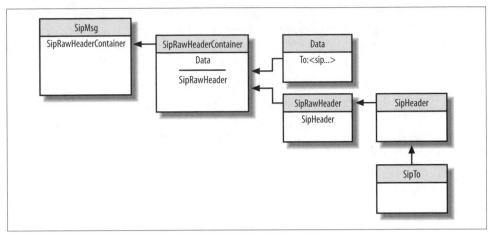

*Figure 8-6. The SipRawHeaderContainer class is derived from the SipMsg class*

Our SIP stack can dynamically resize itself, and the possibility of large amounts of unused memory allocations would seem, well, pointless. With smart pointers, each time the pointer is copied, a reference counter increments by 1. As processes stop using the pointer, the reference counter decrements until it reaches 0. As it reaches 0, the last program to discontinue using the pointer frees the memory allocation.

Reference-counted smart pointers are vulnerable to circular references. If two objects contain smart pointers to each other, and there are no outside smart pointers to either object, the reference counts in these two objects will never decrement to 0. Therefore, the memory in both objects cannot be freed up without outside intervention, such as killing the process and breaking the chain of reference by setting a smart pointer to null. Of course, it is far better to avoid using circular data structures, but if you must use one, be sure to keep the memory freed up by setting the smart pointer to NULL, when you are finished using it, to break the circular reference.

## Parsing Examples

The following three examples show how messages are parsed within the SIP stack.

### Example 1

The SIP stack uses the following process to change the Request URI of an INVITE message that has already been constructed:

```
InviteMsg inviteMsg;
SipRequestLine reqline = inviteMsg.getRequestLine();
Sptr<BaseUrl> url = reqline.getUrl();
```

Unfortunately, the *BaseUrl* class has no method to modify the hostname (as not all URLs have a hostname component). We must recast this *BaseUrl* object into a *SipUrl* object to modify its contents. This *dynamicCast* will attempt to do the conversion. If the URL in the *url* object is not a SIP URI, *sipUrl* will be set to the null pointer. If it is a SIP URI, however, it will be set to a non-null value, so we can alter it as a SIP URI and set it:

```
Sptr<SipUrl> sipUrl;
sipUrl.dynamicCast(url);
if(sipUrl != 0)
{
    // Here, we change the hostname in the URL and set the request line
    // to the new URL.

    sipUrl->setHost(Data("newhost.com"));
    reqline.setUrl(sipUrl);
    inviteMsg.setRequestLine(reqline);
}
```

### Example 2

The stack uses the following process to decode a message:

```
Data message="INVITE sip:...";
        // The preceding line would contain the complete text of a SIP INVITE
        // message.
InviteMsg inviteMsg( message );
```

This calls *SipCommand::decode( Data )*.

### Example 3

If you want to encode the message:

```
InviteMsg inviteMsg;
Data inviteStr =  inviteMsg.encode( );
```

InviteMsg::encode() calls SipMsg::encode( ), which calls the individual Headers:: encode() and MimeEncode( ).

# Transporting

The VOCAL SIP transceiver manages transactions and implements a transport-independent interface for sending and receiving SIP messages by the application layer. The message send operation used by the state machine is asynchronous: transmissions can occur while the receive operation is blocked. In the stateful mode, the transceiver maintains state for transactions as long as there is at least one undeleted message in the transaction database. The delay for deleting messages can be specified as a compile-time parameter. As part of its transaction management functionality, the transceiver takes care of filtering the duplicate messages and retransmitting messages that have timed out before receiving responses.

## State Machines

At any given time, a state machine is in a particular state. When an event occurs, it causes a transition from one state to another. These events are implemented in the form of operators, which perform some action and move the state machine to a new state based upon the event received and the present state.

If a stack or an application is described as *stateful*, it means that past events influence current actions. A state machine is one specific way of implementing or thinking about stateful behavior. The Vovida User Agent and SIP-based servers are state machines. See Chapters 9 and 10 for more information.

## High-Level Design

We implemented the functionality of the SIP stack by creating the following five major components at the transceiver level:

*Sent request database*
: Manages transactions for the client side of the application. In the system-level message flow, this takes care of the downstream interface of the application.

*Sent response database*
: Manages transactions for the server side of the application. In the system-level message flow, this takes care of the upstream interface.

*UDP transport layer*
: Provides a SIP-specific interface with the UDP transport stack. It takes care of retransmitting UDP messages and decoding received messages.

*TCP transport layer*
: Provides a SIP-specific interface with the TCP stack.

*Transaction garbage collector*
: Takes care of deleting transactions as they expire.

## High-Level Message Flow

The high-level message flow is depicted in Figure 8-7: the calling party, at the top, sends an INVITE message to the called party, at the bottom. The called party responds with a 180 Ringing message.

As we have said before, the User Agent (UA) has two parts: the User Agent server (UAS) and the User Agent Client (UAC). The sent request database corresponds to the UAC, and all the outgoing requests and corresponding responses go through it. The sent response database corresponds to the UAS, and all of the incoming requests (and corresponding responses) go through this component. While the SIP UAs and other endpoints use only one of either the UAC or UAS for any given session, the Proxies use both components simultaneously. See Chapter 11 for more information about Proxies.

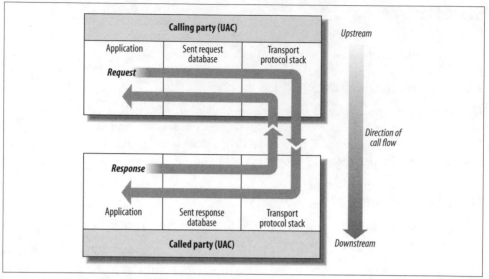

Figure 8-7. High-level message flow through the SIP stack

## Application Interfaces

Figures 8-8 and 8-9 show how SIP messages are sent and received at the transceiver level through the following interface methods:

*sendAsync(sipCommand & sipMessage);*
 Sends SIP request *sipMessage*

*sendReply(statusMsg & sipMessage);*
 Sends SIP response *sipMessage* asynchronously

*sptr(sipMsgQueue) receive(int timeOutMs=-1 );*
 Blocks a specified time for a user (passed as *timeOutMs*, or -1, to block until a message is received) and returns with the next available message from either the UDP or TCP transport stacks

### Illustrations

Figure 8-8 shows the send logic flow through the SIP stack.

Let's look at the interactions shown in Figure 8-8. Starting at the top, the application is running a thread that sends a SIP message by calling either the *SendAsync* or *SendReply* method from the application interface. The two arrows at the top pointing toward the Transaction Management thread represent these methods.

Let's look at the *SendAsync* arrow, which is the path taken by request messages such as INVITE. When the message arrives at the Transaction Management thread, a record of its arrival is entered into the SentRequest database. If the message is tagged

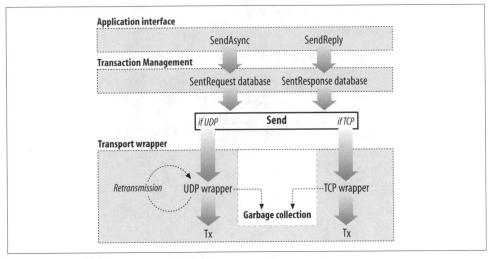

*Figure 8-8. Send logic*

for TCP transmission, it is sent to the TCP wrapper, which sends it out over the network. TCP transmissions are reliable, so there is no provision for retransmissions

If the message is tagged for UDP transmission, then it is sent to the UDP wrapper and a record of this message is kept in the database in case retransmission is required. If this message needs to be retransmitted, the Transaction Management thread knows about it and can queue it up for the Transport Layer thread.

If the message is a reply, such as 200 OK, it follows the *SendReply* path illustrated by the arrow on the right side of the Application interface.

After a period of time, usually 32 seconds, there is no need to keep records of sent messages and the Garbage Collection thread purges the expired message records from the databases.

Figure 8-9 shows the receive logic through the SIP stack.

Let's look at the interactions shown in Figure 8-9. Starting from the bottom of the diagram, a message, arriving from the network over UDP or TCP, is decoded by the transport layer and sent to the received FIFO. At this point, the Transaction Management thread picks up this message and decides what to do with it. If a received message is a status message, this signals the Transaction Management thread to stop retransmitting a request message that is logged in the database. Other messages might be new requests, such as INVITE, that require a reply to be generated and returned. These are just samples of some of the work that the Vovida SIP stack performs at the transaction management level. In our experience, we have not seen this type of transaction management implemented in many other SIP stacks. These other stacks usually push this responsibility up to the application level.

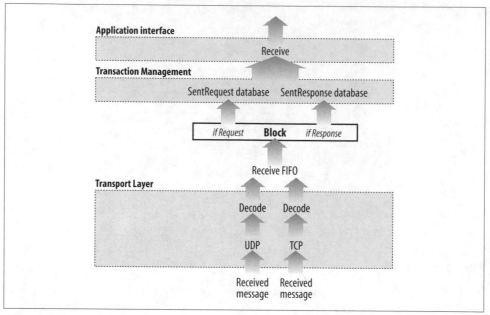

*Figure 8-9. Receive logic*

After traveling through the Transaction Management thread, the message is sent to another FIFO where it waits for the Application interface to retrieve it.

## FIFOs

First In First Out (FIFO) is a computer programming term used to describe a type of queue, which is a place where programs can put objects such as calls or data structures for another process to work on them. A queue of print jobs sent to a printer on the local area network is a good example of a commonly used FIFO queue. When another program needs to retrieve an object from a FIFO queue, it will retrieve the one that was added first. In other words, what went in first will come out first. The program cannot jump to the middle of the queue or retrieve any object other than the one that was inserted first. If there is only one object in the queue and it is removed, the queue becomes empty.

## FIFOs and VoIP

As you have seen from the diagrams in this chapter, there are a number of threads within the stacks and the applications that need to communicate with one another. A really useful method for providing what's known as *threadsafe* processing, which is explained next, has been to provide FIFOs throughout the system. This has helped us tremendously with bug fixes because two threads trying to access the same data simultaneously often led to trouble. By using a single, well-debugged interprocess communication mechanism, we've simplified the job of sharing data.

---

## Threadsafe

A given thread can do only one thing at one time; however, when there are many active threads working with the same types of data, sooner or later a conflict will arise with two or more threads trying to manipulate the same data at the same time. We have chosen two approaches to minimize our exposure to data corruption due to multiple threads manipulating data. First, when possible, a single thread manipulates the data and then passes this data to another thread via a FIFO. Once the thread has passed it off, it no longer manipulates the data. Thus, only one thread can read or write data at a time. If the first approach is insufficient, we use more traditional locking to ensure thread safety.

Thread safety exists where the interactions of multiple threads on a shared resource, such as data structures, do not result in memory corruption or incorrect results.

## Event loops, FIFOs, and timing

One of the most common ways that we use FIFOs is in event loops. This is where a thread waits for the next item in a FIFO. Upon receiving this item, which can be an object (an *event*), the thread takes an action. After completing this action, it again waits for the next item in the FIFO.

This paradigm lends itself to using the event FIFO as a timing mechanism as well as a way to supply work for the thread to do. By adding a time delay to a message to be placed in the FIFO, the thread is "woken up" after a specified length of time. If some other, undelayed message or message with a shorter time delay arrives before the delayed message, they will be received and processed first, but after the time delay has been reached, the original message will arrive.

# Working with TCP

In most of our examples so far, we have been using the UDP for the transport layer. While there are times when it makes sense to use the TCP instead of UDP, there are other times when using TCP makes no sense at all.

## Should you use TCP rather than UDP?

Developers prefer to use UDP for several reasons. One is faster failure detection, and another is a belief that UDP is easier to implement. Let's look at those ideas.

By its nature, a TCP connection takes much longer to detect a failure from the other end than a UDP connection does. TCP can take minutes to time out just from the design of its retransmission algorithm—it was designed to work on networks with packet loss rates that would make them useless for VoIP. On stable network connections, the delay imposed by TCP is considered impractical for most telephony applications.

On the surface, it appears that implementing a UDP stack is much easier than implementing a TCP stack. TCP stacks contain many more lines of code than typical UDP stacks. More code requires more memory. Despite these problems, we have seen many UAs that originally supported only UDP but had added TCP support very soon after their initial release.

This begs the question, "Why would anyone want to use TCP?"

One point in TCP's favor is its capacity for bulk transfers. Imagine two proxies owned by the same service provider that are acting as a bridge between two large networks (maybe one is in North America and the other is in Europe), with hundreds of calls passing between them every second. If these proxies were to use separate UDP messages for each call, they would create an exorbitant amount of overhead and slow down the network. Theoretically, it seems to be much better to collect individual calls and send them as a large, buffered transaction over a TCP stream. This idea remains a theory and has not been tested thoroughly; however, it seems, for the time being, to be entirely plausible. It may be that Stream Control Transmission Protocol (SCTP: RFC 3286) is better suited for this than TCP.

Consider the experience we have gained from working with the Domain Name System (DNS). Originally, bulk transfers in DNS started out using UDP, but eventually people realized that TCP offered a better scheme for providing the same service. This experience has led some people to stop and think about what they're trying to accomplish when they select either TCP or UDP as their transport layer stack. Most VoIP systems are using UDP for the majority of their transmissions. System architects are thinking about where TCP could provide better service, and most of that imagination is being directed toward the problems associated with sending big bulk transfers between large networks.

### How about firewalls?

The statement, "TCP works better through firewalls," often comes up in discussions about transport and security. This is a weak argument, because TCP does not fix the problems associated with transmitting SIP messages through a firewall. The real issue is sending the RTP audio traffic through the firewall, and RTP does not work with TCP. Solving the problem of transmitting RTP packets through firewalls will lead to solving the UDP/SIP problem as well. In practice, firewalls are not a good argument for determining whether to use UDP or TCP with SIP.

### When should TCP be used?

TCP is really useful when it is used with Transport Layer Security (TLS, RFC 2246) for enhanced security. TLS works properly only over a streaming protocol, such as TCP. (TLS is a big topic that is outside the scope of this book.) Also, SIP messages containing Secure Multi-Purpose Internet Mail Extensions (S/MIME) bodies are likely to be too large to easily transfer over UDP.

## When should TCP not be used?

Using TCP as a method for sending many calls over the same connection works well. However, if you use a new TCP connection for every single call or, worse yet, one TCP connection for every single transaction (and we have seen implementations out there that did just this), TCP's performance is absolutely horrendous because the number of extra IP "messages" required just to set up the TCP connection outstrips any advantage that TCP may have presented. We don't recommend using TCP for individual connections because of the negative impact it will impose on your system's performance. We do recommend it as a stepping stone to securing SIP, which will likely require TLS and S/MIME.

# Transaction Databases

The SIP messages in the sent response database and sent request database are organized based on a hierarchy derived from the nature of the SIP transactions. These levels of hierarchy include:

*Level I*
> Consists of the To, From, and Call-ID headings, which identify a *dialog* (formerly known as a *call leg*). This level is the root node of the entire transaction during a call.

*Level II*
> Consists of the command sequence number (CSeq#) and the top Via branch tag. This gives a root node for every sequence, revision, or forking situation that occurs during the call.

*Level III*
> Based on the command sequence METHOD parameter and required to distinguish between responses to INVITEs and CANCELs.

Apart from this hierarchy, the two databases also maintain some state-specific information at the call-leg level (i.e., with the LEVEL-I node):

*SentRespDB*
> Remembers the To tag of the outgoing response, to drop unmatched incoming requests (there is no need to take care of the From tag here, because responses are copied over from requests).

*SentReqDB*
> Remembers the From tag of the outgoing request to match the incoming responses and discards the To tag of the incoming response.

The main functionality of these databases is to take care of filtering duplicate messages (from UDP transport) and to implement a transaction history so that the applications can query and access relevant information about a call's transaction.

# Data Structures

The SIP stack handles data transactions through a series of key structures, threads, and interactions. This section discusses these structures in detail.

## Key structures

We have implemented some special-purpose data structures to achieve high concurrency and to support the overall structure of the SIP stack. The key data structures include:

*Transaction ID*
> Implemented as a compound data type, having one field each for the three levels of hierarchy.

*Transaction hash table*
> A special-purpose implementation of a hash table that is used by the transaction databases. We chose to use an application-specific implementation, rather than a general implementation, to improve performance.

*Transaction-level nodes*
> Compound data types that are implemented for every level in the hierarchical organization. These structures store the information specific to the corresponding level, as well as the direct references to parent node and container objects.

*Transaction list*
> A template implementation of a doubly linked list. This implementation is required because the list items need to store direct reference to their container objects in the list.

## Threads and interaction

When an application instantiates a VOCAL SIP transceiver, it spawns the following five threads:

*UDP Receiver*
> Responsible for listening to the receiving socket interface and, whenever a message arrives, grabbing it, decoding it, and pushing the message into the FIFO.

*UDP Sender*
> Responsible for taking the next message that needs to be transmitted, sending it out on the network, and updating the appropriate databases so that retransmission can happen. If it gets a failure, it receives an Internet Control Message Protocol (ICMP) message saying, "You can't transmit something to there." This is shown in Figure 8-10 as the "Failure" line that is drawn between the UDP sender and the receiver FIFO. If we receive a transmission error, we generate an error message to the application layer, which, in effect, says, "No, you can't do this." This error is passed back to the application as a status message.

*TCP Receiver*

    Works the same way as the UDP Receiver, except that you can't have a failure because it is assumed that the connection worked.

*TCP Sender*

    Works the same way as the UDP Sender.

*Garbage Collector*

    Removes messages that have been in the database for longer than the configured expiry time, which is normally set to 32 seconds.

The interaction between these threads is shown in Figure 8-10.

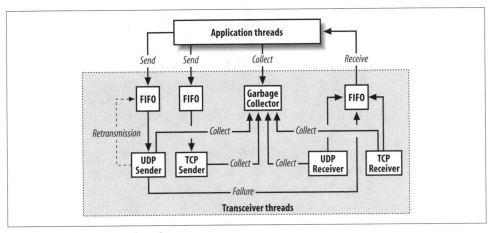

*Figure 8-10. Transceiver threads*

## Internet Control Message Protocol (ICMP) messages

ICMP (RFC 792) is a low-level communication protocol used mostly by routers and gateways. A well-known example of an ICMP message is *ping*. The stack handles ICMP errors, through the *SipUdp_impl* class, by returning the appropriate status codes to the application such as 403 Connection Refused, 404 Host Down, and 408 Host Unreachable.

When you send a UDP message, using a bad address, some router along the way will respond with an ICMP transmission error, such as "Host not reachable," which is the only meaningful ICMP error that we receive. In order to signal that error up to the application, the SIP standard says that a host not reachable error shall be treated the same as a Proxy returning "Destination not reachable." Therefore, we turn ICMP errors into "Destination not reachable" SIP responses and propagate them up to the application. This approach makes it really easy for people coding applications because they don't have to deal with several sets of error messages. Receiving an ICMP error is the same as receiving a 400-class error.

# Advanced Interactions

Besides the basic interactions expected from a SIP stack are several interactions that keep engineers up all night thinking about how to improve their implementations. Here is a sampling.

## Network Address Translation traversal

If Network Address Translation (NAT) is set, the SIP stack puts the address from which the packet was actually received into the RECEIVED parameter of the Via header. This happens if and only if the actual Via address is different from the received packet's address. Also, while sending responses, the packet will be sent to the RECEIVED parameter in the Via field, if the RECEIVED parameter is there. Otherwise, the packet will be sent to the Via address.

Here's an example: A device sends a packet through a NAT firewall to a proxy server. When the proxy receives the packet, the SIP Via header will say it is from the device—because this is how the device created the message. However, the routing information provided by the UDP stack will say that the message actually came from the firewall.

If the NAT flag in the proxy is turned on, the following happens: The proxy sets the SIP Via message to say that it is from the device but adds a tag that says it was received from the firewall. Later, when the reply (200) for this message arrives, and the proxy needs to decide where to send it, the proxy looks at the Via, which lists the device with a note that the actual sender was the firewall. The proxy sends the reply to the firewall, which, hopefully, forwards it on to the device without dropping it.

## DNS SRV support

DNS SRV is a special type of DNS lookup that can request hostnames for services, such as SIP. For example, a DNS SRV lookup for _sip._udp.vovida.org would receive a list of proxies that provide the SIP service in this domain. This method includes priorities and weights that allow for load balancing and backup servers. At the time of this writing, the Redirect server does not support DNS SRV record queries, but there is code around to support this. For more information about DNS SRV, see RFC 2782.

The stack does a DNS SRV or a DNS lookup in cases in which the *requestline* contains a hostname that cannot be resolved. The DNS lookup returns the available list of servers and their port numbers. The stack then does a *getNextRecord*, which returns the first highest-priority entry—the immediate destination of the message.

DNS SRV support assumes that retransmissions should be turned on at all times in the application. If there is no response to the request, even though it was sent to supposedly available servers, the appropriate status code of 403, 404, or 408 is returned to the application. In case of 403, the stack does not do a DNS lookup again, since this is a server-side request-rejection response; it continues with the next request. In all other cases, the process is repeated a configurable number of times.

# Compiling and Running the Stack

From *Vovida.org*, you can download the stacks as individual files, or you can download VOCAL, which contains the stacks, servers, User Agents, translators, and testing tools. If you prefer, we have a Concurrent Versioning server (CVS) available. Log on to our site and go to the Faq-o-matic for more information about CVS. The latest version of VOCAL is intended to use the latest SIP stack. *Vovida.org*, however, also offers older versions of VOCAL and their stacks for those who need them to fulfill special purposes.

## Copying and Untarring the Files

In the following instructions, `1.3.0` refers to the current version number:

1. From V*ovida.org*, copy this file to your hard drive:

   ```
   sip-1.3.0.tar.gz
   ```

2. Untar the SIP stack file by typing the following:

   ```
   tar -xvzf sip-1.3.0.tar.gz
   ```

## Compiling

To compile both the SIP stack and user agent from */sip-1.3.0/sip/*, type:

```
make sip
```

To compile the SIP stack from */sip-1.3.0/sip/sipstack/*, type:

```
make sip
```

To compile the SIP (UA) from */sip-1.3.0/sip/ua/*, type:

```
make ua
```

## Running Tests

To make sure that the SIP stack and UA compile properly on your host, run the following:

```
run ua
```

You can also run the tests and check the outputs for errors:

```
make test
```

or:

```
make verify
```

# Bugs/Limitations

We have added Bugzilla, the Mozilla project's bug-tracking tool, to *Vovida.org*. We encourage all members of the community who have checked out a copy of either VOCAL or one of the protocol stacks to enter bugs, or anything that appears to be a bug, into Bugzilla. As we continue developing our code, we often find better ways to do things and discover the root problems behind some of the long-standing bugs.

Sometimes the root cause of the bug is an open issue in the standard that is up for discussion by the SIP Working Group at the IETF. While it would be impossible to predict which of our current bugs will or will not be fixed between the time of this writing and the time you read this passage, we would like to highlight one bug and one limitation as examples of some of the issues that drive our development process. We encourage you to look through our Bugzilla archives to gain a better understanding of the state of our code. Maybe you can fix some of our bugs!

## The Difficulty with Reducing the Size of the Stack

Occasionally, we see this question posted to the Vovida SIP mailing list: "Why is your stack so big?" Several people in the community are interested in running our SIP stack in either embedded or space-restrictive environments and ask us for tips about reducing the stack's footprint. While we really haven't done a lot of work with respect to embedding our stack, we can comment about the size of the stack in general.

The reason the SIP stack is so big is very simple: we use a large number of C++ templates, and every time we use one, it is instantiated into every single file, resulting in a countless number of copies of the same thing in many different locations. All that we need to do to reduce the size of the SIP stack is to make sure that the compiler is set up to instantiate the templates only once in any given instance of a template. That's hard to do with older versions of the *gcc* compiler, which we normally use for our Linux and Solaris ports. With other compilers, it's much easier. If anyone were to port SIP to another platform that simplified single template instantiation, it would make a huge difference to the size of the stack.

## Performance Issues

Some people on our mailing lists wonder why VOCAL can't process more calls per second. One writer asked, "How come some applications can process hundreds of megabytes per second of video data and do pretty complicated processing on it, yet VOCAL can't receive more than a few kilobytes per second of SIP messages before bringing the network down to its knees? This seems counterintuitive."

Of all the issues involved with processing SIP messages, none of them actually has to do with a network bottleneck or packet loss. It is the applications and how they process the SIP messages that slow down the system.

## Parsing

As we have shown in this chapter and Chapter 7, SIP is a text-based protocol that provides a great deal of flexibility with respect to the construction of the messages. This flexibility comes at a price: the overhead associated with parsing each message. As we expect a variety of permutations and combinations in the message string, we must perform at least some minimal parsing to determine the message's ID and its destination.

When an endpoint initiates a call, it knows an address for its intended calling party but does not know any routing information. If the routing information is available locally, this helps speed up the performance. If the routing must be queried from a remote location server, the internetwork connection time may slow the call processing time.

We chose to make a distributed stack with separate parsing for each SIP header and data type, such as a SIP URI, distributed into the objects that represent these headers and data types. While a monolithic parser may provide better performance, it would also be harder to modify than our current model.

We have addressed some of the parsing issues with our lazy parsing method, described earlier in this chapter. However, there is not much more performance that can be squeezed out of the parser without making fundamental changes to the design of VOCAL.

## Application reliability

We have already discussed the issues surrounding the transport layer protocols TCP and UDP. Our stack includes UDP reliability and transaction matching. Every database lookup, retrieval, and retransmission adds to the processing overhead.

You could elect to remove the reliability from the application by running the messaging over TCP or making the application stateless. As discussed earlier, TCP does not work well for single VoIP calls because it takes too long to detect failures. Alternately, you could choose to remove the UDP transaction and reliability from our stack in order to improve performance. Making the application stateless may improve your performance; however, you must make sure that your application does not need, nor will ever need, state, or else your performance will be dragged down.

## Multithreaded applications

VOCAL's applications are multithreaded, and when many threads are trying to process the same variables, there is contention for ownership, which is also known as *mutex*. The threads must be managed with FIFOs and other methods that ensure the smooth transition of data ownership between the threads.

Also, there is a historical issue that could be improved. When each thread sends an object to a FIFO, it actually creates a copy of the object and sends the copy, then retains the original until the current process has completed. The applications could be rewritten to send the objects as is without making copies, but that work is considered a low priority for the time being.

While this is the end of our discussion about the Vovida SIP stack, it's just the beginning of our discussion about SIP. The following chapters take separate aspects of the protocol specification and provide some further insight into how these aspects can be implemented.

# CHAPTER 9

# Base Code

Most VOCAL components execute state machines, which share a common base code: a common framework that we wrote to help ensure some design consistency between several different server types including the Marshal, Redirect, and CPL Feature servers. This chapter looks at how we implemented the base state machine and the different types of classes found within the base code. As this code is widely used within VOCAL, it is important for you to understand how it has been written if you intend to modify any of the other processes.

## State Machine

The general model of this state machine provides a class for each state. The constructor for each of these classes lists a series of operators that are instantiated at the same time the state class is instantiated. Each of these operators has been written as a separate class, and the events are defined in a general event class. For example, all the events that drive the state machine for the SIP User Agent are defined within *vocal-1.3.0/ sip/ua/DeviceEvent.hxx*.

When a thread sends an event to a state machine, it passes the event to the current state and then to each operator that is listed within the constructor for that state. Each operator contains code that enables it to work with some events and to ignore all others. If the event is one that it is designed to handle, the operator executes and returns a definition of the next state for the state machine to transform itself into. This makes the state machine move to its next class: the next state in the state machine.

To get a better understanding of this, imagine the state machine shown in Figure 9-1.

This machine has two states, A and B, plus three events that cause it to change states—on, off, and reset. When an event happens, it is passed to every available operator. Most operators have no interest in this event and ignore it. The one operator that has been written to anticipate this event acts on it and returns the next state.

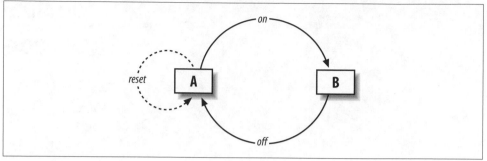

*Figure 9-1. State machine example*

You would implement this machine by deriving two state classes called *StateA* and *StateB*; three operator classes called *OpOn*, *OpOff*, and *OpReset*; and a general class that defines all the events. The constructor for *StateA* instantiates *OpOn* and *OpReset*, while *StateB* instantiates *OpOff*.

The constructor for *StateA* could look like the following code example:

```
StateA::StateA( )
{
    addOperator(new OpOn);
    addOperator(new OpReset);
}
```

The constructor for *StateB* could look like this:

```
StateB::StateB( )
{
    addOperator(new OpOff);
}
```

If the state machine is in *StateA* and an on event happens, it is passed to both of the operators listed in *StateA*'s constructor, *OpOn* and *OpReset*. The reset operator has no interest in the on event and ignores it. The on operator, meanwhile, has been waiting for the on event, and now that this event has happened, it takes the state machine to the next state, *StateB*.

In *StateB*, if another on event happens, there is no on operator listed in the constructor to act on it; therefore, this event is ignored. When the off event happens, the listed operator, *OpOff*, returns the next state, *StateA*. This is not the only way to implement a state machine, but it was the best way for us to make the code simple to understand and change.

## Class Structure

Implementing an application on top of the base code requires writing a collection of states, one for every state in the state machine, which inherit from the base state class (*vocal-1.3.0/proxies/base/state.hxx*). This section discusses the operator, state feature, and builder classes.

## Operator Class

Most of the operators share a common structure in which all of the work is completed by a function called *process*, which processes events. As events are passed into this function, the first task performed is a "relevance check." If the event is not relevant to the operator, the *process* function returns a zero. If the event is relevant, the process function goes to work on the event. If the operator is not concerned with transforming the state machine into a new state, it returns zero after processing the event. If it is concerned with transforming the machine, it returns an enumerate-type value that indicates the next state or simply returns the type of the next state.

## State Class

The state classes also share a common structure. These classes define a name function, their current name, which enables printing out a representation of the state name for debugging purposes. In the constructor, the names of the operators that must be executed in this state must be added by writing an add operator call in which the state creates a new instance of the operator, by creating a new instance of the operator class.

Operators can also be added to run at any time that the state machine enters a given state, or this can also be set up to run when the machine leaves one state and enters another. In the SIP User Agent, this is achieved by using the add-entry operator and add-exit operator methods. We've found this to be a useful shortcut to simplify state machines. The entry and exit operators are added in the constructor. One of the states, the Trying state (*vocal-1.3.0/sip/ua/StateTrying.cxx*), shown here, lists an entry operator followed by several event-handling operators and three exit operators:

```
StateTrying::StateTrying()
{
    addEntryOperator(new OpStartTimer);

    addOperator(new OpOnHook);
    addOperator(new OpStartRingbackTone);
    addOperator(new OpReDirect);
    addOperator(new OpFarEndAnswered);
    addOperator(new OpFarEndBusy);
    addOperator(new OpFarEndError);
    addOperator(new OpTimeout);

    addExitOperator(new OpStopRingbackTone);
    addExitOperator(new OpTerminateTrying);
    addExitOperator(new OpCancelTimer);

}
```

As a simple example, when the SIP UA state machine enters the Trying state, the entry operator starts a timer. When it leaves this state, one of the exit operators, *OpCancelTimer*, cancels the timer. Many of the states within the SIP UA do not require entry or exit operators because there are no active processes that require starting or stopping during the transition into or out of those states.

## Feature Class

When we were white-boarding VOCAL, it seemed logical that developers would want multiple state machines that would see the same messages but perform completely different activities. Each one of these state machines was going to be called a *feature*. So far, we have not implemented an application that has more than one feature, yet there is a *feature* class in the code ready for the day, if it ever comes, when we do implement applications with multiple features. Also, the mechanism for moving from one feature to another has not been defined. All of our applications have one feature: the state machine that they run.

This concept is not to be confused with a Feature Proxy server as described in Chapter 13.

## Builder Class

The classic application has a builder that builds all of its other components: a handful of states and a number of operators linked to each state. Occasionally, there's an application like the Proxy that has only one state, and there are states that have only one operator.

While providing a means for executing the state machine is the main thing that the base code does, it also provides the following functions:

- It provides code for parsing the command line.
- It provides the code that will form a SIP stack that can be started and used.
- It can start heartbeating with other systems within the server, if that is desirable. See Chapter 17 for more information about heartbeating.

The important data structures within the state class include a *CallContainer* class that is useful for keeping track of different types of call information that the state machine might need. The *CallContainer* class provides storage for *CallInfo*, which keeps basic information about the call. You can derive a class from *CallInfo* and then use that class to store any specific data that is required to preserve the state machine as it moves between states. In other words, for storing information specific to your application, use classes derived from *CallInfo* and *CallContainer*.

# High-Level Flow

The two major functional blocks in the base code are the SIP and Worker threads.

## SIP Thread

The SIP thread is a conceptual bundling of the threads that run the SIP stack. As we've seen in Chapter 8, there are actually several threads in the SIP stack, but it is easier to understand how the base code works with the SIP stack if you think of all the SIP threads as one complex thread that sends and receives SIP messages to and from the network. SIP messages are received by the SIP thread from its stack's message queue and placed in its output First In First Out (FIFO) queue. See Chapter 8 for more information about FIFOs.

## Worker Thread

The Worker thread is the application thread that takes messages from the FIFO and processes them using the builder, feature, state, and operator structures. Each class in the structure is a class container for the class below it. The content is filled according to the designer's requirements. For example, the Marshal server (Chapter 11) has one builder, one feature, and one state but multiple operators for each message type, whereas the UA (Chapter 10) has multiple operators and multiple states. In the event that the Worker thread needs to send a response or forward a request, it makes a call to asynchronously send the message. The SIP thread then sends the message over the network.

# Key Data Structures

Figure 9-2 shows the classes and how they encapsulate and inherit from one another.

Let's have a look at the classes that appear in Figure 9-2:

*BasicProxy*
> A derived class from the *HeartLessProxy*. A *BasicProxy* object should be used in a system that supports the heartbeating mechanism, as discussed in Chapter 17. This class adds three heartbeating threads.

*Builder*
> A base class that receives events from the Worker thread for processing. *Builder* is a *Feature* container, which is a *State* container, which is an *Operator* container. *Builder* contains a pointer to the proxy *CallContainer*. *Builder* objects are used to construct *HeartLessProxy* and *BasicProxy* objects.

*CallContainer*
> A base class that is used as the proxy's container for *CallInfo* objects.

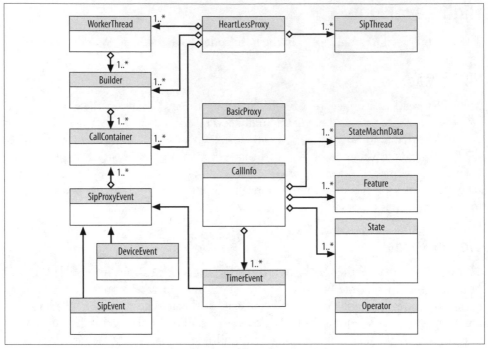

*Figure 9-2. Base code classes*

**CallInfo**

> A base class that stores information about a call. A *SipProxyEvent* object contains *CallInfo*. *CallInfo* data members are thread-safe.

**DeviceEvent**

> Derived from *SipProxyEvent* and a placeholder for events received from a device. These are used in the SIP User Agent for events such as a phone going on- or off-hook.

**Feature**

> Keeps track of running the state machine that is associated with it.

**HeartLessProxy**

> So named because it doesn't heartbeat. This class creates a call container, call processing FIFO, SIP stack, and Worker and SIP threads. The main method is the *run* method, which starts the Worker and SIP threads. Instantiation of the *HeartLessProxy* provides a *Builder* object and a SIP port number.

**Operator**

> A base class for defining operators for the state machine. When deriving operators, the most important function to write is the *process* function, which is where all the code is written that the operator is supposed to use. Generally, when we derive classes from Operator, we name them *Op<name>*.

*SipEvent*

Created by the *SipThread* object and derived from *SipProxyEvent*, when it receives a SIP message. *SipEvent* gets posted to the FIFO. The state machine will read it and pass it into the appropriate operators.

*SipProxyEvent*

Derived from CallContainer, a basic event class that derives the SipEvent, DeviceEvent, and TimerEvent classes. This class does nothing more than provide a common parent for these derived classes.

*SipThread*

A thread that blocks messages received by the SIP stack. On receiving a SIP message, it creates a *SipEvent*, which is then posted to the FIFO.

*State*

A base class for deriving the states in the state machine. It is an *Operator* container. The object represents a state in the state machine. It keeps track of all the operators that are associated with that state and then calls them when appropriate.

*StateMachnData*

A base class that contains data that is static for each state machine. This is really used only by the CPL Feature server. See Chapter 14 for more information.

*TimerEvent*

Derived from *SipProxyEvent* and posted when a timer expires. The User Agent typically uses this class.

*WorkerThread*

A thread that has an input queue on which it is blocked. The input queue contains SIP proxy events.

Looking at the list of base code classes, we can make some general comments about how they work together. The *HeartLessProxy* is the same as the *BasicProxy* except that it doesn't heartbeat. Heartbeating is explained in Chapter 17. Most developers build their applications with either one of these two classes. The *Feature*, *State*, and *Operator* classes are where you can derive your key classes for building applications. Some classes like *CallContainer*, *CallInfo*, and *StateMachnData* are useful for keeping information that you need to persist over transactions. The things that drive the state machine forward are events, and the three most common types are the *SipEvent*, the *DeviceEvent*, and the *TimerEvent*, and all of these derive from a basic event, which is the *SipProxyEvent*.

# Dependencies

The SIP Base Proxy, one of the specific classes included in the SIP base code, requires the following additional software packages within the VOCAL tree:

*SIP*

The SIP stack as described in Chapter 8.

*SDP*

This is the Vovida implementation of RFC 2327 for describing media sessions used in SIP. This is a cascading dependency because some files include files from the SIP stack directory, which in turn includes files from the SDP directory.

*UTIL*

Contains utility classes/functions and exception classes.

*PTHREAD*

The Portable Operating System Interface (POSIX) threads, a Unix standard to enable code portability between the many different flavors of this operating system. PTHREAD is available in our *contrib* directories.

*LIBCEXT*

C programming language extensions that provide compatibility with things that are not derived from Linux. The Linux GNU C Library contains nonstandard extensions. By including LIBCEXT, the code can work with compilers that don't use the Linux C Library.

These packages are required for making the compile build properly. If you want to understand the base code, you will need to understand these other bits of code. If you are trying to decide whether you can use this base code in your application, these are the other things you need to take along with you.

The SIP Base Proxy can be built with or without support for heartbeating. Heartbeating requires the following additional VOCAL packages:

*HEARTBEAT*

Using the Heartbeat Library provides a mechanism to send and retrieve heartbeat events to and from a multicast address. It also keeps track of which servers are active (sending) or inactive (not receiving). There is no dependency on the Heartbeat server, only on the library.

*PSLib*

The *PSLib* is a library that is used to access data from the Provisioning server, making it easier to obtain information without having to do a great deal of coding.

*PSUTILIB*

This directory contains utilities used just by the Provisioning server and *PSLib*.

*LIBXML*

The eXtensible Markup Language (XML) library found in the *contrib* directories.

# CHAPTER 10

# VOCAL User Agent

The Vovida SIP User Agent (UA) is a key component used for testing our SIP-based telephony applications. Besides SIP messaging, the User Agent also has a media component, making it a fully functional softphone. You can make calls between two UAs or between a UA and another SIP endpoint, with or without going through a Proxy or a Redirect server.

## Call Processing

One of the key abstractions of the UA is the state machine, as discussed in Chapter 9, which contains a few paths that describe answered calls and many paths that describe unanswered calls and error conditions. If we were to describe the states for every error condition and nonanswered call in detail, this chapter might exceed the page length of all of the other chapters in this book combined! As in Chapter 7, we have narrowed our discussion about the User Agent to its essential components and behaviors for the sake of clarity.

The state machine is driven from one state to the next by events. An example of an event is a SIP event, which occurs when the UA receives a SIP message. As SIP messages are parsed by the SIP stack and passed to the UA, they are transformed into events and entered into a First In First Out (FIFO) queue from which the state machine retrieves them.

There are other types of events such as timer events that provide time-outs for replies or the entry of phone numbers. You can set up a timer to stop a phone from ringing after a set length of time. There are also hardware events, which are created by interaction with hardware devices.

As these events drive the state machine forward, different threads collect them. The SIP stack thread receives SIP messages, while another thread, running the device manager, whether it's for a Quicknet card (*http://www.quicknet.com*), a sound card, or a test keyboard, puts events into the FIFO. As these events flow into the FIFO, they cause the state machine to move from one state to the next. We implemented our state machine by creating a separate class for each state.

Alongside the classes and events are operators that contain the code to process incoming events and cause the state machine to move from one state to the next. The SIP UA uses the base code that is described in Chapter 9.

Figure 10-1 is a simplified state diagram. We will discuss some of its components and then show how these components work together to make and receive calls.

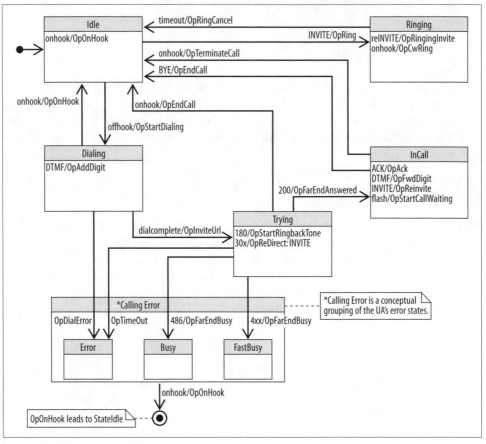

*Figure 10-1. Simple state diagram*

In Figure 10-1, the labels on the arrows, such as *onhook/OpOnHook*, represent hardware events being returned from the FIFO and then being processed by the operator. In the case of *onhook/OpOnHook*, the hardware event, *onhook*, is returned from the FIFO and processed by the operator, *OpOnHook*. See Chapter 9 for more information about events and operators.

## States

Here's a look at some of the states shown in Figure 10-1:

*Idle*
> Nothing is happening.

*Dialing*
> The UA is *offhook*, and the calling party user is entering the phone number.

*Trying*
> The calling party user has finished dialing, and the system is processing the call initiation. The SIP INVITE message is traveling through the network to the called party, and the UA has not yet received a final response.

*Ringing*
> The calling party's SIP INVITE message has reached the called party's UA and has caused it to ring.

*InCall*
> The called party has answered the call, and both parties are speaking.

## Operators

Here is a look at the operators that work when a call is initiated and answered by the called party:

*OpStartDialing*
> Brings the UA from the Idle to the Dialing state.

*OpAddDigit*
> Collects the digits as the user dials the phone without changing the UA's state.

*OpInviteUrl*
> Brings the UA from Dialing to Trying as the completed phone number is routed to the called party.

*OpRing*
> Brings the called party's UA from Idle to Ringing as the INVITE is received.

*OpAnswerCall*
> Brings the called party's UA from Ringing to InCall as its user picks up the ringing UA.

*OpFarEndAnswered*
> Brings the calling party's UA from Trying to InCall as it receives notice that the other party has answered the call. At this point, the session has been established.

*OpEndCall*
> Brings each UA from InCall to Idle when the call ends and each party hangs up.

# Walking Through Two Call Flows

Using the diagram shown in Figure 10-1, let's walk through what happens when a user makes a call and then receives a call.

## Making a call

The first state is Idle. The phone goes *offhook*; the device generates an *offhook* event that is sent to the FIFO. The *OpStartDialing* operator detects this event and moves the state machine to the Dialing state. Now, as the user dials the phone, multiple digit events come in, and each one of those calls the *OpAddDigit* operator. This operator does not move the state machine to another state; the state machine remains in the Dialing state until the user has finished dialing.

With the UA provided with VOCAL, the user is finished dialing if the number dialed matches a dial pattern or if a dial timer expires before a match to a dial pattern is made. This is explained in Chapter 5, and the configuration parameters are shown in Appendix A.

Once the user has finished dialing, there is a *Dialing Complete* event, which calls the *OpInviteURLCall* operator. This operator is what actually sends a SIP INVITE message out to the other side. It also moves the state machine to the Trying state.

Once the machine is in the Trying state, it is waiting for the other end to pick up the phone. When the other end picks up and answers the phone, it sends a SIP 200 message. The *OpFarEndAnswered* operator processes this message and takes the state machine to the InCall state. At this point, the users are speaking to each other until they decide to end the call. The user who phoned hangs up and, by doing so, generates an *onhook* event. The *OpTerminateCall* operator processes this event, sends a SIP BYE message to the other user, and takes the state machine back to the Idle state where it started.

That completes a walk through the state machine and its operators as a user makes a call. Let's look at this process again from the point of view of the user receiving a call.

## Receiving a call

The user's phone, resting in the Idle state, receives a SIP INVITE message. The *OpRing* operator processes this message and takes the state machine to the Ringing state. At this point, the hardware is signaled to start ringing the phone, and the ringing begins. When the user answers the phone, an *offhook* event is generated. The *OpAnswerCall* operator processes this event and takes the state machine to the InCall state. Here, the users are speaking to each other until the conversation ends. The user who phoned hangs up, and in doing so, sends a SIP BYE message. The *OpEndCall* operator processes this event and takes the state machine back to the Idle state.

### New calling parties and called parties

In transfer and conferencing situations, discussed later, we have to talk about a third party. We use the terms *new called party* and *new calling party* to introduce the idea that while two users are engaged in a call, either a new party interrupts the call, or the two users decide to add a new called party to the conversation. We hope you don't find this confusing.

## Error States and Operators

When you start adding error states, such as Busy, the state diagram becomes more complicated even though all of the code remains basic. There are a number of states that represent conditions such as Busy, or Dialing Error, and operators that move you from one state to the next based on the present state and the event that just occurred.

## Early Media

The VOCAL UA supports *early media*, which is an idea that's worth some discussion. Early media enables a caller to hear things like a ringing tone and the called party's voice when the other end is answered. Every country has its own unique ringing tone, and early media helps you hear the ringing tone when you make a local, long-distance, or international phone call over the PSTN. As for the called party's voice, when the far-end phone is answered, you want to hear the entire phrase "Hello" or its localized equivalent. It would be unsettling to hear this greeting clipped due to a delay in setting up the voice channel. Early media starts setting up the voice channel from the called party back to the caller as the phone is ringing.

Let's talk about how this works in the SIP world. When you make a phone call from a SIP UA, you send an INVITE to the far end. When the far end starts ringing, it sends back an 18x message that contains the Session Description Protocol (SDP) headers, which are used by the UA to set up a one-way path of Real-time Transport Protocol (RTP) packets coming toward you. This enables you to hear the ringing from the far end. Once the 200 message is sent, after the far-end user answers the phone, it connects up the second leg of the two-way RTP packet stream, and you have a two-way voice channel.

## Using the INFO Message with MGCP

In order to enable the SIP UA to talk to a voice mail system, it needs a method for sending dual-tone multi-frequency (DTMF) digits. Originally, we were transporting those digits through the SIP INFO message. You encapsulated DTMF information inside an INFO message by inserting a Media Gateway Control Protocol (MGCP) message that contained the digits. This sounded like a great idea at the time, and it was the subject of a working draft of the IETF, but it proved to be a total disaster. Thankfully, that draft expired, and we're changing our approach and looking at new methods such as RFC 2833, which looks at using Audio/Video Transport (AVT) tones over RTP.

## Multicall Processing

Now, adding multicall processing to the state diagram makes it more complex, especially when you add all of the possible error conditions. To simplify the diagram in Figure 10-2, we removed all of the error states and left behind the states that are used for single calls and when the UA is handling two calls at once.

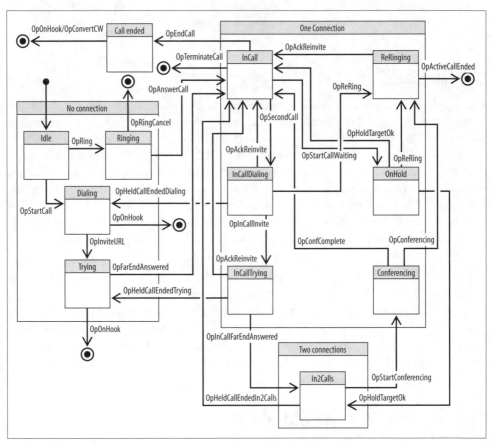

Figure 10-2. A complex state diagram

Let's look at some of the states and operators that are shown in Figure 10-2.

## States

Here is a list of the states shown in Figure 10-2 that enable processing two calls at once:

*Conferencing*

Three users can actually talk to one another over an ad hoc session hosted by one of the User Agents. The conferencing state is shown as one connection because the UA and the two called parties are connected to the conference bridge. The UA does not support local conference mixing.

*In2Calls*

The user who is setting up the ad hoc conference call has made connections with two called parties but has not yet conferenced them together onto the same session. One connection is on hold while the other is active. The ability to transfer and conference calls depends on the availability of a conference bridge and the parameter settings within the *ua.cfg* file. See the later "Conferencing" section.

*InCallDialing*

The calling party has put one called party on hold and is dialing the number for another called party.

*InCallTrying*

The calling party, while having put one called party on hold, has finished dialing a second called party and is waiting for that party to answer.

*OnHold*

This is the state while an incoming call waits while the intended called party is on another call. The intended called party can either toggle between this new call and the original call or transfer both calls to the conference bridge.

*ReRinging*

This is a transition state between the operators that bring the held party out of hold and back into a live session. After the incoming party has been put on hold for a configured length of time, a timer triggers the *OpReRing* operator to ring the local user's phone to reestablish the call.

## Operators

Here is a look at the operators that work when one call is put on hold and another call is initiated for conferencing:

*OpAnswerCall*

Looks at the SDP information, sets up the appropriate media requirements, and starts the real media going. It also sends back appropriate SDP to the far end. If the UA is already in a call, and if the UA doesn't support Call Waiting, OpAnswerCall returns a busy signal. If the UA does support Call Waiting, OpAnswerCall signals that a second call is coming in.

*OpAckReinvite*

If the UA is in a call, sends a RE-INVITE to a third party, and then receive a status message from that party, OpAckReinvite sends an ACK to it, providing the response message from the third party was a 200 or higher.

*OpActiveCallEnded*

If the UA is in a call and it receives a BYE, OpActiveCallEnded sends a 200 to the sender of the BYE and then removes that call and cleans up the processes. If the UA is in two calls, OpActiveCallEnded switches it back into one call. If the UA is in one call, OpActiveCallEnded takes you back to idle state.

*OpInCallInvite*

If the UA is in a call and the user puts the call on hold and then dials a third party, OpInCallInvite takes the dialed digits and sends out an INVITE to that new party

*OpHeldCallEndedin2Calls*

If the UA is in two calls and it receive a BYE from one party, OpHeldCall-Endedin2Calls sends a 200 to that party. This operator is almost the same as OpActiveCallEnded except that it works with two simultaneous calls only.

*OpConfComplete*

Closes the conferenced session and reduces the call from two to one called party.

*OpConferencing*

Brings the called party, who was on hold, from ReRinging to Conferencing. The ability to transfer and conference calls depends on the availability of a conference bridge and the parameter settings within the *ua.cfg* file. See the later "Conferencing" section.

*OpHeldCallEndedTrying*

If the user puts a call on hold and dials another number but hangs up before the new called party answers, this brings the held party back into a live call.

*OpHoldTargetOk*

Brings an incoming call from OnHold (Call Waiting) to InCall.

*OpInCallFarEndAnswered*

Brings the user from InCallTrying to In2Calls when the new called party answers the phone.

*OpReRing*

Brings a called party from OnHold to either ReRinging or In2Calls.

*OpSecondCall*

Brings the calling party from InCall to InCallDialing after the first called party has been put on hold.

*OpStartCallWaiting*

Brings the calling party's phone, while engaged in a call, from InCall to OnHold when a new incoming call comes in.

*OpStartConferencing*

Brings the calling party's phone from In2Calls to Conferencing by transferring both called parties to the conference bridge.

# Conferencing

You may have been to a corporate meeting that involves several off-site participants connecting through a single speakerphone. The technical issue that surrounds this type of call is channel mixing. How do you mix the channels from three or more endpoints into the same call so that all parties can hear one another clearly? A device called a *conference bridge*, also known as a *Multimedia Conference Unit* (MCU), provides the solution.

## Conference bridges

Conference bridges are media servers that enable call channel mixing. Some conference bridges are built into rack-mounted casings that are connected to the phone system via communication cables; others are cards mounted inside private branch exchange (PBX) cabinets. Large-scale conference bridges are normally managed by a phone company operator and are capable of mixing a dozen or more channels in one connection without any perceptible quality degradation. Small bridges, like the ones found inside PBXs, are limited to mixing a few calls at once and are activated by feature buttons on the users' phone sets.

## Scheduled versus ad hoc conference calls

There are two types of conference calls, *scheduled* and *ad hoc*. Scheduled calls normally include a large number of participants that require conference mixing for a formally prearranged meeting. Ad hoc calls are normally limited to three or four participants who get together informally.

On a scheduled conference call, each participant is either invited to call a well-known phone number at a scheduled time to register with the conferencing operator, or the operator calls the participants just before the call is scheduled to take place. The conference bridge used for these calls is normally a large-scale server on the phone company's premises or, in the case of the largest organizations, on the premises of the organization's central phone system.

On an ad hoc call, one of the participants calls two or three of the others and transfers them to a local conference bridge. This bridge could be part of the phone set's firmware, an MCU server attached to the network, or a card inside the PBX. The user who initiates the ad hoc conference either uses a feature button on the phone set or does a *flash hook* (see the next paragraph) to put the first participant on mute/hold while the second participant is called.

## Flash hook

Flash hook was originally a quick depression and release of the phone's hook: depressing the hook for less than a second activated the flash hook. Many users found themselves hanging up on calls when they intended to flash hook. Some modern phone sets provide a flash hook button separate from the hook to help prevent such mistakes.

## Call Waiting

If Call Waiting is active, a user who is currently engaged in a call can receive an incoming call from a third party. VOCAL Version 1.3.0 does not support Call Waiting; however, if someone in the community wanted to enable this function and contribute his source code back to VOCAL, we would appreciate the effort. There are operators and events available in the SIP UA code to support Call Waiting, but changes to the APIs in both the SIP stack and the base code have compromised this functionality. However, to illustrate how Call Waiting is supposed to work, we can look at how the events and operators would interact with the state machine.

Two users are on a call, and one user receives an INVITE from a third party, which activates *OpStartCallWaiting*, bringing the state machine to OnHold. The user decides to accept the incoming call, flash hooks to put the first call on hold, and toggles to the second call. The flash hook event activates *OpHoldTargetOk*, which brings this call from OnHold to InCall. The user can flash hook between the two callers repeatedly until one of the parties hangs up. If, while toggling, the user hangs up while the other two calls are active, the other two calls are transferred to each other. This can be frustrating because it means that the user who has hung up has lost control of the call and cannot break in and tear it down. Not a problem if the other two users are on local extensions; big trouble if they are long-distance connections over the PSTN! For the time being, this functionality is disabled.

## Conferencing with the VOCAL UA

VOCAL does not include a conference bridge, but we have tested our system successfully with a number of different MCUs connected to dedicated Marshal servers. See Chapter 11 for more information about Marshals. The VOCAL UA supports three options for conferencing and call transfer that can be set up in the *ua.cfg* file. Let's look at the passage from that file:

```
# TRANSFER & CONFERENCE
#
# Ua_XferMode--Use this to turn on transfer or ad hoc conferencing.
#             The options are: Off/Transfer/Conference.
#
# Conference_Server--Specify the URI for the Conference server.
#                   The URI consists of the conference bridge number and
#                   the IP address of the proxy server or the conference
#                   bridge itself (if no proxy is being used).
```

```
#
Ua_Xfer_Mode          string          Off
Conference_Server     string          6000@192.168.5.4
```

The possible parameters for *Ua_Xfer_Mode* are described in Table 10-1.

*Table 10-1. Transfer and conference options*

| Parameter | Capability |
|---|---|
| Off | The UA cannot handle more than one call. If the UA is in a call and another call comes in, the new caller will either go to voice mail or receive a busy signal. |
| Transfer | The UA can handle two calls. The first flash hook gives the dial tone, and the user can dial the third party's number. The second flash hook completes the feature. |
| Conference | The UA can handle two calls. If the UA is in a call, the user can flash hook to transfer both calls to the conference bridge. |

At the time of this writing, the SIP standards bodies were finalizing the details of the new REFER message that will be used for conferencing. Our UA is ready to handle conference and transferring as soon as the REFER specification has stabilized.

## Walking Through Two Call Flows

The UA is capable of putting a call on hold, dialing a second call, and keeping track of both calls at once. And it can toggle between the two or conference both onto the same ad hoc session. Here is a walk through two call flows: the first occurs when the *Ua_Xfer_Mode* parameter within the *ua.cfg* file is set to Transfer; the second occurs when this parameter is set to Conference. See Appendix A for more information about the *ua.cfg* file.

### Transferring a call: Ua_Xfer_Mode = Transfer

A user is on a call and flash hooks, taking the state machine from InCall to InCall-Dialing via the *OpSecondCall* operator. The user hears the dial tone, dials a third party, and then hears the ringing tone; the state machine has been taken to InCall-Trying via *OpInCallInvite*. If the user wants to do a blind transfer, she hangs up before the third party answers and the second party is transferred to the third party, taking the state machine back to InCall. If the third party is busy, the user can flash hook back to the second party and inform him.

If the user wants to do a consultation transfer, she waits for the third party to answer, which takes the state machine from InCallTrying to In2Calls, then informs the third party about the incoming call before transferring it. Naturally, if the third party is busy or refuses the call, the user can flash hook again to inform the second party.

### Conferencing call flow: Ua_Xfer_Mode = Conference

Two users are on a call, and they decide to bring a third party into an ad hoc conference call. One user flash hooks, which puts the first call on hold and activates *OpSecondCall*, bringing the state machine to the InCallDialing state. The user dials a new number, and the digits are collected. When the user is finished dialing, *OpInCallInvite* sends an INVITE message out to the network and brings the state machine to the InCallTrying state while the user waits for the new called party to answer. The new called party picks up the phone and answers, activating *OpInCallFarEndAnswered*, which brings the state machine to In2Calls.

After speaking with the new called party, the user flash hooks, activating *OpStartConferencing*, which transfers all parties to the conference bridge and brings the state machine to Conferencing, in which the audio is mixed and everyone is talking. This is considered one connection because the UA is no longer keeping track of two calls; it is engaged with a single connection to the conference bridge.

## Looking Through the Code

From a coding point of view, the UA is constructed from classes and threads.

### Classes

Here is a look at some of the classes behind the VOCAL UA. These classes can be easily divided into base classes and event classes, as shown in the following diagrams. Figure 10-3 shows the UA classes.

The classes in the gray shading are base classes. Here are descriptions of the event classes:

*DeviceThread*
> Works with the hardware devices such as the sound card.

*DigitCollector*
> Receives digit input while the user is dialing and processes the digits according to the dialing plan.

*FeatureThread*
> The main thread that executes the state machine.

*LoadGenThread*
> Tracks statistics when the UA is being run in Load Generation mode. This is used for bulk testing to make large numbers of calls.

*NullHwDevice*
> Used for testing when neither a sound card nor a Quicknet card is available.

*QuickNetDevice*
> A device thread that works with Quicknet cards (http://*www.quicknet.com*).

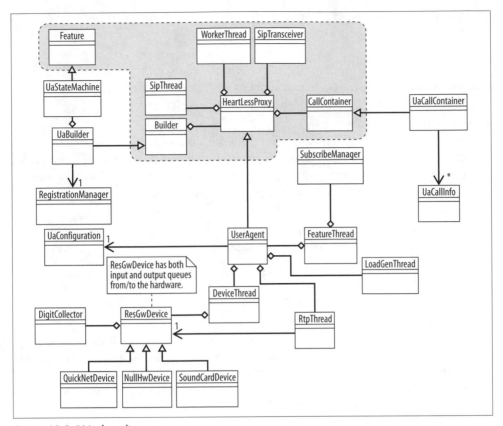

*Figure 10-3. UA class diagram*

*RegistrationManager*
> Keeps track of the UA's registration with the network and its expiry.

*ResGwDevice*
> Residential Gateway Device, a base class. Anything that was a UA that did audio would need to deal with certain operations. We derive certain devices from that, such as the Quicknet card and sound card, or if there is nothing, we derive the *NullHardware* class. If you have a specific audio device that is different than the ones that our classes support, you need to derive a class to make that device work with VOCAL.

*RtpThread*
> Processes Real-time Transport Protocol (RTP) packets.

*SoundCardDevice*
> A device thread that works with the sound card.

*SubscribeManager*
> An experimental thread designed to work with SIP SUBSCRIBE and NOTIFY requests. SUBSCRIBE and NOTIFY are examples of SIP request messages that

are not directly related to call control. You can find out more about these and other request messsages in RFC 3261.

*UaBuilder*

The main control for the UA, which sends events and SIP messages to their intended destinations. *UaBuilder* contains pointers to the SIP stack, *StateMachine*, *CallContainer*, *SubscribeManager*, and *RegistrationManager*.

*UaCallContainer*

Derives from the *CallContainer* class from the base code. This contains the information about the current call that is set up; for example, if an INVITE is received that has certain record route information that should be saved somewhere, so that the correct route header can be put into future messages.

*UaCallInfo*

Inherits from the base code class *CallInfo*, which contains a smart pointer to state. This enables each call to have a different state while sharing the same state machine.

*UaConfiguration*

Contains the data found in the configuration file, *ua.cfg*.

*UaStateMachine*

Derived from the feature base code. Processes events received from *UaBuilder*, by cycling through all of the operators (designated as *Op<name>*) in the current state, calling each in turn. Each operator returns either a new state or a null pointer indicating no change in state. All operators receive the event for processing regardless of how they respond. Each state contains both entry and exit operators.

*UserAgent*

The class in the base code that inherits from *HeartLessProxy*, because our UA does not heartbeat, and initiates all of the other threads. *HeartLessProxy* is in the base code.

### Call info class

There's only one important data structure, the *UaCallInfo* class, which keeps track of the information about the call such as the identity of the called party. The *UaCallInfo* class also includes the current list of who it has already contacted while trying to make this call. If it keeps getting redirected to many places, it has the contact that it is currently trying to reach plus all of the others it's tried to contact. This enables it to detect if the request has looped back to itself.

The *UaCallInfo* class has the session description that describes the media session for itself and the other side. If the UA is in two calls, it needs to have the session descriptions for both calls.

The *UaCallInfo* class also keeps track of any INVITE messages that it has received. The INVITE message contains an address in its Contact header that enables the other side to send a BYE or another SIP request to the originator of the INVITE. The UA reuses the data found within the INVITE message to generate and send a BYE.

## Event classes

Figure 10-4 is a simplified class diagram of the events that drive the UA's state machine. The grayed areas indicate base classes.

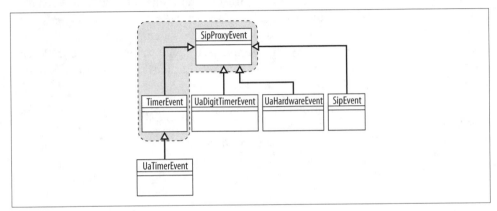

*Figure 10-4. UA event classes*

The UA event classes are as follows:

*SipEvent*
> Occurs when a SIP message is received by the UA.

*SipProxyEvent*
> A common base code type that the others can derive from. This has no specific function other than acting as a superclass that others can derive from.

*TimerEvent*
> A base class that is a generic timer event class.

*UaDigitTimerEvent*
> Related to dialing completion. Local and long-distance calls to countries with fixed dial plans must be mapped in a dial plan. International calls and calls within countries with variable dial plans cannot be mapped as easily because mapping all possible combinations, and maintaining this mapping, would be excessively time consuming. It is much easier to have a timer that expires a configured number of seconds after the user has stopped dialing, thereby triggering the INVITE message. The *vocal1.3.0/sip/ua/ua.cfg* file provides configuration fields for the initial digit time-out (how much time is allowed between the UA going *off-hook* and the user entering the first digit) and the enter digit timeout (how long the UA waits after the last digit is entered before it attempts to dial).

*UaHardwareEvent*

Triggered by data being sent to or received from an audio device such as a Quicknet card or a sound card.

*UaTimerEvent*

A generic event used for queuing *SIPProxyEvents* in the FIFO.

## Threads

The *UserAgent* class spawns a number of threads, including one that processes data from a device such as a Quicknet card or a normal sound card. The UA can also run without a sound device by using a PC keyboard to send signaling information. In this mode, it can't set up a phone call, but it can test signaling and interactions with other network components. It is not unusual to see our developers testing their code without sound cards.

So, once it starts this device thread, it also starts a Real-time Transport Protocol (RTP) thread to process RTP packets. The device thread monitors the card, and whenever the card does something, like go on-hook or off-hook, it creates the appropriate *UaHardware* event and inserts that into the FIFO. And then the state machine pulls it out of the FIFO and executes the state machine. The UA then loads all of the required configuration information from the *ua.cfg* file, and starts the main state machine.

*Device*

Handles the sound card or the Quicknet card and gets information such as the UA going on-hook or off-hook.

*DigitTimeout*

Used when the UA is being dialed and works with dial patterns to indicate when the user has finished dialing. See Chapter 5 and Appendix A for more information.

*Feature*

Very experimental thing right now; used to manage SUBSCRIBE/NOTIFY signaling.

*HeartbeatTxThread*

Used only when the UA is being used in the voice mail system.

*LoadGen*

Used only in load generation to print statistics.

*Main*

Runs the state machine.

*RegistrationManager*

Keeps track of when the time-out is going to happen for the registration and reregistration, causing the UA to send another REGISTER message when appropriate.

*RTP*
> Receives packets and sends them to the sound card and takes packets off the sound card and sends them out over the network.

*SIP stack*
> Discussed in Chapter 8.

# Other UA Processes

There are some other uses for the UA besides making and receiving user calls. For example, we use the UA as a load generator for performance testing.

## Load Generation

When providing load generation for performance testing, the UA automatically dials calls and some extra operators, such as *OpAutoCall* and *OpAutoAnswer*, are added, automatically moving the state machine from Call to Dialing. As the test calls are completed and the UA returns to the Idle state, it waits a few seconds and then automatically generates another call.

In many of these setups, there are several UAs set up in a lab with ingress and egress Marshal servers and a Redirect server. The load-generating UAs are calling other test UAs in the lab. They are not usually calling across to another network or over the PSTN, unless those routes are specifically intended for testing. When lab testing is in progress, there are also a load-generation thread and a monitor that track statistics about how much of the load is being generated.

## Registration

As you read in the SIP chapters, the UA needs to register with the Registration server in a SIP network before it can start sending and receiving phone calls. To register, it sends a REGISTER message, which is handled by the *Registration* class and the *RegistrationManager* class. The registration is good for a limited time only: the UA has to resend its registration to the Registration server repeatedly. The *RegistrationManager* class keeps track of the timers that trigger reregistration.

Having discussed the User Agent, we will now look at the servers that interact with SIP messages sent and received by UAs connected to VOCAL: the Marshal, Redirect, and CPL Feature servers.

# B2BUA

There is a new type of SIP application being developed in the VoIP community that can handle prepaid calls while being able to modify the media and initiate session teardowns during established calls. This functionality is beyond the ability of standard SIP proxy servers; therefore, it is being implemented as a Back-to-Back User Agent (B2BUA).

The B2BUA is defined in the in the SIP standard as follows:

> Back-To-Back User Agent:
>
> Also known as a B2BUA, this is a logical entity that receives an invitation, and acts as a UAS to process it. In order to determine how the request should be answered, it acts as a UAC and initiates a call outwards. Unlike a proxy server, it maintains complete call state and must participate in all requests for a call. Since it is purely a concatenation of other logical functions, no explicit definitions are needed for its behavior.

One way of implementing the B2BUA for providing prepaid call control is to enable it to act as a User Agent server (UAS) in the caller-initiated call leg and have it create another call leg to the destination as a User Agent Client (UAC). After the call is set up, the B2BUA could be designed to send SIP messages to modify the caller's media to convey call duration or billing-related information. It may also use HTTP messages to carry the same information. The B2BUA can be implemented to control the call by tearing it down when the caller's prepaid time has expired.

In some other implementations, the B2BUA is being implemented to provide third-party call control (3PCC) whereby a user (who may be a customer of a web-based business, for example) can initiate a call between two phones by clicking a hyperlink on a web page. The B2BUA enables a series of transactions that invite both phones (for example, a call center's phone and a customer's phone) to a call. The B2BUA also works to tear down the call when the conversation is over. This functionality is similar to what we set out to achieve with the JTAPI Feature server.

A B2BUA is now available on *Vovida.org* for downloading and testing. We welcome your comments on the *b2bua@vovida.org* mailing list.

# CHAPTER 11

# SIP Proxy: Marshal Server

The Proxy server is a creature of the SIP RFC, 3261, and, as its name implies, is intended to act on behalf of other network endpoints rather than on its own. In VOCAL, we implemented two flavors of the SIP Proxy server as distinct entities and called them the Marshal server and the Feature server. There are also SIP Proxy servers built into the MGCP-SIP and H.323-SIP translators. See Chapters 15 and 16 for more information about the translators.

The Marshal server is the first point of contact for all traffic entering the VOCAL system: it authenticates users, provides data that is useful for call detail record collection, and forwards SIP request and response messages to other network entities. The Feature server provides core network features, such as call forwarding, and is the subject of Chapter 13.

## High-Level Design

"Why do we need Marshal servers?" This question was raised repeatedly during our initial white-boarding of VOCAL. Logically, we could have used the Redirect server (RS) to provide authentication and call detail record (CDR) data, but combining those elements into the RS would have negatively impacted our ability to scale the system. While you can add additional RSs to the system to provide redundancy, you can't easily add more hardware to the RS to scale the system because of its requirement to keep all of the registration information together in the same place. To promote scalability, we decided to decentralize the authentication and CDR data collection to another server type based on the SIP Proxy and name these servers Marshal servers.

 We refer to Marshal servers type proxies as *Marshals* and label them as *MS* in diagrams.

Another advantage of having the Marshal is its ability to keep the SIP messages circulating within the VOCAL system consistent with our implementation of the stack, filtering out syntax and content variations within SIP messages coming from third-party endpoints. When the Marshal receives an INVITE message, it rewrites the headers to match our implementation. In the lab, sometimes one manufacturer's SIP IP phone won't talk directly to a different manufacturer's gateway. Having the IP phone send its INVITE through our network to the gateway solves this problem. In that transaction, our Marshal rewrites the message into something that the other VOCAL servers and the gateway can understand.

The Marshal is also useful for providing a ring of security around the outside of the system. This eliminates the need to provide extra security for every server type within VOCAL. This streamlines performance, because once a message is inside the system, it does not require further authentication as it moves from hop to hop.

While the Marshals solve a lot of problems, they also add to the sheer number of servers required to deploy the system. We have found that the previously mentioned advantages of implementing Marshals into VOCAL more than make up for the extra memory and signaling latency required for parsing the messages between the servers. As far as SIP Proxies go, Marshals are basic: they receive SIP messages and forward them to other servers or endpoints. Marshals also have the ability to monitor heartbeats and use this data to assist with load balancing among the active RSs and to avoid forwarding messages to inactive servers.

## Transaction Statefulness

Marshals are transaction-stateful proxies in the sense of a SIP transaction. For example, as shown in Chapter 8, a typical initiation transaction includes INVITE, 100, 180, 200, and ACK messages. The BYE, 200 sequence, used to tear down calls, is a separate transaction. Marshals are not call stateful. You can use one Marshal to start a call (the initiation transaction) and another to tear it down (the teardown transaction). This is useful if the first Marshal crashes during your call and allows a different Marshal to handle the call teardown for load balancing.

 We are using the informal *call* rather than the proper *dialog* in describing SIP-based VoIP sessions to remain consistent with the expression used in Chapter 8.

## Ingress and Egress Signaling

Any given Marshal can be used for ingress and egress signaling. In a typical configuration, one Marshal works with all of the UAs, and another works with the gateways. If there are two users, both with UAs, and one user calls another, the INVITE goes in through the UA Marshal, using it as an ingress proxy, which forwards it to the RS. As the INVITE leaves the network, it goes out through the same UA Marshal, using it as an egress proxy, which forwards it to the called party.

Having messages go through the same Marshal twice during the same call creates a tricky coding issue, because, typically, this condition could be considered a loop that must be stopped. The Marshal is programmed with a functionality called *spiral loop detection* that detects the change in the message's destination from going to the RS on the way in to going to a UA on the way out. This functionality enables ingress and egress through the same Marshal. That sounds simple, but writing a bug-free implementation was not easy.

In our implementation, we look at where the message has been before it arrives at the Marshal. If its prior location isn't a server within our system, we know that it is an incoming call. If the prior location is somewhere within our system, we know it is an outgoing call. When a Marshal receives an INVITE message, it looks for a Proxy-Auth header within the message. If this header exists, the Marshal knows that the user has been authenticated and forwards the message to the next address in the Via field. If the Proxy-Auth header does not exist, then, based on the type of authentication scheme provisioned for the user, the Marshal returns a 407 Proxy Authentication Required message back to the user.

As authenticated SIP messages pass through the VOCAL servers, each server adds a branch to the Via header. As Via is a required header, unlike the Record Route header, it presents an accurate reflection of where the message has been within the system.

The concept of adding branches to a Via header is something that SIP has derived from the HTTP standard. In our implementation, there are potentially four distinguishable branches per SIP transaction:

1. The ingress Marshal server has forwarded the message to the Redirect server.
2. The ingress Marshal has forwarded the message to the next hop within the system after receiving a 302 Moved Temporarily message from the Redirect server.
3. The egress Marshal server has forwarded the message to the Redirect server.
4. The egress Marshal server has forwarded the message to the next hop outside of the system after receiving a 302 Moved Temporarily message from the Redirect server.

The branch appears in this form:

```
Via: SIP/<version>/<transport layer protocol> <IP address>:<port>;branch=1
Via: SIP/<version>/<transport layer protocol> <IP address>:<port>;branch=2
Via: SIP/<version>/<transport layer protocol> <IP address>:<port>;branch=3
Via: SIP/<version>/<transport layer protocol> <IP address>:<port>;branch=4
```

Figure 11-1 illustrates the use of these branches in a message flow. For clarity, the message flow shown in this illustration is not complete: the 100 and ACK messages have been removed. For an illustration of a complete flow, see Chapter 7.

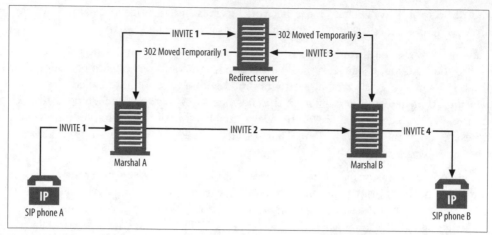

*Figure 11-1. Partial SIP message flow, illustrating the use of branches in the Via field*

Looking at Figure 11-1, when Marshal A receives an authorized INVITE message from an endpoint, it creates an INVITE message with a 1 branch in its Via header, indicating that the message is an incoming redirect, and sends the message to the RS. The RS returns a 302 message, including the 1 branch. Marshal A then creates a new INVITE with a 2 branch to indicate that the message is an incoming contact. When Marshal B processes the INVITE to send to the RS, it creates a new INVITE message with a 3 branch in its Via header to indicate that the message is an outgoing redirection. When the RS receives this message, it returns a 302 message, including the 3 branch. Marshal B then generates an INVITE and sends it to SIP phone B with a 4 branch, indicating that the INVITE is now an outgoing contact message.

Figure 11-2 shows a partial continuation of this message flow. The ACK messages have been removed.

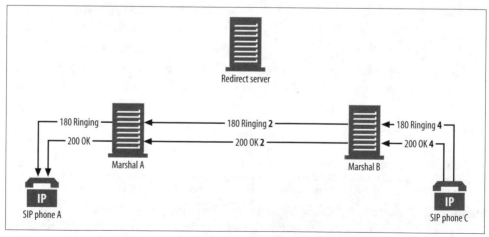

*Figure 11-2. Partial SIP message flow (continued)*

From the messages shown in Figure 11-2, you can see that the SIP phone returns a 180 Ringing message with the 4 branch. Marshal B changes this branch to 2 before forwarding it to Marshal A, which strips the branch out before forwarding the message to SIP phone A. The 200 message is handled the same way.

The INVITE, 180, and 200 messages contain multiple Vias. For example, in an INVITE from Marshal B to phone B, the following is a representation of the headers that would appear in the messages:

```
INVITE sip:sphonB@ipaddress
Via SIP/2.0/UDP <IP address and port of MSB>;branch=4
Via SIP/2.0/UDP <IP address and port of MSA>;branch=2
Via SIP/2.0/UDP <IP address of phoneA>
...
```

The 200 message would look something like this:

```
200 OK
Via SIP/2.0/UDP <IP address and port of MSB>;branch=4
Via SIP/2.0/UDP <IP address and port of MSA>;branch=2
Via SIP/2.0/UDP <IP address of phone A>
...
```

For more information about the structure of SIP messages, see Chapter 8.

Without the branches, if the same Marshal were being used for both ingress and egress, the SIP stack would think that the outgoing INVITEs, 180s, and 200s were retransmissions or loops, would not pass them up to the Marshals at the application layer, and the messages would never reach the SIP IP phones. As far as the SIP stack is concerned, different branches indicate different messages regardless of any similarity in name or content.

## Types of Marshals

Each type of external device interacts with VOCAL through a device-specific Marshal. For scalability and redundancy, there may be more than one Marshal for each type of device. On the other hand, similar devices may be grouped and share a common Marshal. Of course, if you like, you can run multiple Marshals on the same host. The initial device types and the corresponding Marshals are as follows:

*User Agent Marshal*
> Generally used as a proxy server for multiple SIP UAs, the UA Marshal is capable of performing various authentication schemes on a per-user basis. A UA Marshal serves as the link between SIP UA and the rest of VOCAL.

*PSTN Gateway Marshal*
> Designed to interface with any SIP gateway, the PSTN Gateway Marshal does not know or care about individual users. It simply knows the gateway to which it interfaces and acts as the link between it and the rest of VOCAL for all SIP traffic. This gateway may be a SIP-to-PSTN gateway (such as the Cisco AS5300), a 3660 gateway, a 2600 gateway, or another SIP proxy.

*Conference Marshal*

Designed to support ad hoc SIP conference calls, the Conference Marshal performs some Call-ID translation to make ad hoc conferences work with conference bridges that are not capable of supporting this functionality on their own. Similar to the PSTN Gateway Marshal, it interfaces to a single SIP gateway or SIP Proxy. In order to use PSTN conference bridges, the Conference Marshal server may be configured to interface with a SIP-to-PSTN gateway, which in turn interfaces to a PSTN conference bridge.

*Internetwork Marshal*

Designed to authenticate calls through a clearinghouse using the Open Settlement Protocol (OSP) and to request Quality of Service (QoS) for calls through a router using Common Open Policy Service (COPS), the Internetwork Marshal is similar to the PSTN Gateway Marshal. It interfaces to a single SIP gateway or SIP Proxy.

As developers continue building VOCAL, they will probably create additional types of Marshals. Whether an existing Marshal type should handle similar but nonidentical devices or a newly defined type of Marshal should handle them is a judgment call that the developers will have to make. For example, rather than using the User Agent Marshal server for all UAs, there could be a separate Marshal type created to handle each make and model of third-party UAs. It is desirable, however, to aggregate functionally similar devices whenever possible.

## Class Diagram

The Marshal is built on top of the base code and the SIP stack. Figure 11-3 is a class diagram of the Marshal that we created using the Together modeling tool (*http://www.togethersoft.com*), and these files are available for you to download from *CVS./doc/documents/proxies/marshal*. The base classes, which are discussed in Chapter 9, are shaded in gray.

Figure 11-3 illustrates a simplified version of the Marshal. (The Together tool generated the code in this directory as we drew the diagram. It is not the real source code or even a subset of the real source code. A diagram generated from the real source code is very complex and, thereby, of limited use for this discussion.)

The following is a list of the Marshal-specific classes illustrated in Figure 11-3:

*CdrInterface*

Wraps code that interfaces to the CDR server to instruct it to start or stop collecting records.

*Marshal*

Defined in *Marshal.cxx*, this class contains the code for the Marshal instance and the utility methods used by the Marshal. This is where the Marshal is created as a type of base proxy.

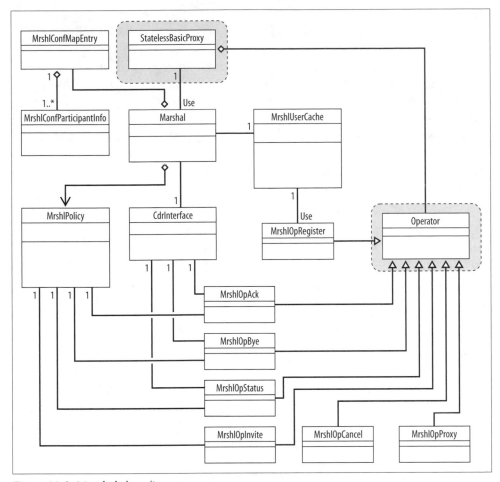

*Figure 11-3. Marshal class diagram*

*MrshlConfMapEntry*
> Comes into play for VOCAL User Agent ad hoc conference scenarios. *MrshlConf-MapEntry* is a class used to store/access data on an individual ad hoc conference.

*MrshlConfParticipantInfo*
> Comes into play for VOCAL User Agent ad hoc conference scenarios. *MrshlConf-ParticipantInfo* is a class used to store/access data on a participant in an ad hoc conference.

*MrshlOpAck*
> The operator that acts upon SIP ACK messages received by the SIP thread.

*MrshlOpBye*
> The operator that acts upon SIP BYE messages received by the SIP thread.

*MrshlOpCancel*
> The operator that acts upon SIP CANCEL messages received by the SIP thread.

*MrshlOpInvite*
> The operator that acts upon SIP INVITE messages received by the SIP thread.

*MrshlPolicy*
> Used for the Internetwork Marshal, this class interfaces with the Policy server through the *pepAgent* as needed for the Open Settlement Protocol (OSP). This class is called from various *MrshlOp* classes.

*MrshlOpProxy*
> Handles all remaining SIP commands including INFO, OPTIONS, TRANSFER, REFER, and unrecognized methods. This is a default handler, which simply forwards the command to the next hop.

*MrshlOpRegister*
> The operator that acts upon SIP REGISTER messages received by the SIP thread.

*MrshlOpStatus*
> Handles all SIP responses. It contains handling for 18x, 200, 3xx, 4xx, 5xx, and 6xx responses.

*MrshlUserCache*
> Called from various *MrshlOp* classes, this class provides a cache of all user data needed by the Marshal to perform various types of user authentication.

## Threads

The base Marshal is derived from the basic proxy, which contains the following threads:

*Heartbeat thread*
> Performs the continuous heartbeat exchange with the RS.

*SIP thread*
> Operates the SIP stack. It receives SIP messages from external devices and from VOCAL, determines if a message is an application-level message that needs to be handled by the Marshal, and if it is, puts it in the input FIFO of the Worker thread. It also accepts SIP messages from the Worker thread and transmits them to VOCAL as instructed by the Worker thread. The SIP thread also performs any necessary retransmission of these messages without additional intervention by the Worker thread. It also filters out duplicate messages on behalf of the application.

*Worker thread*
> The application layer of the Marshal. It continuously takes SIP messages from its FIFO, processes the messages, and puts any appropriate SIP response messages in the FIFO of the SIP thread. The FIFOs of both the Worker thread and of the SIP thread are instantiations of the FIFO class.

## Source Code

The source code for the Marshal is in the CVS directory:

```
./vocal-1.3.0/proxies/marshal/base/
```

The *main( )* method is contained within *ms.cxx*, which parses command-line parameters, sets up provisioning, and creates and runs a Marshal instance.

## Basic Proxy State Machine

The Marshal state machine, as constructed by *MrshlBuilder*, is a variation of the state machine–generation mechanisms offered by the basic proxy. The state machine consists of a single feature, *MrshlFeature*, and a single state, *MrshlState*. All events placed in the Worker thread's FIFO pass from the *MrshlBuilder* to the *MrshlFeature* to the *MrshlState*. From there, the event passes through the ordered set of operators until it reaches the one that knows how to handle it.

### Operators

The Marshal operators are ordered as follows:

1. *MrshlOpAck*
2. *MrshlOpBye*
3. *MrshlOpCancel*
4. *MrshlOpInvite*
5. *MrshlOpRegister*
6. *MrshlOpStatus*
7. *MrshlOpProxy*

The only significance to this ordering is that *MrshlOpProxy( )* must be last because it serves as a default operator to proxy all SIP messages, which are not handled by any of the previous operators. The overhead of chaining through the operators is so small that rearranging the order so that the most frequently used ones are ahead of the others does not significantly affect performance.

### Return values

Each operator returns one of the following return values:

*PROXY_DONE_WITH_EVENT*
> If an operator handles a message to the extent that no other operators need deal with it, it returns PROXY_DONE_WITH_EVENT. This return value causes the process of chaining through the operators to stop before reaching the remaining operators. For example, *MrshlOpBye* returns PROXY_DONE_WITH_EVENT after it finishes handling a BYE message.

*PROXY_CONTINUE*

In all other cases, the operator returns PROXY_CONTINUE, which causes the message to be passed to the next operator in the chain.

Figure 11-4 shows a sample message flow through the different Marshal classes for a BYE message.

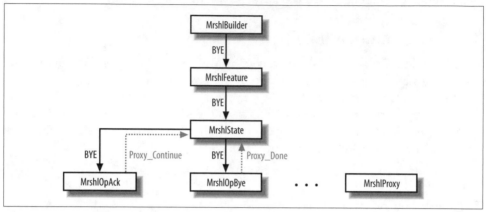

*Figure 11-4. Sample message flow*

## SIP Request Message Handling

Handling SIP messages is a coordinated effort between the SIP and Worker threads. The SIP thread handles SIP messaging at the UDP port level, taking care of message reception, transmission, and retransmission and duplicate message filtering. The Worker thread handles each type of SIP message differently. The following sections highlight the processing performed by the Worker thread and the corresponding operator for each type of SIP message. These sections also state any assumptions made regarding stack-level details abstracted away from the application layer by the SIP thread. In all cases, it is the *process()* method of the operator that does the real work.

### REGISTER

The Marshal handles SIP REGISTER messages with the *MrshlOpRegister* operator. When the Marshal receives a REGISTER message, it performs the following steps:

1. Marshals the incoming command.
2. Checks for a key in the ProxyAuthorization field.

   If the key is missing, that means that the command is an incoming SIP message. Adds the key to the field.

   Else, the command is an outgoing SIP message. Removes the key from the field.

3. Checks the MaxForwards header in the message.

   If it is 0, responds to the request with 483 Too Many Hops.

   Else, decrements the value in the MaxForwards header by 1 and proceeds.

4. If the message is incoming, authenticates the user by the configured authentication method (basic, access list, or digest).

   If the user's authentication fails, responds with 401 Unauthorized.

   Else, proceeds.

5. Checks the Via header for possible loops.

   If the command is incoming and the Marshal appears in Via or if the command is outgoing and the Marshal appears in Via more than once, returns 482 Loop Detected.

   Else, proceeds.

6. Sends a 100 Trying response.

7. Adds itself to the Via header with the branch using this syntax:

   ```
   SipCommand::computeBranch( ).1
   ```

8. Adds itself to the Contact header.

9. Forwards the message to the RS.

The Marshal uses the ProxyAuthorization header in a proprietary manner. Other VOCAL proxies do not look for this field to determine if the message is authorized for our system.

## INVITE

The Marshal handles the SIP INVITE message with the *MrshlOpInvite* operator in a different way for each type of Marshal server. User Agent Marshals forward the INVITE to an RS, unless there is a Route header, in which case they forward the INVITE based on the route. All other Marshals send the INVITE directly to their corresponding gateway, unless there is a Route header, in which case they forward the INVITE based on the route.

When the Marshal receives an INVITE message, it performs the following steps:

1. Marshals the incoming command (same as Step 1 in "REGISTER").

2. Sends a 100 Trying response back to the sender.

3. Adds its own Via with appropriate branch.

   If a Route header exists or if the Marshal is forwarding the message directly to a gateway, the branch has this form:

   ```
   <SipCommand::computeBranch( )>.2
   ```

   or:

   ```
   <SipCommand::computeBranch( )>.4
   ```

Otherwise, the INVITE is forwarded to an RS and the branch has this form:

```
<SipCommand::computeBranch()>.1
```

or:

```
<SipCommand::computeBranch()>.3
```

4. Adds its own RecordRoute.

   This is *not* done if a Route header already exists.

5. Sets RequestURI to the RS, gateway, or URL specified in the Route.

6. If the Marshal is an Internetwork Marshal, checks with the Policy server if the user is allowed to make calls.

   If not, returns 401 Unauthorized to the user.

   Else, proceeds.

7. Sends the message through the SIP stack.

## ACK

The Marshal handles the SIP ACK message with the *MrshlOpAck* operator in two types of cases. The simple case is receiving an ACK for a 4xx response generated by the Marshal. In this case, the ACK contains no Route header, and the Marshal needs to do nothing with the ACK, so it ignores it.

The other case is receiving an ACK for an INVITE/200 sequence. In this case, the Marshal starts billing and policy usage for the call and forwards the ACK to the called party based on the Route header.

When the Marshal receives an ACK message, it performs the following steps:

1. Checks the ACK message for a Route header.

   If a Route header does not exist, that means that the ACK was intended for this Marshal. Simply returns.

   Else, proceeds.

2. Marshals the incoming command (same as Step 1 in "REGISTER").

3. Starts billing for the call.

4. If this is an Internetwork Marshal, starts policy usage for the call.

   Else, skips this step.

5. Adds its own Via with a branch of this form:

```
<SipCommand::computeBranch()>.2
```

or:

```
<SipCommand::computeBranch()>.4
```

6. Forwards the message through the SIP stack based on its route.

## BYE

The Marshal handles SIP BYE messages with the *MrshlOpBye* operator in two types of cases. In the normal case, the Marshal receives a BYE with a Route header, which means that the BYE is intended for an active call and the Marshal needs to stop billing and policy usage and forward the BYE based on its route.

There is also an abnormal case in which there is no Route header. This happens if the caller sends a BYE instead of a CANCEL to cancel an INVITE prior to receiving a final response. See a description of this problem at the end of Chapter 8. Although this use of the BYE message is not recommended by the RFC, some vendors have incorrectly implemented it and the Marshal needs to support it. In the absence of the Route header, the Marshal uses the transaction state information in the SIP stack to route the BYE as it would a CANCEL.

When the Marshal receives a BYE message, it performs the following steps:

1. Marshals incoming command (same as Step 1 in "REGISTER").

   If the message is incoming and has no Route header, handles it as a CANCEL message, discussed later.

   Else, proceeds.

2. Stops the billing for the call.

3. If the Marshal is an Internetwork Marshal, stops the policy usage and QoS for the call.

   Else, skips this step.

4. If the Marshal is a Conference Marshal, removes the participant from the conference bridge, and if this is the last participant, tears down the conference call.

   Else, skips this step.

5. Add its own Via with a branch of this form:

   ```
   <SipCommand::computeBranch( )>.2
   ```

   or:

   ```
   <SipCommand::computeBranch( )>.4
   ```

6. Forwards the message through the SIP stack based on its route.

## CANCEL

The Marshal handles SIP CANCEL messages with the *MrshlOpCancel* operator in two types of cases. In the normal case, the Marshal receives a BYE with a Route header, which means that the BYE is intended for an active call. The Marshal needs to stop billing and policy usage and forward the BYE based on the route.

In the abnormal case, there is no Route header. This happens when the caller sends a BYE instead of a CANCEL to cancel an INVITE prior to receiving a final response. In the absence of the Route header, the Marshal uses the transaction state information in the SIP stack to route the BYE as it would a CANCEL.

When the Marshal receives a CANCEL message, it performs the following steps:

1. Marshals the incoming command (same as Step 1 in "REGISTER").
2. Checks for a pending command with matching To, From, CSeq number, and Call-ID fields.

   If there is a match, sends a 200 OK reply and proceeds.

   Else, sends a 481 Transaction Does Not Exist reply and exits.
3. If the Marshal is an Internetwork Marshal, aborts policy usage and QoS for the call.

   Else, skips this step.
4. If the Marshal is a Conference Marshal, removes the participant from the conference bridge. If this is the last participant, tears down the conference call.

   Else, skips this step.
5. Creates a new CANCEL from the pending command with the same To, From, Call-ID, and CSeq number but changes the method to CANCEL.

 For cases where this logic is applied to handle a BYE as a CANCEL, change the method to BYE.

6. Sets Request URI to that of the pending command.
7. Removes any existing Vias and adds its own Via with this form:

   ```
   <SipCommand::computeBranch( )>.2
   ```
   or:
   ```
   <SipCommand::computeBranch( )>.4
   ```
8. Sends the message through the SIP stack.

## SIP Status Message Handling

The Marshal handles SIP status messages with the *MrshlOpStatus* operator. Similar responses are grouped together and handled as follows.

### 100 Trying

There is no processing required for SIP 100 messages at the Marshal, but the SIP stack uses 100 messages to stop retransmissions.

### 181 Call Is Being Forwarded /182 Queued

The *MrshlOpStatus* operator handles SIP 181 and 182 messages as follows:

1. Forwards the message based on data found in the Via field.

2. Checks that there are at least two addresses in the Via field, a local system address and at least one more.

   If there are at least two addresses, proceeds.

   Else, exits.

3. Removes its own address from the Via field.

4. Sends the message through the SIP stack, which sets the Request URI based on the address in the Via field.

### 180 Ringing /183 Session Progress

The *MrshlOpStatus* operator handles SIP 180 and 183 messages as follows:

1. Starts billing for ring time.

   At this point, the Marshal makes sure that it is able to connect to at least one CDR server for billing purposes. If it is unable to connect, the Marshal checks whether it is provisioned to allow or to block unbillable calls and either allows the call to pass through or rejects it with a 402 Billing Required, accordingly.

2. If the Marshal is an Internetwork Marshal, enables QoS.

   Else, skips this step.

3. Forwards based on the address found in the Via field, same as Step 1 in "181 Call Is Being Forwarded /182 Queued."

   If the Marshal is a Conference Marshal, does not forward these responses.

### 2xx (success status messages)

The *MrshlOpStatus* operator handles SIP 2xx messages as follows:

1. Examines the CSeq method to determine the required processing.

   If CANCEL, no processing is required. Exits.

   If REGISTER, removes its own address from the Contact field and proceeds.

   Else, proceeds.

2. If the Marshal is a Conference Marshal, updates the conference map.

   Else, skips this step.

3. Forwards based on the address found in the Via field, same as Step 1 in "181 Call Is Being Forwarded /182 Queued."

### 302 Moved Temporarily

The *MrshlOpStatus* operator handles SIP 302 messages as follows:

1. Sends an ACK message based on the previous INVITE.

2. Searches the SIP stack's previous messages for a corresponding INVITE.

3. If no INVITE is found, forwards a 302 message based on the address found in the Via field (see the earlier section "181 Call Is Being Forwarded /182 Queued").

   Else, creates an ACK with the same Request URI as the corresponding INVITE.

4. Sends the ACK message through the SIP stack.

5. Checks whether the corresponding INVITE was sent to the RS; if not, forwards 302 based on the address found in the Via field (see "181 Call Is Being Forwarded /182 Queued"), and exits.

6. Replaces the Marshal's own address in the Via field of the original INVITE with another that has the appropriate branch.

   If a ProxyAuthorization header exists, the branch has this form:

   ```
   <SipCommand::computeBranch( )>.2
   ```

   Else, the branch has this form:

   ```
   <SipCommand::computeBranch( )>.4
   ```

7. Checks the Contact header in the 302 message; if none exists, returns 404 Not Found to the sender of INVITE and exits.

8. If the Marshal is a Conference Marshal, creates a new conference or adds participant to existing conference.

   Else, skips this step.

9. Sets the Request URI of the INVITE to that of the address found in the Contact field.

10. If the Marshal is an Internetwork Marshal, checks for an OSP token in the INVITE. If it does not exist or is invalid, rejects with a 401 message and exits.

    Else, skips this step.

11. Sends the INVITE message through the SIP stack.

### 4xx (Client Error), 5xx (Server Error), 6xx (Global Failure)

The *MrshlOpStatus* operator handles SIP 4xx, 5xx, and 6xx messages as follows:

1. Sends an ACK based on the corresponding INVITE (see Step 1 in "302 Moved Temporarily").

2. Forwards 302 based on the address found in the Via field (see "181 Call Is Being Forwarded /182 Queued").

# Functionality

Some of the Marshal's functionality has been alluded to in the previous section. Primarily, the Marshal provides authentication services and CDR data that can be used for billing purposes. Marshals also monitor heartbeats from the RS and can be set up to process OSP tokens or conference calls. Security and authentication are big topics and are discussed in separate sections later. Let's look at some of the other functionality provided by the Marshals.

## Billing

Each Marshal is responsible for constructing CDRs for all calls that pass through VOCAL. The Marshals send these records to the CDR server, which can forward them to a third-party billing system if such a system has been attached to the network. If there is no billing system, the CDR server can store these records to provide historical usage statistics. Each Marshal reports the ring time, start, and end of each call by making the appropriate calls into the CDR API provided by the CDR library. When two Marshals are involved in the same call (for example, when a call is routed through a UA Marshal and Gateway Marshal), both Marshals provide separate CDR records to the CDR server.

Marshals are responsible for collecting records that describe only the ring time, start, and end of basic calls. If there are separately billable features available in the system, CDRs for those features are provided by the Feature servers.

Marshals invoke the CDR API of the CDR library as follows:

- Start of ring time, identified by the 180 Ringing reply for INVITE messages
- Start of call, identified by the ACK returned for INVITE messages
- End of call, identified by the BYE message

The Marshal invokes the corresponding CDR API call with the required parameters. The actual messaging, which occurs as a result of these calls and the actual construction of the CDR records, is discussed in Chapter 18.

## Quality of Service

The Internetwork Marshal uses the COPS library to request QoS for all calls placed through it. It calls *enableQos()* whenever it receives a 183 and *disableQos()* whenever it receives a BYE. These routines make the appropriate calls into the policy enforcement point (PEP) library. The actual messaging that occurs as a result of these calls is discussed in Chapter 18.

## Heartbeat with Redirect Server

The base SIP proxy contains a Heartbeat thread that handles the basic heartbeat mechanism for the Marshal. No additional code is needed within the Marshal itself. See Chapter 9 for more information about heartbeating.

## Redundancy

In the event that a Marshal goes down, the ability of external UAs and proxies to send and receive messages through VOCAL depends on both the ability of VOCAL to provide another Marshal and the ability of the UAs and proxies to connect to it. There are two types of redundancy configurations:

*Nonredundant configuration*

In a nonredundant configuration, if a Marshal goes down, there is no replacement Marshal to work with the affected network components. When this happens, these components are not capable of sending or receiving subsequent messages or calls through VOCAL. This service outage remains in effect until the Marshal comes back into service.

When service resumes, the components will be able to send requests to VOCAL; however, if the affected components are UAs, which register, they will not be able to receive calls from VOCAL until after reregistering with the RS. This constraint does not apply to proxies and gateways because they don't register.

*Redundant configuration*

In a redundant configuration, if a Marshal goes down, one or more other Marshals exist to work with the affected components. When this happens, these components receive subsequent messages or calls from VOCAL automatically through the new Marshal; however, affected UAs and proxies cannot send messages to VOCAL until they switch over to the new Marshal. To facilitate redundancy, configure the potentially affected UAs and proxies to use more than one Marshal. Reregistration is not required in redundant schemes.

The Marshal uses the round-robin server code to listen to heartbeats from all RSs. It uses information within the headers of the request to deterministically forward INVITE requests among the active Redirect servers, which provides the following benefits:

- Load balancing: the Marshal load-balances its requests among all active RSs so that all active RSs are utilized equally, and the same request is rerouted to the same RS (provided it is still active) in order to take advantage of any cached information.

- Redundancy: if one or more RSs go down, the Marshal chooses from the RSs that are still active.

## Security

Security mechanisms are built into the Marshals' SIP stacks. All security measures taken by the Marshals, to ensure secure communication between all external SIP entities and VOCAL, are the result of using the appropriate SIP stack security mechanisms. Here is a look at some possible security mechanisms.

# Pretty Good Privacy

Pretty Good Privacy (PGP, *http://www.pgp.com*) is an end-to-end mechanism that seemed like a good idea at the time the architects were writing some of the early drafts of the SIP standard, but it was not popular with VoIP developers and has since been deprecated. Part of the reason for the demise of PGP was its requirement that full public key encryption be used for every transaction. This is problematic from the server side because, while processing hundreds of calls per second, it's difficult to encrypt and decrypt messages fast enough to avoid a traffic bottleneck. If PGP allowed for a session key that was valid for the entire call, this problem might have been avoidable.

# IPsec

In terms of encryption for this type of system, Internet Protocol security (IPsec, *http://www.ietf.org/html.charters/ipsec-charter.html*) is a hop-by-hop mechanism that offers a better solution than PGP. IPsec operates under the SIP layer and permits data transportation over TCP or UDP by setting up a security association between two SIP devices. Once this security association and a set of keys have been set up, IPsec encrypts all traffic associated with the same call to match these keys. Check out a project called Linux FreeS/WAN (*http://www.freeswan.org*), which is an open source Linux implementation of IPsec and Internet Key Exchange (IKE, RFC 2409).

As we have discussed in Chapter 8, SIP can be transported over TCP. There are many developers who would like to be able, instead of just using TCP, to set up an authenticating encryption connection using Transport Layer Security (TLS, RFC 2246). TLS is the IETF version of Secure Socket Layer (SSL). At the time of this writing, VOCAL and the Vovida SIP stack do not support TLS. Development of this will probably be built upon the OpenSSL (*http://www.openssl.org*) stack.

# Firewalls and NAT

A firewall is a set of programs running on a gateway, router, or other device that protects private networks from unauthorized intruders hacking their way in and snooping around. Network Address Translation (NAT) is a device that maps the IP addresses assigned to devices in one network to a different set of addresses in another network. On SIP-based networks, this type of translation works fine for signaling over UDP or TCP connections. The problems with NAT firewalls lie with translating the source and destination of the streaming RTP channel. Let's look at some proposals for solving this problem with both non-NAT and NAT firewalls.

## Non-NAT firewalls

The problem, as discussed before, can be more finely defined as looking at how the firewall allows traffic in and out and how the different protocol stacks used for signaling and streaming media need to be set up. Typically, SIP traffic uses port 5060, and RTP and its control protocol RTCP normally use neighboring even- and odd-numbered ports, respectively; for example, ports 9000 and 9001. The question becomes, "Is this UA known and trusted?" This can be scary from a security point of view because it implies extending the network's security beyond the perimeter of the Marshal servers.

Figure 11-5 illustrates the basic requirement of sending SIP messages through a firewall and setting up an RTP path.

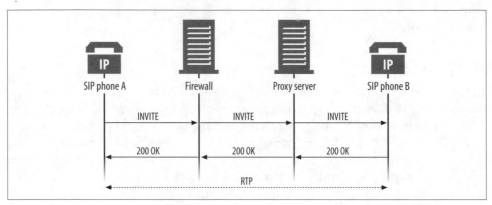

*Figure 11-5. Firewalls*

Alternatively, if your UA always received on the RTP on the same port, you could just enable that incoming port on the firewall and allow anything outgoing. The Cisco analog telephone adapter (ATA) uses port 4000. The Cisco 2600 gateway uses ports starting at 16,768. The Cisco 7960 IP phone uses different ports each time, incrementing the port numbers by 2 for each new call. The VOCAL UA uses whatever port has been configured for it in the *ua.cfg* file.

### SIP ALGs for NATs and firewalls

The first NAT to support a SIP application-level gateway (ALG) was a Linux-based firewall using a module developed by Canadian software engineer, Billy Biggs. Figure 11-6 illustrates a network topology featuring this firewall. In this setup, the NAT firewall understands SIP and uses the data within the SIP header fields to route the messages properly.

Let's look at the message flow illustrated in Figure 11-6. The phone sends an INVITE with the address, 192.168.0.2, written into the *Via* field. The NAT receives this packet and forwards it to the proxy, which appends the address 24.1.1.1 to the *Via* field. When the response is returned, the proxy retrieves the 24.1.1.1 address from the *Via* field and sends the reply to that address. The firewall looks up 24.1.1.1 on its mapping table of IP addresses and firewall appendages, which tells it to forward the message to the phones' true address, 192.168.0.2.

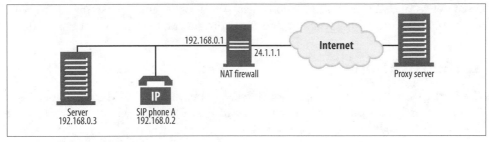

*Figure 11-6. Using a SIP ALG-NAT firewall*

This method works fine for outgoing call signaling but provides no solution for several important issues including:

- RTP (voice channel)
- Incoming calls
- NAT forwarding to the internal SIP Proxy server
- Dynamic Host Configuration Protocol (DHCP, RFC 2131) address assignments and dynamic reassignments without notifying the UA or proxy

### Preconfigured NAT firewalls

Let's talk about a solution to some these issues. One example is a single Cisco ATA behind a Linksys firewall (*http://www.linksys.com*), as illustrated in Figure 11-7.

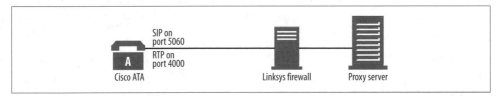

*Figure 11-7. Using the Linksys firewall with a Cisco ATA*

Configure the Linksys firewall so that any incoming traffic on ports 5060, 4000, and 4001 goes to the Cisco ATA. The proxy ideally could accept a REGISTER from a web page that advertises reaching the caller at IP address 24.1.1.1:5060, which is the IP address of the firewall, not the phone. If a DNS name for a Linksys firewall interface is used instead of 24.1.1.1, it will work with DHCP changes. Check out *http://www.dyndns.org* for more solutions; they are supported by many different products, including Netgear firewalls.

If the proxy was set up to take the address of the user on the far end of the connection from the SDP info attached to the INVITE message, the far end would send it to the firewall, which would know to resend it to the UA. If a proxy like this is not available, it is sometimes possible to tell the UA not to put its own IP address into the SDP fields, and to write the firewall's IP address instead. This is possible with the

Cisco ATAs and Cisco 7960 IP phone sets. We have found this setup useful with individual IP phones, but it is painful to configure several phones over the same unit.

# Authentication

The authentication provided by the Marshal depends on the Marshal type. User Agent Marshals perform user-level authentication. PSTN Gateway, Conference, and Internetwork Marshals perform gateway-level authentication.

## What Is Authentication?

Authentication is the process that protects the system from unauthorized users. The Marshal servers authenticate each call by checking the calling party's IP address against a master file. If the Marshal does not have the calling party's address on its list, it requests verification from the Provisioning server. If the Provisioning server does not verify the address, the Marshal refuses to authenticate the call. The authentication method can be either access list or digest.

That takes care of the high-level picture. The lower-level picture has to do with the actual security of a single call.

## User-Level Authentication

All User Agent Marshals perform user-level authentication. Each user within a given VOCAL system is configured with its own authentication scheme and Marshal group.

The authentication scheme is one of the following:

*Marshal group*
> The Marshal group is the group of one or more Marshals through which the user communicates with the VOCAL system. When a User Agent Marshal receives an incoming SIP request, it reads the user portion of the From header to determine the user. Then it gets, and caches, the user's information from the Provisioning server to determine which type of authentication to enforce.

*None*
> If a user is configured with None as its authentication type, all requests for the user are allowed into VOCAL.

*access list*
> If a user is configured with access list authentication, the source address of the IP packet containing the request is matched against that provisioned for the user, and the request is rejected with a 401 Unauthorized message if the addresses do not match. This is true for all requests except BYE.

*digest*

> If a user is configured with digest authentication, the Marshal authenticates only the user's REGISTER and INVITE requests due to requirements outlined in RFC 3261.

For digest authentication, the Marshal acts on behalf of the registrar of the VOCAL system. It performs the authentication of the REGISTER message on behalf of the RS, which stores the contact information specified in the REGISTER message. The Marshal authenticates users by challenging REGISTER requests with a 401 reply.

The Marshal does not act as a User Agent server; therefore, it cannot use the same logic to challenge other requests with a 401. It could, and quite arguably should, respond with a 407 Proxy Authentication Required message, but to do this requires additional functionality in both the SIP stack and the Marshal, which has not been implemented for either.

## Basic

The other method that SIP provides is the basic scheme, in which the user puts a username and a password into the file. This is also incredibly lame because it results in passwords being sent unencrypted over the network. We can't imagine anyone in his right mind ever using this and have no idea why it was included in RFC. 2543. Basic has been deprecated from RFC 3261.

## Access list

If the User Agent Marshal authenticates the user agent, it forwards the user agent's message through to the RS. If the message is REGISTER, the Redirect server registers the user agent and returns a confirmation message back through the UA Marshal to the UA.

Figure 11-8 shows a SIP phone registering with the RS.

Table 11-1 describes the messages illustrated in Figure 11-8.

*Table 11-1. Interactions shown in Figure 11-8*

| Interaction | Step | Description |
| --- | --- | --- |
| SIP phone to UAMS | 1 | The SIP phone is connected to the network and immediately sends a REGISTER message to the User Agent Marshal server (UAMS). |
| UAMS to PS | 2–3 | The UAMS does not have a record of the SIP phone's IP address in its database, and it retrieves data from the Provisioning server (PS) to validate the request. The UAMS adds the SIP phone to its list of authorized users. |
| UAMS to RS | 4 | The UAMS forwards the REGISTER message to the Redirect server (RS). |
| OK returned | 5–6 | The UAMS forwards the OK to the SIP phone. The phone is registered in the system, and it will reregister every few minutes. |

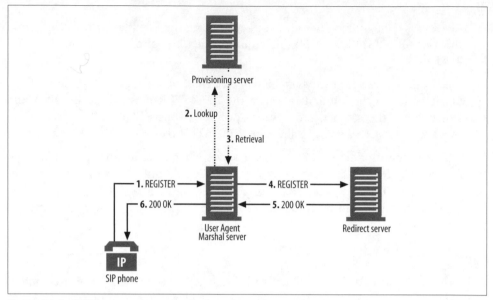

*Figure 11-8. Calling party registration: access list*

## Digest

The next sort of level is digest authentication, and this works the same way as digest authentication in HTTP. The UA sends an INVITE that is effectively unauthenticated to the Marshal and the Marshal returns a 407 Authentication Required message, which is the same as saying, "No, I'm not going to accept that; you have to authenticate." It also includes a *nonce*, a one-time random number that may or may not be public, to help ensure that the security can't be broken by replay attacks at a later time. In a *replay attack*, an intruder replays messages recorded from earlier, authorized transmissions.

> We prefer to say *intruder* rather than *hacker* to describe an individual who illegally breaks into networks. Years ago, the term *hackers* was used to describe creative people who could build simple computers out of spare parts. Often in our normal work, it is common to legally "hack" one of our programs to create a workaround for a persistent bug, and by extension, that makes most of us hackers. How the public media managed to twist this term to suggest something criminal is beyond our knowledge.
>
> Another twisted term is *showstopper*. In our shop, a showstopper is a bug that is severe enough to delay a release. In the popular press, it is a breakthrough feature, a showstopping accomplishment. Go figure.

The UA takes its username, password, this nonce, and some other information to create a unique key. Basically, it hashes all of this data together through a one-way hash, which is the MD5 algorithm (a unique one-way hash, RFC 1321), to come up with its key. The UA then sends the INVITE again, including this authorization information, the encrypted key, inside the message.

The receiving Marshal does the same thing as the UA. It knows the username and the password for this user and does all of the same things to form a key. If the keys match, then clearly the user is using the correct password, and the Marshal can proceed with the authentication. The nice part about this process is that it doesn't result in sending the password across the network. Both sides knew the password. It wasn't a public key encryption scheme in which people didn't need to know one another's passwords. Both sides need to know the same password, but they can confidently authenticate the password without sending it in the network.

Figure 11-9 shows a SIP phone using digest authentication to register with the RS.

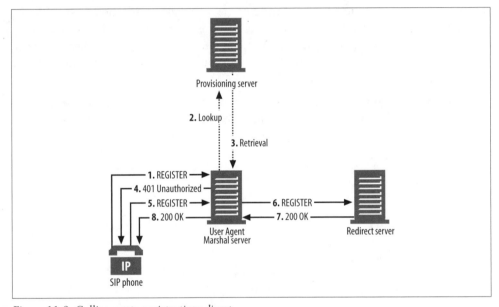

*Figure 11-9. Calling party registration: digest*

Table 11-2 describes the messages illustrated in Figure 11-9.

*Table 11-2. Interactions shown in Figure 11-9*

| Interaction | Step | Description |
| --- | --- | --- |
| SIP phone to UAMS | 1 | The SIP phone is connected to the network and immediately sends a REGISTER message to the User Agent Marshal server (UAMS). |
| UAMS to PS | 2–3 | The UAMS does not have a record of the SIP phone's IP address in its database, and it retrieves data from the Provisioning server (PS) to validate the request. The UAMS adds the SIP phone to its list of authorized users. |

*Table 11-2. Interactions shown in Figure 11-9 (continued)*

| Interaction | Step | Description |
|---|---|---|
| Unauthorized | 4 | The UAMS returns a 401 Unauthorized message to the SIP phone requesting a password. This message includes the nonce to use. |
| New REGISTER message | 5 | The SIP sends a new REGISTER message that includes a password. |
| UAMS to RS | 6 | The UAMS authenticates the calling party and forwards the REGISTER to the RS. |
| OK returned | 7–8 | The RS replies with a 200 OK message, which is forwarded to the SIP phone by the UAMS. The phone is registered in the system, and it will reregister every few minutes. |

## Gateway-Level Authentication

All PSTN Gateway, Conference, and Internetwork Marshals perform gateway-level authentication for all requests. When provisioning these types of Marshals, the technician is prompted for the address of the gateway associated with the Marshal. When the Marshal starts up, it receives and caches this address from the Provisioning server. For all incoming requests, the Marshal checks the source address of the IP packet containing the request. If it does not match the Marshal's gateway, the request is rejected with a 401 message.

Having looked at how the Marshal server has been developed, let's look at the next hop in the SIP call flow, as the INVITE message winds its way through VOCAL, the Redirect server.

# CHAPTER 12

# Redirect Server

The Redirect server (RS) is very simple and performs only a few basic tasks. It receives SIP REGISTER messages from User Agents (UAs), keeps track of registered users and their locations, and provides routing information for SIP INVITE messages.

## High-Level Design

The RS knows about each user in the system. When an ingress Marshal forwards a new call request to the RS, the RS decides where the call needs to go by looking at the calling party, the intended called party, and any registration information it may have about either party. If the Request URI is an external phone number, the RS looks in the dial plan for meaningful routing information.

The signaling message also contains information that tells the RS where the call has already been, which is needed to figure out where it needs to go next. For example, the call may require routing to a Feature server before reaching the egress Marshal. Before forwarding the call, the RS checks to see if the intended servers are up and running. If the next intended server is down, the RS forwards the message to a substitute server if one is available.

## Class Diagram

Key data structures are the subscriber container, IP and phone dial plans, and the server container. Figure 12-1 shows a class diagram of the structures found within the RS. The gray background highlights the base classes, as discussed in Chapter 9. The classes marked with asterisks are single instances of these objects.

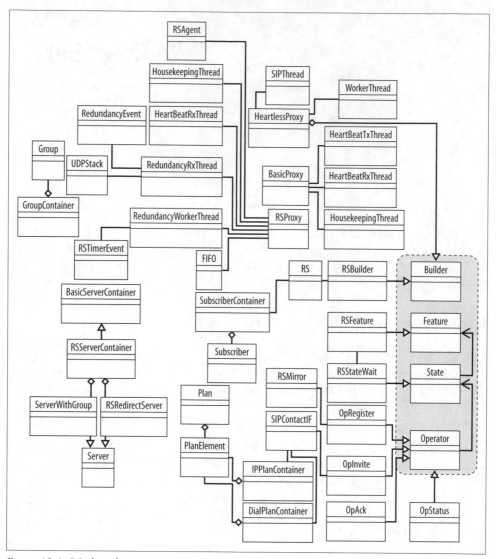

*Figure 12-1. RS class diagram*

## Key classes

The following is a list of the RS-specific classes illustrated in Figure 12-1:

*DialPlanContainer*
An ordered list of pairs of regular expressions and vectors of destinations used to provide routing information for phone numbers.

*Group*
Servers within the same group can perform the same functions.

*GroupContainer*
Encapsulates *Group*.

*IPPlanContainer*
An ordered list of pairs of regular expressions and vectors of destinations used to provide routing information for IP addresses.

*OpAck*
Responds to SIP ACK messages received from the SIP thread.

*OpInvite*
Responds to SIP INVITE messages received from the SIP thread.

*OpOptions*
Responds to SIP OPTIONS messages received from the SIP thread.

*OpRegister*
Responds to SIP REGISTER messages received from the SIP thread.

*OpStatus*
Responds to numbered SIP status messages, such as 200 OK, received from the SIP thread.

*Plan*
Encapsulates *DialPlanContainer* and *IPPlanContainer*.

*PlanElement*
Encapsulates *Plan*.

*RedundancyEvent*
A *SipProxy Event* is posted when a redundancy message is received from a server.

*RedundancyRxThread*
Defines a thread whose job is to process RS redundancy events from the UDP stack.

*RedundancyWorkerThread*
A *ThreadIf* (thread interface) has an input queue that contains events, on which it is blocked.

*RSMirror*
To support system redundancy, mirrors a REGISTER message from the Redirect server to all other RSs. If the REGISTER message is from another RS, it does nothing.

*RSProxy*
Creates a call container, callprocessing FIFO, SIPstack, and UDPstack, as well as Worker, Sip, and Heartbeat threads. The RSProxy::run method starts the Worker, Sip, and Heartbeat threads. A builder object and SIP and UDP port numbers are given upon instantiation.

*RSRedirectServer*

Defines an RS, which derives from the *Server* base class. The added difference is a sync port. This is a UDP port, which the server uses to send synchronization messages back and forth between other servers of this type.

*RSServerContainer*

Used to determine active servers and select substitutes for inactive servers.

*RSTimerEvent*

A Timer event that is posted when a timer goes off.

*ServerWithGroup*

Expands on the base class of *Server* by adding the concept of groups. Each Server can belong to a group or groups. These group names are stored within this class as a list.

*ServerWithoutGroup*

Not in use.

*SipContactIf*

Processes SIP messages.

*Subscriber*

Keeps track of information about individual subscribers.

*SubscriberContainer*

Keeps track of all users in the system and the data that the RS needs to know about them. Encapsulates *Subscriber*.

### The dial plan container

A dial plan is a table that contains patterns that represent phone numbers and routing information for each pattern. See Chapter 5 for examples of dial plans. The *DialPlanContainer* is an ordered list of paired regular expressions and vectors of destinations. If a called party's number matches a regular expression, the corresponding vector of destinations is used for routing the call to the correct servers and the correct endpoint. The ordering of entries in the dial plan is important because the first match is used for routing. Special destination patterns—such as *$USER*, which is replaced with the name of the person being called—can be used. Dial plans are discussed thoroughly in Chapters 2 and 5.

Figure 12-2 shows how plans can flow out of either the dial plan or IP plan containers.

There are two dial plan containers: IP and digit. When the user part of the Request URI is a telephone number, the digit plan is used. Otherwise, the IP plan is used. Each plan has the capability of containing a number of dial plans for complex installations in which several organizations using the same switch require separate plans; however, the code required to make this possible has not been fully activated.

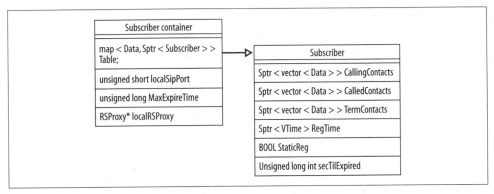

*Figure 12-2. Plan containers*

## The server container

*RSServerContainer* is used to keep track of active servers and select substitutes when servers go down. *RSServerContainer* maps an IP address to a structure that provides information about the type of server and its group. The group container maps a group name to the list of servers in that group that are currently active.

Figure 12-3 shows the relationships among the server objects.

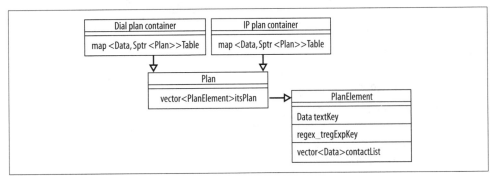

*Figure 12-3. Server objects*

The RS uses Server objects to keep track of the status of all configured servers in the system by heartbeating with the Marshal, Feature, Java Telephony Application Programming Interface (JTAPI), and other Redirect servers. All VOCAL servers belong to a group except for the RS.

Although a JTAPI server was written and included with the VOCAL code, it is a partial implementation of the standard, which was intended as a proof of concept or developer's starting point rather than a fully functioning server.

## Server groups

All servers within the same group can perform the same functions and can become substitutes for one another should one or more fail. This server substitution is enabled by the Heartbeat and HouseKeeping threads, which can be found in the CVS tree under Heartbeat.

Figure 12-4 shows the relationships among the group container and groups.

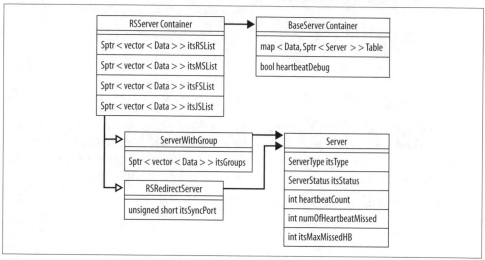

*Figure 12-4. Group containers and groups*

## The subscriber container

*SubscriberContainer* keeps track of the system's users as hash values that map to their phone numbers. *SubscriberContainer* contains objects, called *subscribers*, and each subscriber contains vectors of contacts for a specific user. Figure 12-5 shows the relationships among subscriber containers and subscribers.

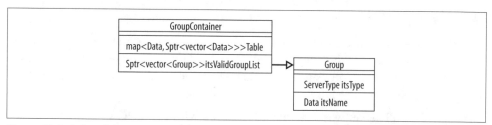

*Figure 12-5. Subscriber container and subscribers*

The calling contacts are a list of Feature servers that need to be accessed when a call originates from a subscriber. The called contacts are the features that need to be accessed when a call terminates on this subscriber. Terminating contacts are obtained from the REGISTER message and tell the RS where the user is currently located. An expiry time is also stored for the term servers so that the system can time out a registration at the appropriate time. It is the responsibility of the UA to reregister with the system before its current registration expires.

## Startup

On startup, before the RS becomes active, it connects to the Provisioning server (PS) to download data and register for updates with specific directories and files. By registering with the PS, the RS ensures that it is notified about new or modified data as it is entered into provisioning by the system administrator.

Provisioning provides the RS with a multicast address for heartbeating, an expiry setting for user registrations, dial plans, a list of all provisioned users on the local network, and a list of all configured servers and their groups. Initially, the RS stores only the usernames, flagging each as either a master or an alias. The RS creates a blank master subscriber object and a blank alias subscriber object that are placed in the subscriber container. Each user record on the RS initially points to these objects until the corresponding user registers with the system. Upon registration, the subscriber record is accessed, and as it is blank, the RS retrieves the correct record from provisioning and creates a unique subscriber object for that user. Starting up with a blank subscriber object lets the system start up more quickly and reduces the startup avalanche load on the provisioning system. For more information about this container, see the earlier section, "The subscriber container."

After it is finished with its initial connection to provisioning, the RS starts the RSProxy thread and the Heartbeat, Receive, and Housekeeping threads. The RS needs to discover if it is the first RS to come up or if there is another RS already active in the system. The RS listens to the heartbeats for 2 seconds; if it does not hear another RS, then it assumes that it is the first RS to start up and becomes active by starting the Heartbeat Transmit thread. If it hears another RS, it begins synchronization with the other RS so that it can act as a redundant backup server.

## Redundancy

Upon startup, if the new RS discovers another RS in the system, it needs to synchronize user registration data with the existing RS to enable redundancy. To synchronize, the new RS sends a *syncReq* message to the sync port on the existing RS and waits for it to reply with an acknowledgment. Provisioning provides the sync port address. Once this exchange has completed, the existing RS sends the registration information for each user to the new RS on its SIP port as separate SIP REGISTER messages. Once it is finished sending this data, the existing RS will notify the new RS on its sync port. This interaction is shown in Figure 12-6.

The existing RS finds this registration information by looking at each user in the subscriber container. If a given user's term information is not blank and has not expired,

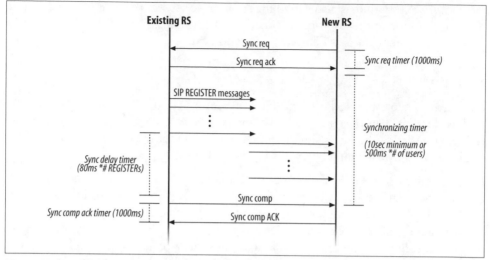

*Figure 12-6. Synchronization messaging*

the RS constructs a REGISTER containing the user's term and time remaining before expiry and sends it to the new RS. When the existing RS has sent all of its user data, it sends a Sync Complete message to the new RS, which returns an acknowledgment. At this point, the new RS can become active and start sending heartbeats, enabling other servers in the system to find it and start using it.

If there is more than one existing RS, it doesn't matter which one the new RS chooses to synchronize with, because all existing RSs are supposed to contain the same user registration information. Multiple RSs allow the system to handle more calls, but redundancy requires a synchronization of all state information between all RSs. The only state information that the RS stores, aside from the provisioning data, is the registration data. The Redundancy Worker Thread makes synchronization possible.

 In these next sections, when the text refers to "the technician," it is refering to someone who is signed onto the Java Provisioning Server Configuration screens as described in Chapters 5 and 6.

## Synchronization (RedundancyWorkerThread.cxx)

The RS uses a UDP port called the *synchronization port* to synchronize registration data with other RSs in the system. The technician must configure this port in the provisioning GUI when maintaining the RS's data. Without the sync port, the RSs would get out of sync. Were this to happen, the registration data could be different or nonexistent for a user on one RS versus another. A call to that user may not terminate correctly, depending upon which RS was accessed during the call. Active RSs stay in sync by mirroring REGISTER messages that come in to a single RS.

---

### Mirroring (RSMirror.cxx)

Servers send REGISTER messages to only one RS. In a system with multiple RSs, a received REGISTER message must be sent to all other RSs regardless of heartbeat status. An RS will mirror a received REGISTER message to all other configured RSs by copying the REGISTER message and modifying the From and Req URI fields before sending it over the SIP port. The receiving RSs will see that it came from another RS and will not try to mirror it again.

### Heartbeat

Heartbeating allows servers to keep track of the status of other servers within the system. A heartbeat is a UDP message sent periodically from a server to a preconfigured multicast address on the network. Any server listening to this multicast address is able to process the message and keep track of the server's active or inactive status.

Several threads compose the heartbeat library. Figure 12-7 illustrates the operation of these threads.

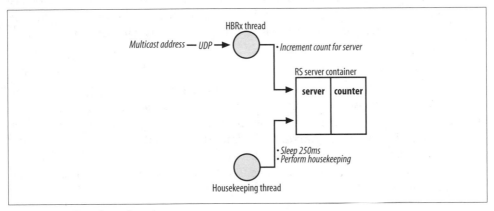

*Figure 12-7. Heartbeat threads*

*HeartbeatTxThread* (HBTx) periodically sends a heartbeat, a multicast UDP packet, onto the network, announcing to the other servers that it is active. The interval between heartbeats is configured in the Provisioning server. The *HeartbeatRxThread* (HBRx) receives all heartbeat packets and looks up the sender's IP address in the server container's index. If the sender is a server that the RS knows about, the RS looks up the server's group name. It then finds the server's record in the group container and sets the flag to indicate that a heartbeat was received. If the heartbeat is coming from an unknown server, *HeartbeatRxThread* ignores it.

The *Housekeeping* thread runs periodically, using the same interval as the heartbeats. Each time this thread runs, it looks through all the servers listed in the server containers to see if any heartbeats have been received since the last time it checked. For every server that has sent a heartbeat, the Housekeeping thread clears the missed heartbeat counter to zero. For every server that has not sent a heartbeat, it increments the missed heartbeat count upward by 1. If this count exceeds a configured threshold, which is usually set between 4 and 8, the Housekeeping thread concludes that this server is down and marks it as inactive. This scheme allows the RS to maintain a list of active servers for each group, enabling it to switch to an active server if one server goes down.

## Registration

When a SIP User Agent registers, it generates and sends a REGISTER message to the Marshal. If it passes authentication, the Marshal forwards the REGISTER message to the RS. The RS, through its *OpRegister* class, checks for an entry in its subscriber container that corresponds to the user's address provided in the To field of the REGISTER message. The Provisioning server provides this entry from the data entered by the system administrator. If the RS cannot find an entry for the user, this means that the user has not been entered into provisioning, and the RS rejects the registration. For those users with valid provisioning entries, the Contact field or fields (as there may be more than one in the REGISTER message) contain the information that the RS stores as the user's termination contact list. This list usually contains only the addresses for the UA Marshal and the UA.

For this user's registration, there is an expiry time set up to keep the registration as current as practically possible. When the RS processes the REGISTER message, it stores the time of registration. Whenever the subscriber data is accessed for call processing, the RS compares the registration timestamp against the current time. If the delta is longer than the configured expiry time, the registration is considered timed out, and this user's termination contact list is removed from the RS's database. The UA must send a new REGISTER message to the RS before calls can be processed to or from that user.

 At the time this was written, the code that removes the terminating contacts upon registration timeout had been implemented but was not in use. Check the *readme* file for the latest release of VOCAL to see if this code is being used.

Initially, there are no calling or called feature contact lists for a user in the RS database. When the user sends its first REGISTER, and as long as the RS is active, it will request the information from the PS. Subsequent calls involving the registered user do not need to query the PS for this information: This information is cached in the RS. The RS also registers its interest in this user with the PS so that any changes to the user at the PS level are passed on to the RS.

The reason the RS waits until the initial REGISTER before requesting the feature contact lists for a user is because an inactive RS may be synchronizing with an active RS and receiving a large number of REGISTER messages, one for each registered user. (VOCAL was originally designed to handle over 100,000 users.) Requesting each user's contact lists from the PS during synchronization would flood the system.

## Calling and called features

The distinction between calling features and called features is not a creation of the packet telephony world; these terms have been used in the Public Switched Telephone Network (PSTN) world for years. A calling feature is one that is activated when a UA places a call, and these can include Call Blocking and Call Return. A called feature is activated when a UA is the intended receiver of a call, and these can include (incoming) Call Screening and Call Forward Busy. See Chapter 13 for more information about features.

## Static registration

Another registration scheme, called *static registration*, always considers users to be registered. No SIP REGISTER messages are expected from statically registered users and, if any are sent, they are discarded. The user's Marshal group and static IP address are configured in provisioning. The RS will obtain this information from the PS along with the user's calling and called contact lists when the user is first involved in a call.

# Call Processing

The RS takes several stages to determine where to redirect the call. First, a contact list is compiled, and then the next contact is determined. Finally, the contact is validated and returned in a status message.

The RS, through its *OpInvite* class, compiles a contact list using the INVITE's From and Req URI fields. First, the RS searches for the From in the Subscriber database. Next, it tries to match the Req URI's username in the Subscriber database and, failing that, will try to match the whole Req URI in either the digital plan or the IP plan. The ;user=ip or ;user=phone portion of the Request URI will determine whether the RS uses the IP plan or digital plan, respectively. If the user tag does not exist in the URI, the default is ;user=ip. The list will be compiled differently for calls terminating on a subscriber within the system than for calls going through the dial plans.

Two examples of the types of contact lists that the RS generates when it receives a SIP INVITE message follow.

For calls to subscribers, the list is compiled as follows:

- The calling contacts found in the From field (obtained from provisioning)
- The called contacts found in the Req URI field (obtained from provisioning)
- The terminating contacts found in the Req URI field (obtained from registration)

Or for calls to something in a dial plan, the list is compiled as follows:

- The calling contacts found in the From field of the SIP request message (obtained from provisioning)
- The contacts found in the dial plan that match the Req URI field

Contacts are made up of either SIP URIs, server addresses, or group names. Terminating contacts are SIP URIs. The calling and called feature contacts are stored as group names to allow the RS to pick an active Feature server when one is required. Dial plan entries can be any of the three forms depending on what the technician enters into provisioning.

The three types of contacts are defined as follows:

*Calling contacts*
> Feature servers related to the originator's calling features. There are few calling features such as call blocking within VOCAL.

*Called contacts*
> Similar to the calling contacts; Feature servers related to the destination number's features. If the call is intended for a subscribed user, any feature turned on for that user will correspond to a Feature server, and a call will need to include this server in its routing through the system. If the call is intended for a user on the PSTN or on another system, the called features will be unknown and not a factor in the call routing.

*Terminating contacts*
> Usually the Marshal server and IP address of the registered user. These contacts are received in a REGISTER message from the user.

A feature that takes place when a number is called (like Call Forward All Calls) is handled before the terminating contacts (address of the set itself). This is because the call may be forwarded to voice mail or another number or may be blocked and will never reach the terminating contacts.

Once the contact list has been generated, the RS is ready to determine which contact should be passed back in the status response message. The RS must determine the call's current place in the contact list by using the INVITE's Via fields, which show a history of the call's path through the system. Usually, the RS searches the contact list for the address contained in the first Via field. Once the address has been found, the next contact on the list is passed back as the next hop in the call.

If the address in the first Via field is a Marshal, the RS generates the contact list differently. When this happens, the RS looks at the second Via in the list. If it exists as a server in our system, we can assume that the call is leaving our system and that it is safe to search for the top Via within the contact list. If the second Via is not in our system, we can assume that the call is entering our system and we should take the first contact in the compiled list and return that in the status response message.

This method can be expressed in the following if/else statement:

- If the top Via is a Marshal and the second Via is not a server in our system, return the first contact in the list.
- Else, search for the top Via in the contact list and return the next contact in the list.

Once the contact has been found, it is verified as active through heartbeating. If the contact is a group name, the RS can choose an active server from those in that group. If the contact is found to be inactive, the RS attempts to obtain an active server from the same group. If none is available, either the next contact is taken (in the case of a Feature server) or the call is failed (in the case of a Marshal). If the contact is not recognized as a server in our system, it will be assumed to be active and the contact will be returned as is in the 302 Moved Temporarily message.

# Routing

The RS handles routing differently for SIP requests, responses, and subsequent messages.

## Initial SIP Requests

When the RS receives a SIP request message, such as INVITE, it can use the Route field to send the message to its next destination. In general, this field is not present in the original message, so there are several other options available for routing. The message can route on the Request URI, or it can use DNS SRV or telephone number mapping (ENUM) to look up the routing. DNS SRV and ENUM are discussed later in this chapter.

## SIP Responses

When the RS receives a SIP response message, such as 200 OK, it can use the address found in the Via field for routing, paying attention to the ingress address received from the tags. This method is explained in Chapter 11.

## Aliases

Aliases are alternate identities for users in the system. One aspiration of packet telephony is giving its users multiple addresses for calling one another—by phone number, email address, or an ID of the user's invention. In traditional phone systems, there are internal dialing schemes that permit corporate users to call others within the enterprise through a truncated version of the local phone numbers. A corporate user in California might call a coworker in North Carolina by dialing a five-digit internal number. In many cases, this type of dialing bypasses the PSTN and transmits the call as packets over a managed backbone. We have prepared VOCAL to handle aliases, but the code has not been fully implemented.

## Least-Cost Routing

An order of priority can be expressed within a dial plan to be used to suggest alternate gateways if the desired gateway is busy. To implement this idea, the RS could return a list of gateways that are capable of providing the desired routing, with the intention of using the first one that is available.

In Version 1.3.0 of VOCAL, we don't have a way of providing advanced parallel and sequential forking that would enable the RS to route a call through several Feature servers and then the first available gateway in a prepared list. Currently, all lists are approached sequentially regardless of the availability of the gateways or entities that make up the list.

Also using external routing servers that use Telephony Routing over IP (TRIP), another routing protocol, would make this type of routing easier. If you had another routing engine, such as Signaling System 7 (SS7) lookups of 1-800 numbers, that knew how to make decisions, the RS would be the right place to make a modification to VOCAL. Right before you look in the dial plan, you would do a translation of the number based on the information provided by the external routing server.

## TRIP

Telephony Routing over IP (TRIP) provides a method for routing calls from an IP system to the PSTN. In a TRIP scheme, a Location server (LS) obtains information about the gateways in its Internet Telephony Administrative Domain (ITAD) and shares that information with other LSs in other ITADs by exchanging TRIP messages. This information can be transferred to the RS to enable routing services.

TRIP has not been implemented in VOCAL, however there is an open source TRIP stack available on *Vovida.org* (*http://www.vovida.org*) that you can download and play around with. At the time of this writing, we were considering two options for implementing full TRIP services into VOCAL.

### Option 1: dial plan in RS

In this option, a Location server (LS) modifies the dial plans as it receives new information. In order to implement this option, several things must be done:

- The RS needs to keep track of heartbeats from the LS.
- Whenever an LS begins heartbeating, the RS must register for updates with a callback function as the LS currently does with the Provisioning server.
- Upon initial connection, the LS needs to download all relevant route data to the connecting RS (probably through the same callback function).

- When relevant data changes on an LS, it calls the RS's callback function to insert the new data as an entry in the appropriate dial plan.

- The RS's dial plan scheme must be modified to allow automated insertion of entries. Currently, when a change is made, the whole plan is discarded and replaced by a new plan. This is due to the fact that first match does the searching. The dial plan must be changed to accept searching by the longest match to allow autoinsertion of entries.

### Option 2: dial plan in LS

In this option, the LS maintains the dial plan, and the RS queries the LS for dial plan info as required. Upon startup, the LS obtains the configured dial plan information from the Provisioning server. The RS will most likely need to download the same information as a backup measure should all LSs go down. It is also possible that there will be an LS/RS pairing so that, if one goes down, the other does as well.

If the LS were incorporated into the RS, the following issues would have to be considered:

- What key will the LS use for an entry in the dial plan? Will it be by prefix or something different? Currently, the RS needs the key to be a regular expression that will be used on SIP URIs. Will the RS compose the key or will the LS? How will this work for countries that don't have the same area code scheme as North America and the Caribbean?

- TRIP does not support load balancing: it simply gives the best route. The RS may have an entry in the dial plan that it obtained from the Provisioning server. How should the static data, obtained from the PS, coexist with the dynamic data obtained from the LS? Should it be configurable from PS GUI as to whether the RS gives preference to static or dynamic routes?

## ENUM

ENUM (not an acronym, more of an abbreviation for *E-number*) allows DNS lookup of an E.164 phone number (a normal phone number in full international format) to see if there is an IP phone associated with it. If there is, the call can go straight to the IP phone, without going to the PSTN. Examples of numbers in full international format are +1(408)555-1212 and +44 55551212.

ENUM turns these phone numbers into URIs, in which the numbers appear in reverse order. For example, an ENUM lookup for +1(408)555-1212 becomes a DNS lookup for 2.1.2.1.5.5.5.8.0.4.1.e164.arpa. The top levels of *e164.arpa* are contentious and should be made configurable. See the E.164 and DNS RFC, 2916, for more information.

ENUM defines the process of taking an E.164 telephone number and mapping it to a particular Internet service (such as a SIP URI) by performing a DNS-type lookup to an ENUM server. The RS could query a preconfigured set of ENUM servers after failing to find the dialed number in the Subscriber database. If the query succeeds, the resulting SIP contact will be used. If the query fails, the dial plan will be searched as usual.

Upon receiving a valid SIP URI from an ENUM query, the RS needs to insert a gateway Marshal before the resulting contact. Currently, the gateway group name is hardcoded, but it will need to be made configurable.

In order to enable ENUM queries on the dialed number, several things must be done:

1. Have BIND installed on your machine.
2. Modify *vocal/build/Makefile.pkg* and set the following paths to the place where BIND is located on your machine:

   ```
   BIND_INCLUDEDIRS
   BIND_LIBDIRS
   ENUM_LIBDIRS
   ```
3. Modify ENUM Makefile within *vocal/contrib/enum* by setting the correct BIND source directory variable BIND_BASE_DIR.
4. Turn on the ENUM compile-time flag in *vocal/build/Makefile.opt* by setting:

   ```
   VOCAL_HAS_ENUM=1
   ```
5. Turn on the ENUM runtime flag in *vocal/proxies/rs/SipContactIf.cxx* by setting:

   ```
   activateEnum = true
   ```
6. Sign up for an ENUM service at a web site such as *www.enumworld.com*, *www.netnumber.com*, or *www.neustar.com*.
7. If you use a different service, you will need to modify *vocal/proxies/rs/SipContactIf.cxx* and add the new ENUM service domain name to *client_default_domains*.

## TRIP and ENUM

Many people seem to be confused about the differences between ENUM and TRIP. Let's try to clarify these differences.

ENUM is an address resolution protocol that lets you translate a phone number into a URI. TRIP is a routing protocol that lets you obtain routing information for a particular telephone number prefix. With TRIP, you discover where the next hop should be in your routing of telephone numbers. With ENUM, you discover a mapping between a phone number and an IP entity.

ENUM answers the question, "Is there a URI (for example, an IP phone) associated with the number that the user is dialing?" The answer to this question helps avoid routing a call from an IP phone over the PSTN to another IP phone. If the system knows that the far end has a URI, it can do a DNS lookup for that URI and contact it directly over IP.

If the number doesn't have an associated URI, TRIP allows you to find the best rate to get to that phone number. You use ENUM first to find out if there is a URI; if there is no match, then you use TRIP, which is a mechanism used to communicate that a specific gateway handles a specific PSTN prefix or a specific range of PSTN numbers.

Neither protocol has any use if you're dialing *user_name@vovida.org*. These protocols have to do with phone numbers, not URI dialing.

# Ongoing Development

VOCAL is not a shrink-wrapped product; it's a project, and we encourage you to help us with its development. Here are some areas of the RS that could be enhanced by members of our community.

## Simplified Routing

One proposed project is redesigning VOCAL to eliminate the need for each server to send an INVITE to the RS in order for it to determine the next hop in the call.

In this new design, the first server to handle a call sends an INVITE to the RS for routing information. The RS generates a list of contacts according to the calling party and the called party. The call's current place within the list is determined by using the Via fields. The remaining contacts are validated and prepared for return in a 302 Moved Temporarily response message.

The first contact in the list to be returned is placed in the Contact header's SIP URI. The remaining contacts are added to the Contact header as embedded routes. The server extracts the contacts and places the first address in the Request URI of the INVITE message and the remaining addresses in the Route header before sending it to the next hop in the system. Thereafter, servers remove themselves from the top of the Route list and add themselves to the route record before passing the INVITE to the next server in the Route list.

If a server changes the Req URI or From fields of the INVITE message or finds that the next server to receive the INVITE is unavailable, it queries the RS again for a new route list as if it were the first server to handle this call. The RS returns the contacts for the remainder of the call's path through the system.

If a server cannot handle the embedded routes within the 302's Contact header, it can send the call to the next hop. The receiving server can then query the RS for the following hop:

```
OpINVITE::process( )
```

The code needs to be modified to call a new function, *SipContactIf::validateRemainingContacts()*, and place the routes correctly into the SIP Contact header. The SIP stack requires additional modification to complete this task.

As for each server placing the list of contacts in the Route header of the INVITE message before sending it to its next hop, the SIP specification as of bis-06 of RFC 2543 added support for "loose routing." VOCAL needs to be updated to handle loose routing. If you are using older third-party Feature servers, this scheme may not work.

## Dial Plan

The following are the issues surrounding development of the dial plan.

- Modify to match on the username portion instead of whole Request URI. This modification would require that administrators reprovision their existing dial plans.
- Change number matching to a longest-match scheme. This may require changing structure to *Trie* (a retrieval object). Some *Trie* objects have already been implemented for use in the LS. These may be reused for the dial plan.
- Should keys remain as regular expressions?
- Should we keep $USER?
- Should we add $RESOLVE_HOST? As in looking past the at sign (@) to the domain part of a SIP URI when the address requires IP dial plan routing. Currently, the RS cannot distinguish SIP URIs by their domain parts.

## Customer Partitioning

Currently, VOCAL uses a design that works for a single customer. All provisioned data (subscribers, servers, etc.) can be viewed and modified from the GUI. The same dial plans apply to all calls.

Multiple customers on the same soft switch would require separate dial plans and subscribers belonging to a particular customer would be given access only to that customer's dial plan. Customers would have access to only their customer-specific information from the GUI, but a system administrator would have access to everything.

In order to make this work from the RS point of view, the following is needed:

*Global subscriber list*
Requires that subscribers are stored as full E.164 numbers so that they can be combined in the same container and still remain unique. Subscribers are marked as belonging to a particular customer.

*Global dial plan*
Catchall that applies to all customers and calls passing through the system.

*Customer-level dial plans*

Allow customers to specify which servers to use for particular calls.

*Digit manipulation tables*

May need to expand a dialed number before searching the subscriber container or dial plans.

*Dial plan variables*

Replace $USER with User field of the Request URI and $AC with customer's or subscriber's area code. We need a way to identify a customer so that both the PS and the RS know which customer a subscriber or plan belongs to.

The following assumptions can be made about call processing:

- Full E.164 numbers are used to identify the calling party. Gateways can be configured to do this.
- Subscribers are stored as full E.164 numbers.

A call originating from within a particular customer can terminate within the same customer, to another customer, or outside the system entirely.

INVITE's From field is used to determine the identity of the customer. If From is a subscriber in the global database, the customer's digit manipulation table and dial plans can be accessed. If From is not in the database, the global database and global dial plan are used.

This process goes through several phases depending on where the call originated.

If the call originated internally:

1. Locate the From user in the Subscriber database.
2. Use the From in the customer's digit manipulation table on the dialed number.

   Normalize the Request URI into a full international number. For example, if the number is 914085551212, strip off the 9; if the number is 65551212, strip off the 6, and prepend +1(408); if the number is 751212, prepend +1(408)55.
3. Attempt to find the dialed number in the Subscriber database.
4. If the number cannot be found in the database, attempt to resolve it by looking for the number of the customer listed in the From field of the INVITE in this customer's dial plan.
5. If the number is not found in this customer's dial plan, attempt to resolve the number in the global dial plan.

If the call originated externally and the number listed in the From field is not found in the Subscriber database:

1. Attempt to find the dialed number in the Subscriber database.
2. If this number is not found, attempt to resolve it in the global dial plan.

The Subscriber database could contain names of, or pointers to, the customer's digit manipulation table and plans. That would reduce the need to link to a customer to find the table and plans.

The following open issues exist:

- Will the other servers in the system be customer-specific?
- How should subscribers who don't have real E.164 numbers be handled?

## Here's a Project for You

At times we have talked about combining the RS, MS, and all other servers together into one large server to provide a simple, nonscalable system that is not highly reliable but is easy to set up and use. This combined server would merge the authentication and billing functions of the Marshal into the RS code to produce a single, multifunctional SIP server.

# CHAPTER 13

# CPL Feature Server

This chapter looks at the role of the Call Processing Language (CPL) Feature server, including the different types of features that it supports and how you can write your own CPL features. In our development of VOCAL, we implemented features such as Call Forward, Call Screening, and Call Blocking. The Feature server (FS) can also execute arbitrary CPL scripts written by users, but it is difficult to link new features to the original provisioning screens without having the in-depth knowledge and experience of Vovida's developers. The new, not yet implemented, provisioning system, discussed in Chapter 19, offers a more user-friendly approach to linking new features to GUI screens.

## What Are Features?

Features are enhanced functions that enable customers to do more than simply call and receive phone calls. Core system features such as Call Forward No Answer are provided by the network. Other set-based features such as Transfer are dependent on the design of the phone set. Set-based features are also known as *phone-based* or *user-based features*. Another way of grouping features is determining whether they are calling features or called features. Calling features such as Call Blocking are activated when a user makes a call, which may prevent the user from dialing a 1-900 toll number. Called features such as Call Screening are activated when a user receives a call, which may prevent someone from calling the user. Before going into detail about how these features were implemented in VOCAL, it would be useful to see how they were traditionally built into the PSTN.

## Feature Development in the PSTN

The PSTN has been built on monolithic servers that implement call control, feature servers, and everything else in a single box. In this environment, anyone who wants to add a new feature needs to perform regression testing. As there are potentially hundreds of features that need to interoperate with every other feature on the box, this testing is time consuming and arduous, making the costs of implementing new features in the PSTN an expensive proposition.

The PSTN is evolving, but its monolithic nature has endured. Despite the advent of deregulation and competition, traditional phone system customers have very little control over their service and must adapt their needs to what the vendor is willing to provide. While large, multinational corporations can influence the development of new features, the majority of traditional phone customers cannot.

## Feature Development in VoIP

The VoIP world provides a less-complicated implementation environment than the PSTN. For example, much of the regression testing has been removed by its distributed feature implementation onto separate servers, which limits the interaction between features to the sequence of the features being executed. You don't have to test every feature against every other feature; you have to test a new feature only against those that come directly before and after it in the execution sequence.

Writing new features is relatively straightforward in the VoIP environment, and the tools are getting better every day. Most business analysts who follow telecom say that the future of VoIP resides in its ability to receive implementations of new features on a variety of different scales. VoIP engineers have been known to poke fun at the "disruptive, revolutionary, cool new application that nobody has thought of yet" mantra being touted by some marketing people, but just as the practicality of email and web pages has made the Internet welcome in homes and businesses, smart endpoints and simple feature implementation will usher VoIP into the same environments.

## Accessing Features in VoIP

Anyone who has worked in an office with a private branch exchange (PBX) has been trained to access features either by entering feature codes on their phone sets or by pressing buttons that are programmed to launch specific features. In the VoIP world, you can mimic this interaction by writing a dial plan that maps feature codes to different Feature servers in the system. Dial plans are discussed in detail in Chapter 5.

# Core Features

Core features are network features that operate independently of the User Agent appliance used by the customer. These features include Calling Line Identification features, Call Forwarding, Call Blocking, and Call Screening.

You might ask yourself, "Why implement features on a Feature server; why not just add them to the functionality of the phone?" While some features belong on the endpoint, there are several reasons it is desirable to implement some of your features within the network. For example, in the case of the Call Forward No Answer feature, what happens if your phone is down? And your phone may be down for various reasons such as a power failure or a connectivity problem with the local network. Having a dead phone doesn't eliminate your need to have calls forwarded to voice mail.

As for call blocking, you can't trust the phone to block the calls properly, and some users can't be trusted with 1-900 numbers or whatever numbers are being blocked. The same thing goes for Calling Party ID Blocking: you don't want to send the ID to the phone and then hope that the phone doesn't display it to the end user. Also, if the user changes her set, the new phone may not support the same features as the old phone did. For stability and consistency, it is better to have some core system features implemented on a server within the system.

## Calling Line Information Features

Calling Line Information features are the calling features that appear in a call display device:

*Calling Number Delivery (CND)*
> Also known as Calling Line Identification (CLID); provides information to the terminating line about the directory number where the call originated as well as the date and time of the call.

*Calling Name Delivery (CNAM)*
> Also known as Calling Party Name Delivery (CPND); provides information to the terminating line about the calling party's name as well as the date and time of the call.

*Calling Party Identity Blocking (CIDB)*
> Allows a subscriber to control whether his number (CND) or name (CNAM) is delivered when he places an outgoing call.

## Call Forwarding

Call Forwarding is an enhanced method of working with unanswered phone calls. These calls can be forwarded to another phone set or to a service such as voice mail or a paging device. These called features include:

*Call Forward All Calls (CFA)*
> Allows a customer to reroute all calls to an alternate number. When CFA is activated, a call to the listed number is redirected to a user-selected alternative number or a voice messaging system.

*Call Forward No Answer (CFNA)*
> Allows a customer to specify where an unanswered call should be routed. When CFNA is activated, a call to the listed number that does not answer in a specified

number of ringing cycles will be forwarded to an alternative number selected by the user.

*Call Forward Busy (CFB)*
Allows a customer to specify where a call should be routed when the listed number is in use. When CFB is activated, a call to the listed number, while it is in use, will be redirected to another number.

## Call Blocking

Call Blocking is a calling feature. It prevents the customer from establishing connections to specified parties such as 1-900 numbers. For Version 1.3.0 of VOCAL, long-distance call blocking works only for calls in the North American Numbering Plan (NANP), which includes Canada, the United States, and many Caribbean nations. Calls cannot be blocked if they originate from Europe, Asia, or other locations that are not part of the NANP. For more information about the NANP, see *http://www.nanpa.com*.

## Call Screening

A called feature, Call Screening prevents incoming calls from specified parties to establish connections with the customer. For Version 1.3.0 of VOCAL, phone numbers entered for Call Screening must include the area code, regardless of whether they are local or long-distance phone numbers. Call Processing Language (CPL) does not provide a pattern-matching method that differentiates 7-digit (local) phone numbers from 10-digit (long-distance) numbers.

# Set Features

Set features are features that depend on the User Agent appliance. The VOCAL system supports Call Transfer, Call Return, and Call Waiting features.

## Call Transfer

Call Transfer allows a user, on any existing two-party call, to place the existing call on hold and originate another call to a third party. The user may consult privately or connect the original call to the third party.

## Call Return

Call Return allows the subscriber to place a call back to the last number that called her. Call Return can be either a core system feature, with the user required to dial a code such as *69, or a set-based feature allowing the user to select from a list of callers.

## Call Waiting (CW)

Call Waiting notifies a telephone user, who is on an established call, that an additional external call has been presented and is waiting to be answered. The waiting call receives normal ringing until it is answered, the incoming calling party abandons the call, or the ringing cycle timer expires and the call is given Call Forward No Answer treatment (if applicable). On some phones, only one Call Waiting call can be present at a time. Additional calls are handled with Busy treatment (CFB, if applicable). Implementation of Call Waiting requires support from the phone sets.

## Cancel Call Waiting (CCW)

Otherwise known as *Do Not Disturb*, Cancel Call Waiting allows the subscriber to dial a feature activation code prior to making a call. For the duration of the subsequent call, the Call Waiting feature will be disabled for that line. The Cancel Call Waiting feature lasts only for the duration of one call, and when the subscriber goes on-hook again, his Call Waiting feature is reenabled.

# New Features

Here is a brief discussion about some features that are not yet implemented into VOCAL but have been the subject of enthusiastic conversations among VoIP developers.

## Call Park, Call Pickup

Call Park and Call Pickup is like Call Transfer. When a call is received, a user may want to park it on the system and then pick it up at another phone that is on the same network. A user puts a call on hold, and then the same user or a different user picks up the call from a different phone. When the call is picked up, the original phone that is holding the call has to be contacted to transfer the call to the user's new location. From a technical point of view, this is not a major issue and has been discussed as being a transfer mechanism that would work with the SIP REFER message. From a human point of view, it is a different story.

The basic problem is that, when one user parks a call for someone else, she needs to tell the other user which call to pick up. In many organizations, the user who is picking up the call enters the phone number for the extension where the call was parked. The call that was parked likely came in on the main number, 1000; therefore, the user wants to pick up whatever call is parked on extension 1000.

However, if a call is waiting on 1000, it may be preventing other calls from coming into the switchboard. This problem is circumvented by including parking locations within the system to enable the receptionist, or any other user, to transfer a call to a parking location where it can wait for pickup without tying up any of the system's extensions. If the receptionist is parking the call, he normally pages the intended receiver to notify her of the call and its parking location. "Ms. Smith, call parked on 3" is an example of this type of page. Call Park, Call Pickup is not implemented in VOCAL.

## Ring Again

In the world of the PSTN, a busy signal may lead to a voice prompt telling you that if you press a certain key sequence on your phone, the network will call you back, or ring again, when the called party is available.

In the VoIP world, the hope is that presence might be a way of achieving a Ring Again feature. One user phones another, and the called party is busy. The calling party then sends a SUBSCRIBE message to his phone to say, "Tell me when the other user gets off her current call." Then when the called party hangs up the current call, it returns a NOTIFY message to the calling party's phone, which makes an indication. If the user agrees to call, the call is put through.

This scenario works between IP phones; we're not too sure how it would work between an IP phone and the PSTN. It seems possible that the IP phone would send the SUBSCRIBE to the gateway and the gateway would do the appropriate PSTN signaling to make Ring Again happen.

## Multiline Appearance

Multiline Appearance is when an incoming call rings more than one phone set simultaneously. This is useful for small call centers in which the caller needs to speak to a member of a team of people, and it doesn't matter which individual answers the call. This is really easy to implement in SIP. There should be a proxy that forks the call to every user in a group simultaneously. This is known as *parallel forking*. See Chapter 7 for more information, including details on call flows.

# SIP Messages and Feature Servers

The VOCAL system routes calls to Feature servers by using SIP messages. When the servers first come online, they download a register from the Provisioning server but do not download the file that controls the feature. It is not until the first time that the server runs a script that it downloads the controlling file from the Provisioning server.

## SIP Messages to the Feature Servers

Figure 13-1 shows a Feature server receiving a message from a Marshal and then requesting routing information from the Redirect server. It is possible that a call signal may be routed to several Feature servers before leaving the VOCAL system. Some calls may not be routed to any Feature servers before going to the outbound Marshal.

The call flow shown in Figure 13-1 is continued in Figure 13-2.

We left the 100 Trying messages out of these figures to make them less cluttered.

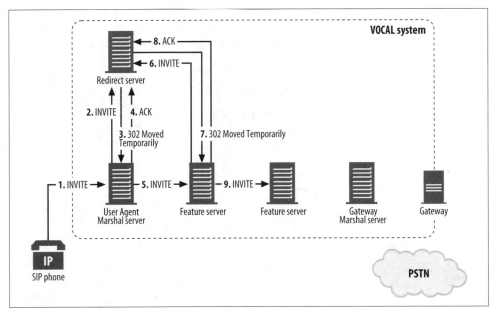

*Figure 13-1. SIP message flow to the Feature servers*

Table 13-1 describes the messages illustrated in Figure 13-1.

*Table 13-1. SIP message flow to the Feature servers*

| Interaction | Step | Description |
|---|---|---|
| SIP phone to UAMS | 1 | A call is initiated at one SIP phone to call a party attached to the PSTN. The SIP phone sends an INVITE message to the User Agent Marshal server (UAMS). |
| | 2 | The UAMS authenticates the user and forwards the INVITE message to the Redirect server (RS). |
| | 3–4 | The RS generates a new packet with routing information and sends it to the UAMS as a 302 Moved Temporarily message. The routing information includes a call blocking feature. |
| UAMS to an FS via the RS | 5 | The UAMS generates a new INVITE message and sends it to the Call Blocking FS. |

Table 13-1. SIP message flow to the Feature servers (continued)

| Interaction | Step | Description |
|---|---|---|
| The message is redirected to a second FS | 6 | The Call Blocking FS generates a new INVITE message and sends it to the RS. |
| | 7–8 | The RS generates a new packet with routing information and sends it to the Call Blocking FS as a 302 Moved Temporarily message. The routing information includes a calling party ID blocking feature. |
| | 9 | The Call Blocking FS generates a new INVITE message and sends it to the Calling Party ID Blocking FS. |

## Feature Servers to the PSTN

Figure 13-2 shows the INVITE message being redirected from the Feature servers to the PSTN.

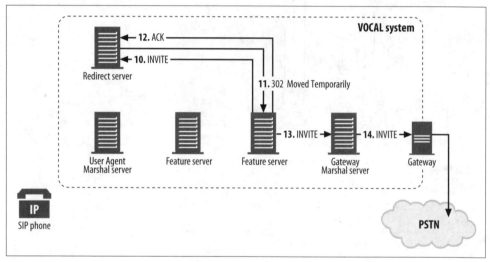

Figure 13-2. Feature servers to the PSTN

Table 13-2 describes the messages illustrated in Figure 13-2.

Table 13-2. Message flow of Feature servers to PSTN

| Interaction | Step | Description |
|---|---|---|
| FS to GMS | 10 | The Calling Party ID Blocking FS generates a new INVITE message with calling party ID blocking instructions and sends it to the Redirect server (RS) for routing. |
| | 11–12 | The RS recognizes that the INVITE is destined for the PSTN and generates a 302 Moved Temporarily message with routing instructions. |
| | 13 | The Calling Party ID Blocking FS receives the Redirect message, generates a new INVITE message, and sends it to the Gateway Marshal server (GMS). |
| GMS out to the PSTN | 14 | The GMS forwards the INVITE to the gateway, which sends it to the PSTN. |

# Scriptable Feature Development

*Scriptable features* are features that can be expressed in a scripting language such as CPL. The VOCAL system supports new-feature development through CPL scripts.

## What Is CPL?

The Call Processing Language (CPL) is a scripting language that describes and controls Internet telephony services that have been implemented on either network servers or User Agents. CPL scripts are normally simple and easy to edit. For more information about CPL, see RFC 2824.

CPL is limited because it was originally intended for end users to create their own scripts, in the same way that many people create web pages with HTML. The intention behind limiting CPL was to prevent regular users from overwhelming or crashing a shared server; however, the problem with this limitation is that it also prevents professional engineers from expanding out of the boundaries that CPL defines. The list of limitations is long—for example, there are no variables and no loops.

From the six features that we designed, only four of them include standard CPL language. For features such as Caller ID Blocking, we created our own proprietary tags. If we took these features and ported them to another softswitch that supported standard CPL, the scripts would fail.

So why did we choose to work with CPL rather than VoiceXML (*http://www.w3.org/TR/voicexml*), Tcl (*http://www.sun.com/960710/cover*), or some other scripting language? First of all, CPL and VoiceXML are not mutually interchangeable languages. CPL provides functionality for processing features on demand; VoiceXML is meant as a direct user interface for people to select items from menus. VoiceXML servers provide the same sort of functionality as interactive voice response (IVR) systems except that VoiceXML is designed to expect voice commands as well as dual-tone multifrequency (DTMF) tones.

In general, SIP Feature Proxy servers should support any standard CPL script written by anyone. Our specification for VOCAL stated that it would support the basic, core system features that are described earlier in this chapter. We aspire to create a fully functional CPL Feature server and, as of this writing, that server could be completed with 2 or 3 months of effort.

Our Provisioning server also had some limitations with respect to feature implementation: enabling the GUI for provisioning new servers required the direct intervention of the Provisioning server development team. With the new Provisioning system, as discussed in Chapter 19, anyone can write a new GUI on the fly without months of research and effort beforehand.

## What Is VoiceXML?

VoiceXML is a web-based markup language, like HTML, but, instead of being designed for supporting graphical interaction, like web pages, VoiceXML was designed to support voice interaction. VoiceXML enables voice browsing based on audio output and audio input, which can be a human voice, DTMF tones, or a combination of the two. We are thinking seriously about implementing VoiceXML into a future voice mail system.

Check out the Open VXI VoiceXML interpreter (*http://sourceforge.net/projects/openvxi*): an open source library designed to interpret the VoiceXML dialog markup language scripts.

## CPL Definitions

CPL is defined in RFC 2824. However, at the time we started working with CPL, the second working group draft had just been published, and we made a decision to work with this draft until there was a compelling reason to switch. This implementation was an experimental proof of concept and has not been maintained to incorporate the latest changes from the IETF working group.

This is a summary of all the CPL tags, including attributes and acceptable output:

*CPL <cpl>*
> Top-level tag
>
> Attribute: None
>
> Output: <timezone>* <subaction>* <outgoing>? <incoming>?

*OUTGOING <outgoing> (%Node)*
> Do this *if* script owner equals originating party of the call (calling feature).
>
> Attribute: None
>
> Output: Any

*INCOMING <incoming> (%Node)*
> Do this *if* script owner equals destination party of the call (called feature).
>
> Attribute: None
>
> Output: Any

*ADDRESS SWITCH <address-switch>*
> Allows matching on one of the addresses in the original call request.
>
> Attribute: field="origin | destination | original-destination"
>
> Attribute: subfield="address-type | user | host | port | tel | display | password | asn1"
>
> Standard output: (<address> | <not-present>)+ <otherwise>?
>
> Vovida output: <not-present>? <address>+ <otherwise>?

---

*ADDRESS <address> (%Node)*

    Attribute: is=""

    Attribute: contains="" used only if subfield=display

    Attribute: subdomain-of="" used only if subfield=host | tel

    Output: Any

*STRING-SWITCH <string-switch>*

    Allows string matching on a field.

    Attribute: field="subject | organization | user-agent"

    Output: (<string>|<not-present>)+ <otherwise>?

*STRING <string> (%Node)*

    Attribute: is=""

    Attribute: contains=""

    Output: Any

*TIME-SWITCH <time-switch>*

    Allows matching on the time/date the triggering call was placed.

    Attribute: timezone="local | utc | {customized timezone name}"

    Output: (<time> | <not-present>)+ <otherwise>?

*TIME <time> (%Node)*

    Attribute: year="",month="",date="",weekday="",timeofday=""

    Output: Any

*PRIORITY SWITCH <priority-switch>*

    Make decision based on the priority specified for the original call.

    Attribute: None

    Output: ( <priority> | <not-present> )+ <otherwise>?

*PRIORITY <priority> (%Node)*

    Attribute: less=", greater=", equal=" emergency | urgent | normal | nonurgent"

    Output: Any

*LOCATION SET <location> (%Node)*

    Add or clear literal locations from CPL's set of specified locations.

    Attribute: url="{supported signaling protocol}"

    Attribute: clear="yes | [no] "

    Output: Any

*LOOKUP URL <lookup>*

    (1) Look for locations through an external server that returns "application/url."

    (2) Look for locations by querying registration servers.

    Attribute: source="{URL} | registration"

    Attribute: timeout="{ [30] sec }"

Attribute: use="{comma-delimited list of caller preference fields to use}"

Attribute: ignore="{comma-delimited list of caller preference fields to ignore}"

Attribute: clear="yes | [no]"

Output: <success> | <notfound>? | <failure>?

REMOVE LOCATION <remove-location> (%Node)

Filter addresses out of the location set.

Attribute: location="{caller preference locations addr-spec field of Reject- Contact hdr}"

Attribute: param="{comma-delimited list of caller preferences parameters}"

Attribute: value="{comma-delimited list of caller param values}" Must be same number as param

Output: Any

PROXY <proxy>

Causes the triggering call to be forwarded on to the currently specified set of locations. After a proxy action has completed, the CPL server chooses the "best" response to the call attempt, as defined by the signaling protocol or the server's administrative configuration rules.

Attribute: timeout="{ [20 if noanswer defined] sec}"

Attribute: recurse=" [yes] | no " {handle call attempts to redirection responses that were returned from the initial server}

Attribute: ordering="first-only | sequential | [parallel]"

Output: {success} <busy>? <noanswer>? <failure>?

REDIRECT <redirect>

Causes the server to direct the calling party to attempt to place its call to the currently specified set of locations. REDIRECT immediately terminates execution of the CPL script.

Attribute: None

Output: None

REJECT <reject>

Causes the server to reject the call attempt.

Attribute: status=" busy | notfound | reject | error | {SIP 4xx-6xx err}"

Attribute: reason=""

Output: None

MAIL <mail> (%Node)

Email a user the status of the CPL script. Additional params specified in mailto:.

Attribute: url=" {mailto: url}"

Output: Any

*LOG <log> (%Node)*
> Log info about call to nonvolatile storage.
>
> Attribute: name=""
>
> Attribute: comment=""
>
> Output: Any

*SUBACTION <subaction> (%Node)*
> Defines a subaction to be used by top-level or later-defined subactions.
>
> Attribute: id=""
>
> Output: Any

*SUB <sub>*
> References a subaction to execute.
>
> Attribute: ref=""
>
> Output: None

*TIMEZONE <timezone>*
> The time zone where the script will be executed.
>
> Attribute: name=""
>
> Output: <standard> <daylight>?

*STANDARD <standard>*
> Attribute: offset="{UTC offset}"
>
> Attribute: abbr="{abbreviation of this timezone}"
>
> Attribute: year="", month="", date="", weekday="", timeofday="" (std active when)
>
> Output: None

*DAYLIGHT <daylight>*
> Attribute: offset="{UTC offset}"
>
> Attribute: abbr="{abbreviation of this timezone}"
>
> Attribute: year="", month="", date="", weekday="", timeofday="" (day active when)
>
> Output: None

*BUSY <busy> (%Node)*
> Used by: <proxy>

*FAILURE <failure> (%Node)*
> Used by: <lookup>, <proxy>

*OTHERWISE <otherwise> (%Node)*
> Used by: <address-switch>, <string-switch>, <time-switch>, <priority-switch>

*NOANSWER <noanswer> (%Node)*
> Used by: <proxy>

*NOTFOUND <notfound> (%Node)*
   Used by: <lookup>

*NOT PRESENT <not-present> (%Node)*
   Used by: <address-switch>, <string-switch>, <time-switch>, <priority-switch>

*SUCCESS <success> (%Node)*
   Used by: <lookup>

   <success> = used only 1 time

   <failure>? = used 0 or 1 time

   <string>+ = used once or many times

   <node>* = used zero or many times

   [no] = default value

### Standard default actions

When a CPL action reaches an unspecified output, the action it takes depends on the current state of the script execution. This section summarizes the default actions:

- If *incoming* and *location*, *lookup*, *remove location*, *proxy*, *redirect*, or *reject* were not done, then treat the call as if there were no CPL script.

- If *outgoing* and *location*, *lookup*, *remove location*, *proxy*, *redirect*, or *reject* were not done, then proxy calls to the addresses in the location set.

- If *location*, *lookup*, or *remove location* were done, but no *proxy*, *redirect*, or *reject*, then proxy/redirect the call to the address or addresses in the location set. If the set is empty, reject the call.

- In Proxy, if there is no *time-out* and no <noanswer> ring for the maximum length of time allowed by the server.

- Proxy action previously taken—Return the best of all accumulated responses to the call at this point.

## CPL Supplements

Additional content added to CPL that was not defined elsewhere is documented here. Unless the following are added to future standard CPL definitions, these supplements remain proprietary within the VOCAL system.

The following is a new tag:

*<vmproxy>*
   Redirects a call to the Voice Mail FS

The following tags are described by the IETF working group and have been modified for this implementation:

*<address-switch>*

> npa-area-code, npa-prefix, and npa-line-number
>
> These fields are actually subsubfields of tel and are available only if the URI has signaling protocol, "tel:" or if the URL parameter "user=phone" is defined.
>
> These fields make sense only for telephone numbers that follow the North American Numbering Plan Administration's formats which is usually Nxx-Nxx-xxxx. The 'npa-area-code' fetches the first set of three digits. The "npa-prefix" fetches the second set of three digits. The 'npa-line-number' fetches the last four digits.

*<lookup>*

> Supports source SIP URL's ONLY (source="sip:6398@vovida.com").
>
> "use=" and "ignore=" support "media,language,class,duplex,scheme,feature,media,mobility"

*<log>*

> Does not define (success, failure) as done in the latest *cpl.dtd* at the time the FS code was written.

VOCAL's Feature server handles these default actions:

- If a user isn't configured on the Feature server, reject the call (default script available).
- For incoming scripts:

  If the location set is empty, reject the call.

  If the location set is nonempty, redirect to the first location.
- For outgoing scripts:

  Look up the next hop for To: Proxy to that location.

Verify that *CPLOpDfltAction* is added to <address-switch> and other switch CPL tags *if* <otherwise> was not defined in the script.

# How CPL Script Converts to a C++ State Machine

The Feature server executes CPL scripts by parsing them into eXtensible Markup Language (XML) document object model (DOM) trees, turning these trees into state machines and then executing them. The following diagrams show how CPL scripts become features and DOM trees.

Figure 13-3 shows how a script becomes a feature.

The FS turns a script, brought in from Provisioning, into a state machine by calling *CPLParser* to build the tree. The parser takes a long string provided by the Provisioning server and turns it into a state machine.

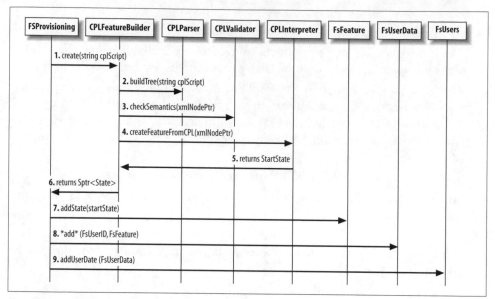

*Figure 13-3. How a script becomes a feature*

Figure 13-4 shows how a script becomes a DOM tree.

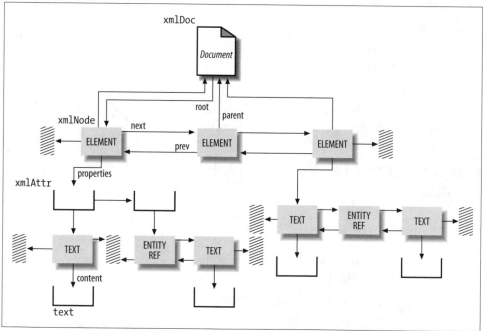

*Figure 13-4. CPLParser: script becomes a DOM tree*

The FS doesn't really care about this, but when it receives a string of characters, it calls the *libxml* library, which creates this data structure. The FS uses this structure, but it has to understand a few things, such as how to traverse it and how to get the data it needs. For more information about *libxml*, see *http://xmlsoft.org*.

## Types of States and Operators

In the FS state machine are a state object and three different types of operators. One major difference between the FS state machine and the UA's state machine is that while the UA's machine never needs more than one operator to change states, in many cases, the Feature server's machine may need to process several operators between states.

The State object has multiple operators that execute their processes until a nonzero (next state) is returned. Each state has zero or more CPL operators including *ProxyOp* and *EndOp*. The FS states include Start, End, and Proxy.

The three types of operators include:

- Operator objects that contain a pointer to one operator, including Address, String, Time, Priority, Location, Null, remove-location, mail, log, otherwise, and not-present.
- Operator objects that contain pointers to one or more operators. These operators execute until a nonzero (next state) is returned. These operators include Address-Switch, String-Switch, Time-Switch, Priority-Switch, and nonAnswer.
- Operator objects that point to a state, including FwdCall (proxy), Lookup, Reject, Proxy, and End.

The interaction of these operators is shown in the following set of examples.

## Changing States with the State Machine

Simple states call the *Simple operator*, which calls another until the next state is reached, as shown in Figure 13-5.

*Figure 13-5. Calling the Simple operator*

Simple states call the *Compound operator*, which decides which operator to call to reach its next state, as shown in Figure 13-6.

*End operators* may return the same state as that where the operator is being held, as shown in Figure 13-7.

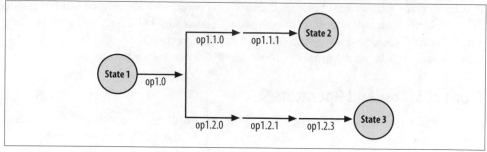

Figure 13-6. Calling the Compound operator

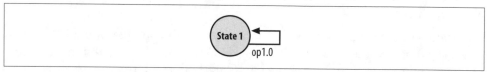

Figure 13-7. An End operator returning the same state

## How the State Machine Works

In general, *CPLStateX* will call each of its operators in its list until one of them returns a nonzero state. The returned state is saved in *CallInfo*. If End state is returned, *CallInfo* is deleted.

*CPLOperators* calls its list of operators until one of them either returns a nonzero state or a state.

- Objects created for every state machine are *StartState*, *StartOp*, *EndOp*, *EndState*, *ProxyState*, and *ProxyOp*.
- Each state has a *ProxyOp*, which loops back to the original state.
- Each state has an *EndOp*, which must always be executed last. *EndOp* always goes to *EndState*.
- Before we rewrite the To: field in the message, we must query the RS for the contact list. We use the contact list to forward the call.
- If we encounter a <proxy> and it has no outputs, *FwdCallOp* points to *ProxyState*.

## From DOM Tree to FS State Machine

Once *CPLFeatureBuilder* has successfully parsed and validated the CPL script, it is time to generate a state machine for the script:

```
CPLInterpreter::createFeatureFromCPL( xmlNodePtr )
```

Starting at the root node, call the recursive *traverse()* to process each node:

```
traverse( xmlNodePtr )
  nodeToI()  //Gets an ID tag for the node based on its CPL type
```

```
nodeStart() //processes the node (retrieve attributes and create operators and
  states)
traverse( children ); //node one level deep
traverse( next ); //node at same level
nodeExit() //do any cleanup
```

*CPLInterpreter* has a member variable, *Sptr<State> startState*, which points to the Start state. The entire state machine is linked by smart pointers to the Start state. If the smart pointer to the Start state decrements to zero, the Start state will be deleted, causing the rest of the state machine to be deleted.

## Attaching States/Operators to the State Machine

Every operator and state class has *::setNextOperator( Sptr<Operator>, int prevNode)*.

*::setNextOperator()* does one of the following, depending on the needs of that class:

- Pushes the new operator to its Operator list (for example, *allOperators.push_back(op)*)
- If the operator points to a *nextState*, calls the state's *setNextOperator()* (for example, *cplNextState->setNextOperator(op)*)
- Adds the operator to one of its special pointers (for example, *busyReceivedOperator = op;*)

## How CPLBuildStatus Tracks Where to Add New Operators

In four scenarios, the *CPLBuildStatus* class must add new operators:

- Adding a Simple operator
- Adding a Compound operator
- Using the Compound operator list
- Linking subactions to the state machine

### Adding a Simple operator

This process is illustrated in Figure 13-8.

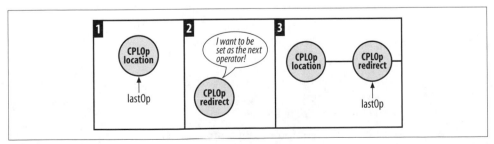

*Figure 13-8. Adding a Simple operator*

The following process is used to add simple operators; the list numbers correspond to the numbered panels in Figure 13-8:

1. *CPLBuildStatus* has *lastOp* pointing to the operator that was most recently added to the state machine.

2. When a new operator is to be added to the state machine...

3. ...*lastOp's setNextOperator()* is called to add the new operator; *lastOp* points to the most recently added operator.

### Adding a Compound operator

This process is illustrated in Figure 13-9.

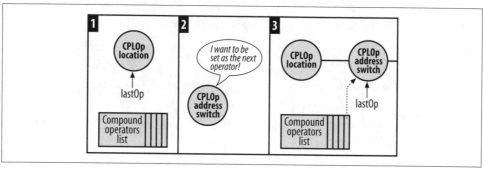

*Figure 13-9. Adding a Compound operator*

The following process is used to add compound operators; the list numbers correspond to the numbered panels in Figure 13-9:

1. *lastOp* points to the most recent operator added to the state machine.

2. When a compound operator needs to be added...

3. ...a pointer to the compound operator is pushed onto the *CompoundOperators* list; *lastOp->setNextOperator(compoundOperator)* is called, and *lastOp* points to the most recently added operator.

### Using the Compound operators list

This process is illustrated in Figure 13-10.

The following process uses the compound operators list; the list numbers correspond to the numbered panels in Figure 13-10:

1. When a compound operator is added to the state machine, the *Compound-Operators* list stores a pointer to the compound operator. Other operators are continually added to the state machine.

2. When an output of the compound operator needs to be added to the state machine...

---

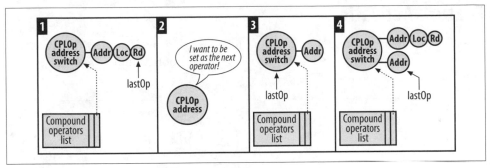

*Figure 13-10. Using the Compound operators list*

3. ...*CPLBuildStatus* uses the pointer from the back of the compound operators list to reinitialize *lastOp*.

4. The new operator is added to *lastOp* using *setNextOperator()*.

## Linking subactions to the state machine

This process is illustrated in Figure 13-11.

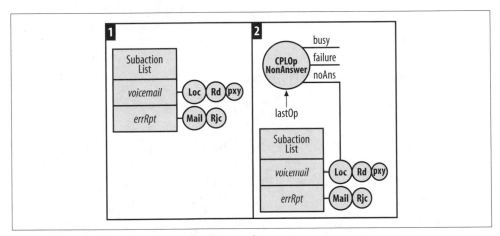

*Figure 13-11. Linking subactions to the state machine*

The following process links subactions to the state machine; the list numbers correspond to the numbered panels in Figure 13-11:

1. Subactions consist of operators and states that are not automatically included into the feature's state machine until a *<sub ref="subaction name">* is encountered. In the interim, subaction state machines are found in the subaction list.

2. When a *<sub ref="name">* is encountered, the subaction list is searched for "*name*". If found, the subaction's operator is linked to the state machine by calling *lastOp->setNextOperator()*.

# Feature Activation

The FS state machine model is the same for executing one feature. This model is illustrated in Figure 13-12. This model does not support multiple features on the same server.

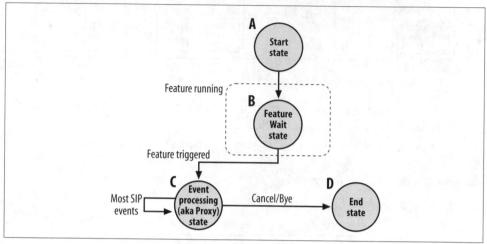

Figure 13-12. Generic state machine on FS

## Walkthrough of Feature Activation

The following process describes what happens when the FS creates a new *CallInfo*:

1. When an INVITE message arrives, the FS does the following:

   a. Creates a new CallInfo, which holds info about the call leg, current state, timers, and feature data (for example, last caller's number, possible proxy locations, proxy status, and anonymous blocking data).

   b. Pulls the user port from the RequestUrl and From fields.

   c. Searches for called or calling features:

      i. Searches FsUsers map for RequestUrl:-user.

         If not found, goes to Step ii.

         If found, retrieves the called feature and the state machine begins at START (illustrated as circle A in Figure 13-11). See Step 2.

      ii. Searches FsUsers map for From:-user.

         If not found, deletes CallInfo and the message processing is finished.

         If found, retrieves the calling feature and the state machine begins at START. See Step 2.

2. At START, the feature has a choice of what to do with the INVITE message:

    a. The feature executes one or more operators until it cannot continue because it needs a response message. Eventually, the SIP stack provides an automatically generated response to allow this process to complete.

    b. The feature transitions to a Feature Wait state (illustrated as circle B in Figure 13-11). This allows the feature to continue processing subsequent messages that arrive at the FS.

3. Triggering a feature transitions the feature to the Event Processing or Proxy state (illustrated as circle C in Figure 13-11): the Proxy state has default handling mechanisms for most messages. It will handle all subsequent messages until a CANCEL or BYE arrives.

4. When the CANCEL or BYE message arrives, the feature transitions to the End state (illustrated as circle D in Figure 13-11). The CallInfo is deleted and the message processing is finished.

## State Machine for Call Forward Busy/No Answer

The Call Forward Busy/No Answer (CFB/NA) feature is illustrated in Figure 13-13:

- Circles = states that wait for events to arrive.
- Rectangles = operators that execute another operator or return the next state.

The following process occurs when the CFB/NA feature is activated:

1. INVITE event arrives.
2. Queries Redirect server for next contact.
3. Waits for response from Redirect server.
4. When message arrives, acknowledges RS.
5. Forwards message to the location returned from the RS.
6. Waits for the next event at *AwaitAnswer* state.

    When the message arrives, and it does not indicate that the call was answered, the following occurs:

    a. Saves specified location to *locations Set*.

    b. Redirects to location in *locationSet*.

    c. Next state is *endState*, meaning call can be safely torn down.

## State Machine for Call Forward All Calls

The CFA feature is illustrated in Figure 13-14.

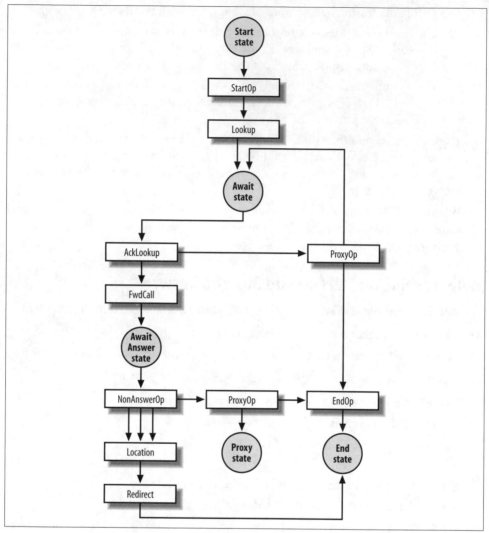

*Figure 13-13. Call Forward Busy/No Answer Process*

The following process occurs when the CFA feature is activated:

1. When INVITE message arrives, calls *StartState::process()*.

2. StartState calls *StartOp::process()*.

3. *StartOp* calls *Location::process()*; *Location* saves the URL to a *CPLUrlData* object.

4. *Location* calls *Redirect::process()*; *Redirect* uses the URL in the *CPLUrlData* object to send a 302 back to the originator. *RedirectOp* returns *endState*.

5. *EndState* is detected and the call info is deleted.

6. Any ACK messages that arrive will be discarded.

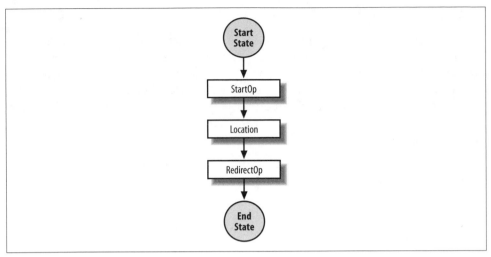

*Figure 13-14. Call Forward All process*

# How to Develop a Feature

To develop a feature using the CPL FS, write the feature as a CPL script. If the desired node and outputs are not available, you may extend CPL and the FS. The next section provides an example of adding some new CPL tags.

## Desired CPL

As a tutorial lesson, let's add new CPL functionality.

The first functionality is:

```
<im source="somebody@somewhere.com" msg="Come quick. I need you." >
```

This is the instant messaging functionality. Using this CPL node tag instructs the Feature server to send an instant message to the source with the specified message. *<im>* has no outputs.

The second functionality is:

```
<addressBkLookup name="Alan Smith" book="company.db" timeout="10" >
    <found> ....
    <not-found> ...
```

This tag instructs the Feature server to query a database file for user information. The Feature server waits for a configured maximum time to hear from the database. Depending on the success or failure of the database search, the CPL script can create different outputs.

# Sample CPL Scripts

The following represents a sample of the CPL scripts used to create the features provided with the VOCAL system. Some of these scripts have adjustable values; others include custom tags or custom uses of tags and attributes that are not compliant with the current CPL standard. These are called *VOCAL keywords*. The text following each example provides more information.

## Example 1: Call Blocking

The phone number prefixes, highlighted in bold italics, can be changed to match your requirements.

```
<?xml version="1.0" ?>
<!DOCTYPE cpl SYSTEM "cpl.dtd">
<!-- Feature to Block 1-900 and 9-1-(xxx)xxx-xxxx calls assuming -->
<!-- those are long-distance calls. -->
<cpl>
<subaction id="rejectcall">
    <reject reason="feature activated" status="reject"></reject>
</subaction>
<outgoing>
  <address-switch field="original-destination" subfield="user">
      <address subdomain-of="91900"><sub ref="rejectcall"></sub>
      </address>
      <address subdomain-of="912"><sub ref="rejectcall"></sub>
      </address>
      <address subdomain-of=";"><sub ref="rejectcall"></sub>
      </address>
      <otherwise>
        <lookup clear="yes" source="registration" timeout="2">
                <success>
                    <proxy ordering="first-only"></proxy>
                </success>
                <notfound><sub ref="rejectcall"></sub>
                </notfound>
        </lookup>
      </otherwise>
  </address-switch>
</outgoing>
</cpl>
```

The script makes an assumption regarding the dial plan that:

- 9+1+(xxx) xxx-xxxx = long-distance calls.
- 9+01+............. = international long-distance calls.
- 9+011+ .......... = international long-distance calls.
- 9+xxx-xxxx = local calls.

This script can be modified to support other dial plans.

91900 represents the first five digits of a 1-900 phone number (including the initial 9 dialed to send calls to the PSTN). 912 represents the first three digits of phone numbers in the 2xx area code. In this example, calls to area codes starting with 8xx are not blocked, to permit toll-free calling.

### Example 2: Caller ID Blocking: complete

The location attribute, highlighted in bold italics, represents a custom use of the tag.

```
<?xml version="1.0" ?>
<!DOCTYPE cpl SYSTEM "cpl.dtd">

<cpl>
  <outgoing>
    <lookup clear="yes" source="registration">
      <success>
        <remove-location location="complete_calleridblock">
          <proxy ordering="first-only"/>
        </remove-location>
      </success>
      <notfound>
        <reject status="reject"/>
      </notfound>
    </lookup>
  </outgoing>
</cpl>
```

complete_calleridblock is a VOCAL keyword. This function removes the username and number from the From URL of the call and replaces it with Anonymous & "-".

### Example 3: Call Forward All Calls

These values should not be changed.

```
<?xml version="1.0" ?>
<!DOCTYPE cpl SYSTEM "cpl.dtd">
<!-- Call Forward Unconditionally/All Calls -->
<cpl>
  <incoming>
    <location url="sip:6399@192.168.26.229:6399" clear="yes">
      <redirect/>
    </location>
  </incoming>
</cpl>
```

### Example 4: Call Forward Busy/No Answer

The timeout value, highlighted in bold italics, can be changed to match your requirements.

```
<?xml version="1.0" ?>
<!DOCTYPE cpl SYSTEM "cpl.dtd">
<!--Feature to call forward if called party does not answer, is busy, -->
```

```
<!--or returns a failure code -->
<cpl>
<incoming>
  <lookup clear="yes" source="sip:6389@vovida.com:5060;user=phone">
   <success>
    <!-- timeout of 15 seconds approximates to 4 rings -->
    <proxy timeout="15" ordering="first-only">
      <busy>
        <location clear="yes" url="sip:918773817287@192.168.6.26:5060;user=phone">
          <redirect></redirect>
        </location>
      </busy>
      <noanswer>
        <location clear="yes" url="sip:918773817287@192.168.6.26:5060;user=phone">
          <proxy ordering="first-only"></proxy>
        </location>
      </noanswer>
      <failure>
        <location clear="yes" url="sip:918773817287@192.168.6.26:5060;user=phone">
          <redirect/>
        </location>
      </failure>
    </proxy>
   </success>
   <notfound>
      <reject status="reject" />
   </notfound>
  </lookup>
</incoming>
</cpl>
```

The timeout, `<proxy timeout="15"...>`, is measured in seconds and can be changed. The time required for four rings (usually 15 seconds) is recommended.

### Example 5: Call Return

The `name` attribute highlighted in bold italics (`lastcaller`) represents a custom use of the tag.

```
<?xml version="1.0" ?>
<!DOCTYPE cpl SYSTEM "cpl.dtd">
<cpl>
  <incoming>
    <log name="lastcaller" comment="">
      <lookup clear="yes" source="registration">
        <success>
          <proxy ordering="first-only"/>
        </success>
        <notfound>
          <reject status="reject"/>
        </notfound>
      </lookup>
    </log>
  </incoming>
</cpl>
```

lastcaller is a VOCAL keyword. This function saves the originator's URL, overwriting any existing data. This data is retrieved when the destination party activates Call Return.

### Example 6: Call Screening

The subdomain-of attribute, highlighted in bold italics, can match a phone number prefix or act as a placeholder.

```
<?xml version="1.0" ?>
<!DOCTYPE cpl SYSTEM "cpl.dtd">
<!-- Block calls from phone numbers that begin with prefix X -->
<cpl>
<subaction id="rejectcall">
  <reject reason="feature activated" status="reject"></reject>
</subaction>
<incoming>
<address-switch field="origin" subfield="user">
    <address subdomain-of="6389"><sub ref="rejectcall"></sub>
    </address>
    <address subdomain-of=";"><sub ref="rejectcall"></sub>
    </address>
    <otherwise>
        <lookup clear="yes" source="registration" timeout="2">
            <success>
                <proxy ordering="first-only"></proxy>
            </success>
            <notfound><sub ref="rejectcall"></sub>
            </notfound>
        </lookup>
    </otherwise>
</address-switch>
</incoming>
</cpl>
```

The subdomain-of attribute can be used to match phone number prefixes.

The subdomain-of attribute can be substituted with contains or is. The contains attribute indicates that the attribute's value is found as a substring. The is attribute indicates an identical match between the attribute's value and string from a field/subfield.

Table 13-3 describes how the attributes of the address tag affect call screening.

*Table 13-3. Call screening CPL tags: address attributes*

| If the address attribute defines | And the originating party's number is | The call will be |
|---|---|---|
| subdomain-of =" 6" | 6389 | Rejected |
| is = "6" | 6389 | Accepted |
| contains = "89" | 6389 | Rejected |
| contains = "98" | 6389 | Accepted |
| is = "6389" | 6389 | Rejected |

**Example 7: Voice Mail**

The <vmredirect/> tag, highlighted in bold italics, is a custom Vovida tag.

```
<?xml version="1.0" ?>
<!DOCTYPE cpl SYSTEM "cpl.dtd">
<cpl>
  <incoming>
     <vmredirect/>
  </incoming>
</cpl>
```

searches for an available Voice Mail User Agent (UAVM) and redirects the message to that host/port.

# Feature Server Files

The files found within in the FS directories can be divided into the base files, those files that enable CPL-to-C++ conversion, and related classes.

## Base Files

The *fs/base* directory contains the following files:

*fs.cxx*
> Contains the FS's *main( )*

*FsConfigFile*
> Contains a function for parsing the execution command line and global variables required by the FS

*FsBuilder*
> Derived from *Builder*; spawns threads to do the event processing and heartbeats; has a *CallContainer* to track current active calls

*FsProvisioning*
> Called to retrieve system data, subscriber list, and CPL features from the Vocal Provisioning server

*FsFeature*
> Points to the Start state of a state machine that was generated by parsing a CPL script

*FsUserData*
> Stores information about a feature subscriber (*incl/ FsFeature*)

*FsUsers*
> A map holding 0 or more *FsUserData*s; this is a key data structure for the FS to look up its subscribers

*CPLUrlData*
> Instantiated during runtime if it needs to track possible locations to proxy the call; also known as the *locations set*

*FsStateMachnData*
> Created for each call and stored in *CallInfo*; stores any information required for the duration of the call (for example, data pertaining to a feature or the state of a proxying)

## CPL Conversion Files

The *fs/cpl* directory contains the following files necessary to convert a CPL script into a C++ state machine:

*CPLFeatureBuilder*
> Used by applications to interface to CPL. This calls *CPLParser*, *CPLValidator*, and *CPLInterpreter*.

*CPLParser*
> Given a CPL script stored as a string, calls *libxml* to build an XML DOM tree.

*CPLValidator*
> Traverses the DOM tree verifying that each node's attributes contain acceptable values (to be implemented in the future).

*CallProcessingLanguage*
> A file of arrays and constants to define the CPL keywords.

*CPLBuildStatus*
> Used by *CPLInterpreter* to keep track of how operators and states are added to a script's state machine.

*CPLListElement*
> Used by *CPLBuildStatus* to store information required during the construction of a state machine.

*CPLStartStateHolder*
> When a state machine is constructed, the *startState* is stored in this class (to be returned to the application to store in *FsFeature*). This class can hold two *start-States* (in case one incoming and one outgoing feature are located in the same CPL script).

## Classes

Figure 13-15 is a class diagram for the FS. The classes shaded in grey are base classes.

The */usr/local/vocal1.3.0/fs/cpl* directory contains the following classes used in state machines:

*CPLOperator*
> Derived from *Operator* and used as base class for all CPL operators; holds 0 or more operators; processes each operator until one of them returns a state.

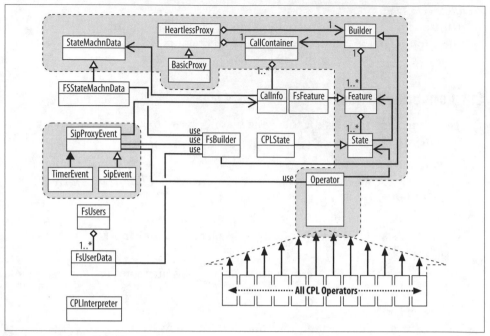

*Figure 13-15. FS class diagram*

*CPLState*

    Derived from *State*; holds one or more operators; processes each operator until one of them returns a state.

*CPLOpProxy*

    Used for default SIP event handling when other operators don't need to process the event. This class does not hold operators. It will return either 0 or the next state.

*CPLOpStart*

    The first operator to be processed when an INVITE arrives for the first time; this is useful for any feature initialization (for example, saving info to the *CallInfo*).

*CPLOpNull*

    Nonfunctional operator; useful as a placeholder.

*CPLOpEnd*

    The last operator to be executed before a call is terminated; this class returns the *endState*, so *FsFeature* knows to clean up the call.

*CPLOpDfltAction*

    Operator used when there's an unspecified output in the CPL script. This checks for a location; if found, redirects the call to that location; otherwise, reject the call.

*CPLOpAddress*
Implements *&lt;address&gt;*.

*CPLOpAddressSwitch*
Implements *&lt;address-switch&gt;*.

*CPLOpFwdCall, CPLOpNonAnswer*
*FwdCall* implements *&lt;proxy ordering="first-only"&gt;*'s proxying while *NonAnswer* waits for the proxy's result.

*CPLOpFork, CPLOpBusy, CPLOpNoAnswer, CPLOpFailure, CPLOpTransition*
Used to implement *&lt;proxy&gt;* when the ordering attribute is equal to parallel or sequential.

*CPLOpLocation*
Implements *&lt;location&gt;*.

*CPLOpLog*
Implements *&lt;log&gt;*.

*CPLOpLookup, CPLOpAckLookup*
*OpLookup* contacts a remote server for the next contact (for example, Redirect server); *OpAckLookup* acknowledges the remote server and saves the contact information into the location set.

*CPLOpMail, MailMan, MailThread, MailThreadMgr*
Implement *&lt;mail&gt;*; a separate thread is used, so sending mail and event processing are done asynchronously.

*CPLOpPriority*
Implements *&lt;priority&gt;*.

*CPLOpPrioritySwitch*
Implements *&lt;priority-switch&gt;*.

*CPLOpRedirect*
Implements *&lt;redirect&gt;*.

*CPLOpReject*
Implements *&lt;reject&gt;*.

*CPLOpRemoveLocation*
Implements *&lt;remove-location&gt;*.

*CPLOpString*
Implements *&lt;string&gt;*.

*CPLOpStringSwitch*
Implements *&lt;string-switch&gt;*.

*CPLOpTime, CPLOpTimeSwitch, CPLOpTimeZone, CPLOpStandard, MyTime, Range, CPLOpDaylight, CPLTimeRange, CPLOffset*
Implement *&lt;time&gt;* and *&lt;time-switch&gt;*.

*CPLOpVmRedirect, VmCallInfo, VmPortContainer*
> Implement proprietary *<vmredirect>*.

*CPLInterpreter*
> Creates a feature state machine based on the tags in a CPL script.

*CallInfo*
> Holds information about a call. For example, reference to the current feature (state machine), state, active timer, and state machine data.

*FSStateMachnData*
> Derived from "StateMachnData"; holds feature-specific state machine data of a call, for example, feature activation prefix and the caller's URI for the Caller ID Blocking feature.

*FsFeature*
> Derived from "Feature"; holds CPL script name; cleans up call related data when the state machine transitions into the End state.

*FsBuilder*
> Derived from "Builder"; handles incoming SIP messages and timer events for the feature server; associates feature state machine and user data with the received event; passes the event to the feature state machine for further processing.

*FsUsers*
> Singleton container class for user data of this feature.

*FsUserData*
> Holds feature specific data of a user; for example, calling and called feature state machines.

## Code Changes

The following is a tutorial that takes you through the steps required to add new CPL extensions. The Feature server code is largely compliant with the output from the 1999 IETF working group. Check the *readme* file for the outstanding issues that remain.

### cpl.dtd

The new CPL node tags must be defined in the *cpl.dtd* for CPL validation to succeed during the parsing of the CPL script. The *cpl.dtd* is defined in *createCplDtd()* in *CPLFeatureBuilder.cxx*.

In this tutorial, add the following:

```
oFile << "<!ELEMENT im ( %Node; ) >" << endl;
oFile << "<!ATTLIST im" << endl;
oFile << "   source        CDATA    #IMPLIED" << endl;
oFile << "   msg        CDATA    #IMPLIED" << endl;
oFile << ">" << endl;
```

```
oFile << endl;
oFile << "<!ELEMENT addressBkLookup ( found,not-found ) >" << endl;
oFile << "<!ATTLIST addressBkLookup" << endl;
oFile << "  name          CDATA    #REQUIRED" << endl;
oFile << "  book          CDATA    #IMPLIED" << endl;
oFile << "  timeout       CDATA    '30'" << endl;
oFile << ">" << endl;
oFile << endl;
oFile << "<!ELEMENT found ( %Node; ) >" << endl;
oFile << "<!ELEMENT not-found ( %Node; ) >" << endl;
oFile << endl;
```

## CallProcessingLanguage Class

In *CallProcessingLanguage.hxx*, add strings for all the CPL node tags in the *CPLNodeStr* array. Note the position in the array where you've placed your new strings!

```
:
"remove-location",  //existing code
"im",
"addressBkLookup",
"found",
"not-found",
"comment",  //existing code
:
```

In *CallProcessingLanguage*, add some constants to represent the new CPL node tags. Place them in the exact same position in the *CPLNode* array as they were positioned in the *CPLNodeStr* array!

```
:
CPLNodeRemoveLocation,    //existing code
CPLNodeIM,
CPLNodeAddressBkLookup,
CPLNodeFound,
CPLNodeNot_Found,
CPLNodeComments,    //existing code
:
```

In *CallProcessingLanguage.cxx*, add an array of strings representing each attribute in the CPL node. Also add a constant indicating the number of elements in that array.

```
const unsigned int AddressBkLookupAttrTableSize = 3;
const char* AddressBkLookupAttributesTable [] =
{ "name", "book", "timeout"   };

const unsigned int imAttrTableSize = 2;
const char* imAttributesTable [] =
{ "source", "msg"   };
```

In *CallProcessingLanguage.hxx*, export each constant and array.

## CPLOpIM Class

Create the operator that will do the instant messaging functionality and hold the values of *<im>*'s attributes. Create a *CPLOpIM* class derived from *CPLOperator*:

```
#ifndef _CPLOPIM_HXX
#define _CPLOPIM_HXX
#include "CPLOperator.hxx"
class CPLOpIM: public CPLOperator
{
    public:
        CPLOpIM( );
        ~CPLOpIM( );
        bool setAttributes( const char* type, const char* value );
        const Sptr < State > process(const Sptr <SipProxyEvent> anEvent );
        bool setNextOperator(Sptr <Operator> anOperator,
                             int nodeId = CPLNodeEndofNodes);
        const char* const name( ) const;
        void logData(void) const;
    private:
        string mySource;
        string myMsg;
};
#endif
```

In *CPLOpIM.cxx*, implement setAttributes( ), which assigns the value of the attribute to a member variable:

```
bool
CPLOpIM::setAttributes( const char* type, const char* value )
{
    if ( strcmp( type, "source" ) == 0 )
            mySource = value;
    else if ( strcmp( type, "msg" ) == 0 )
            myMsg = value;
    else
       return false;   //Unknown attribute
return true;
}
```

In *CPLOpIM.cxx*, implement name( ) and logData( ) if you want the class' debugging information:

```
const char* const
CPLOpIM::name( ) const
{
        return "CPLOpIM";
}
void
CPLOpIM::logData(void) const
{
    cpLog( LOG_DEBUG, "%s - source(%s) msg(%s)",
                    name(), mySource.c_str(), myMsg.c_str() );
}
```

In *CPLOpIM.cxx*, implement setNextOperator(). This function tells the state machine generator (*CPLBuildStatus*) where to add the next operator. Since the <im> has no specific outputs, it just calls a base function.

```
bool
CPLOpIM::setNextOperator(Sptr <Operator> anOperator,
                         int nodeId = CPLNodeEndofNodes)
{
    addOperator( anOperator );
    return true;
}
```

Finally, implement the process( ) function to do the functionality of the node tag:

```
const Sptr < State >
CPLOpIM::process(const Sptr <SipProxyEvent> anEvent )
{
    //Code that sends an instant message

    /* Do the next operator or CPL node tag */
    return CPLOperator::process( anevent );
}
```

Add *CPLOpIM.cxx* to the *makefile*.

## CPLOpAddressBkLookup and More

To implement the *<addressBkLookup>* node tag, two operators are required. *CPLOpAddressBkLookup.cxx* sends the query to the database file. *CPLOp-AddressBkResults.cxx* takes the result of the query and decides whether to do the not-found or found output.

*CPLOpAddressBkLookup.hxx* is defined as follows:

```
ifndef _CPLOPADDRESSBKLOOKUP_HXX
#define _CPLOPADDRESSBKLOOKUP_HXX

#include "CPLOperator.hxx"

class CPLOpAddressBkLookup: public CPLOperator
{
    public:
        CPLOpAddressBkLookup();
        ~CPLOpAddressBkLookup();
        bool setAttributes( const char* type, const char* value );
        const Sptr < State > process(const Sptr <SipProxyEvent> anEvent );
        bool setNextOperator(Sptr <Operator> anOperator,
                             int nodeId = CPLNodeEndofNodes);
        const char* const name() const;
        void logData(void) const;
    private:
        string myName;
        string myBook;
```

```
            unsigned int myTimeout;
    };

    #endif
```

*CPLOpAddressBkResults.hxx* is defined as follows:

```
    #ifndef _CPLOPADDRESSBKRESULTS_HXX
    #define _CPLOPADDRESSBKRESULTS_HXX

    #include "CPLOperator.hxx"

    class CPLOpAddressBkResults: public CPLOperator
    {
        public:
            CPLOpAddressBkResults();
            ~CPLOpAddressBkResults();
            bool setAttributes( const char* type, const char* value );
            const Sptr < State > process(const Sptr <SipProxyEvent> anEvent );
            bool setNextOperator(Sptr <Operator> anOperator,
                                 int nodeId = CPLNodeEndofNodes);
            const char* const name( ) const;
            void logData(void) const;
        private:
            Sptr< Operator > myFound;
            Sptr< Operator > myNot_Found;

    };

    #endif
```

In *CPLOpAddressBkLookup* and *CPLOpAddressBkResults*, name( ), logData( ), and setAttributes( ) need to be implemented. There is nothing special about these functions; see the earlier section "CPLOpIM Class" for sample code. *CPLOp-AddressBkResults* can have a skeleton setAttributes( ) function since it has no attributes.

In *CPLOpAddressBkLookup*, its setNextOperator( ) can be empty since the next step is a state to wait for results (a *sipProxyEvent*) of the lookup:

```
    bool
    CPLOpAddressBkLookup::setNextOperator(Sptr <Operator> anOperator,
                                 int nodeId = CPLNodeEndofNodes)
    {
            //Unnecessary to call this function

            return false;
    }
```

In *CPLOpAddressBkResults*, setNextOperator( ) must know which output to add to the next operator:

```
    bool
    CPLOpAddressBkResults::setNextOperator(Sptr <Operator> anOperator,
                                 int nodeId = CPLNodeEndofNodes)
```

```
    {
        switch( nodeId )
        {
                case CPLNodeFound:
                        myFound = anOperator;
                        break;
                case CPLNodeNot_Found:
                        myNot_Found = anOperator;
                        break;

                default:
                        cpLog( LOG_ERR, "Unknown output.");
                        return false;
        }

        return true;
    }
```

For the *CPLOpAddressBkLookup*'s process( ), the functionality of the
<addressBkLookup> is implemented here. Since the specification of this tag requires a
wait for the results (another event), the operator must return the next state. A state
instructs the FS to stop processing the CPL until another event arrives:

```
    const Sptr < State >
    CPLOpAddressBkLookup::process(const Sptr <SipProxyEvent> anEvent )
    {
        //Code to send query to database

        return myNextState;

    }
```

For the *CPLOpAddressBkResults'* process( ), there has to be a decision made on
which path of operators to execute next:

```
    const Sptr < State >
    CPLOpAddressBkResults::process(const Sptr <SipProxyEvent> anEvent )
    {
        //Code to determine the result of the lookup

        if ( result_deemed_as_found )
                return myFound->process( anEvent );
        else
                return myNot_Found->process(anEvent );
    }
```

Add the two source filenames to the *makefile*.

## CPLInterpreter Class

In *CPLInterpreter.cxx*, add a function to create some state machine (one or more
operators, zero or one state) for the CPL nodes:

```
    bool CPLInterpreter::processAddressBkLookup( xmlNodePtr ptr )
    {
```

```
        Sptr < CPLOpAddressBkResults > resultsNode = new CPLOpAddressBkResults;

        Sptr < CPLOpAddressBkLookup > lookupNode = new CPLOpAddressBkLookup;

        processNodeAttributes( lookupNode,
                               ptr,
                               AddressBkLookupAttributesTable,
                               AddressBkLookupAttrTableSize );

        Sptr < CPLState > waitState = new CPLState("CPLStateAwaitLookup");

        lookupNode->setState( waitState );

        //Add the default-handling operators to the state object.
        //ProxyOp must be added before EndOp, so it can execute first.
        Sptr < CPLOpProxy > proxyOperator = new CPLOpProxy;
        proxyOperator->setState( waitState );
        waitState->addOperator( myEndOp );
        waitState->addOperator( proxyOperator );

        //Append lookupNode and waitState to state machine
        if (myCplBuildStatus->nodeEntry( lookupNode, CPLNodeUndefined ) == false )
            return false;

        //Append resultsNode to state machine
        return ( myCplBuildStatus->nodeEntry( resultsNode,
                                              CPLBuildStatus::CompoundOp ) );

}

bool CPLInterpreter::processIM( xmlNodePtr ptr )
{
        Sptr < CPLOpIM > aNode = new CPLOpIM;

        processNodeAttributes( aNode, ptr,
                               imAttributesTable,
                               imAttrTableSize );

        //Add log operator to current state machine
        bool result = myCplBuildStatus->nodeEntry( aNode,
                                                   CPLBuildStatus::SimpleOp );

        //Add default action if no node follows
        if ( (result != false) && (ptr->children == NULL) )
            return processDefaultAction( ptr );
        else
            return result;
}

bool CPLInterpreter::processFound( xmlNodePtr ptr )
{
        bool result = myCplBuildStatus->nodeEntry( 0, CPLBuildStatus::NoOpChild);

        //Add default action if no node follows
```

```
        if ( (result != false) && (ptr->children == NULL) )
            return processDefaultAction( ptr );
        else
            return result;
    }

    bool CPLInterpreter::processNot_Found( xmlNodePtr ptr )
    {
        bool result = myCplBuildStatus->nodeEntry( 0, CPLBuildStatus::NoOpChild );

        //Add default action if no node follows
        if ( (result != false) && (ptr->children == NULL) )
            return processDefaultAction( ptr );
        else
            return result;
    }
```

In nodeStart( ), create cases to find the new tags in the switch statement:

```
case CPLNodeIM :
returnValue = processIM( ptr );
break;
case CPLNodeAddressBkLookup :
returnValue = processAddressBkLookup( ptr );
break;
case CPLNodeFound:
returnValue = processFound( ptr );
break;
case CPLNodeNot_Found:
returnValue = processNot_Found( ptr );
break;
```

This concludes the steps on how to add code to the FS. You can now parse a CPL script with the new tags and generate a state machine with the new operators.

## Arguments for CPLBuildStatus

In the previous section, "CPLInterpreter Class," were examples of code using *CPLBuildStatus*, such as:

```
myCplBuildStatus->nodeEntry( 0, CPLBuildStatus::NoOpChild );
```

The purpose of *CPLBuildStatus* is to control how state machines created by each node tag are joined together to form one state machine for the entire script. *CPLBuildStatus* keeps track of outputs (node tags that depend on the result of a decision or action) and nodes that have outputs. *CPLBuildStatus* also keeps a list of the state machines generated by subactions.

When the nodeEntry( ) function is called, the arguments passed are:

- The operator(s) and state(s) to be added to the whole state machine
- A descriptor of the type of node tag

Most CPL nodes fit under one of four categories of descriptors:

*SimpleOp*
> CPL tags that are neither outputs themselves nor expect a specific output (for example, <log>, <location>, <reject>, <redirect> ). *CPLBuildStatus* will not save any information regarding this tag because the next node will be independent of the current node.

*CompoundOp*
> CPL tags that may have more than one output (for example, <proxy>, <lookup>, <string-switch> ). *CPLBuildStatus* will save this node so that the next time it sees an output node, it'll know it belongs to this current node.

*NoOpChild*
> CPL tags that are outputs of another node for which *CPLInterpreter* does not create any operators (for example, <busy>, <noanswer>, <failure>). *CPLBuildStatus* retrieves the parent node and remembers this node.

*CompoundOpChild*
> CPL tags that are outputs of another node for which *CPLInterpreter* creates an operator (for example, <address>, <otherwise>, <time>). *CPLBuildStatus* retrieves the parent node to add this child node.

Some CPL nodes do not fit under these four categories. *CPLBuildStatus* can accommodate uncategorizable nodes. Additionally, even though <noanswer>, <failure>, and <busy> for <proxy> fit the *NoOpChild* descriptor, a different implementation for <proxy ordering=sequential/parallel> makes the same outputs categorized as *CompoundOpChild* in this forking implementation.

# Writing Your Next Killer Feature

Stop thinking about the phone as a phone, and start thinking of it as an Internet device. Don't try to reproduce the PSTN on a VoIP system, throw away your concepts of what a phone can or can't do, and think about what would be a cool way to communicate through some kind of handheld, desk-mounted, or ethereal device. Maybe your idea ties into a game or into bringing someone who is located in another region on Earth into your space through an enhanced form of contact. We don't know what the next killer application will be; however, it is just as likely to be born out of an accident or for an individual's convenience as designed in a professional research laboratory. We'd like to think that with the tools available on the Internet, along with open source software, that our future will be made better through the creativity of scattered individuals.

# CHAPTER 14

# Unified Voice Mail Server

When we first started demonstrating VOCAL at trade shows and other promotional meetings, we realized that we needed an attention getter. Nobody was going to get excited about authentication and security, never mind making and receiving phone calls with IP phones. We needed a feature that was simple to operate and immediately useful, as well as being a novel, fun thing to play around with. Voice Mail seemed to satisfy these requirements.

There was only one small problem: we didn't have a Voice Mail system. Writing a system from scratch was too much of a distraction. So, we did what any self-respecting organization would do: we hired a contractor. Basically, between the contractor and ourselves, we hacked together this little Voice Mail system without ever intending to use it in a real-life situation. We assumed that other companies were building unified messaging systems, and eventually, systems that were vastly superior to anything that we could build in six months would be on the market.

As we wait for a superior system to appear, we keep fixing and improving our system. And even though it was never meant to be a commercial-grade, usable Voice Mail system, it works, and we have deployed it in our own offices with about 70 users. Before you get too excited, it is our duty to inform you that the code for this system is held together by the proverbial chewing gum and baling wire; in other words, *it ain't pretty*. By including this system within VOCAL, our aspiration is to provide developers with a starting point for building other, more robust Voice Mail systems or similar applications.

## High-Level Design

The VOCAL Voice Mail system has three components:

*Voice Mail Feature server (FSVM)*
> Provides a pool of Voice Mail User Agents (UAVMs) for all users to share on a per-call basis, thus avoiding the requirement for a separate UAVM for each user.

*Voice Mail User Agent (UAVM)*

> Takes care of SIP messaging and the RTP stream for a single connected call at a time. This is what the caller talks to when leaving a message.

*Voice Mail server (VMS)*

> Contains the intelligence for the Voice Mail system including the state machine and control of the play and record commands.

The UAVM and VMS were implemented as separate entities to provide for better scalability and to save some development time. We already had a SIP UA, and it was easy to modify this code into a UAVM. It also made sense, from a maintenance point of view, to insulate the VMS from future changes to the SIP or RTP stacks. The FSVM was a late addition to the architecture that enabled us to pool the available UAVMs on a per-call basis. The early design, requiring a dedicated UAVM for every configured user, made sense only for testing or demonstrating phones at a trade show booth.

Eventually, we would like to replace the UAVM with our Real-Time Streaming Protocol (RTSP) server and replace the VMS with a VoiceXML interpreter. Maybe someone in the community will help us link VOCAL to the OpenVXI project (*http://sourceforge.net/projects/openvxi*). These replacements would provide a much more flexible environment for developers to express enhanced functionality.

The UAVM and the VMS communicate through a simple proprietary text protocol called *Voice Mail control protocol* (VMCP), which is explained later in this chapter.

 The Voice Mail User Agent was shortened to UAVM rather than VMUA because the developers wanted to emphasize its relationship to the UA code. The Voice Mail Feature server is closely related to the Feature server code, hence its abbreviation, FSVM.

Figure 14-1 shows a typical call flow for SIP messages coming into the Voice Mail system.

The FSVM has a map that keeps track of all UAVM server and port combinations that are available to receive calls. When the FSVM receives an INVITE message, it finds an available UAVM, changes the Req URI within the INVITE, and proxies the call to the UAVM. Then, the UAVM contacts the VMS using the VMCP protocol.

# Voice Mail Feature Server

The FSVM is basically a proxy server running a unique CPL script that enables it to redirect calls to available UAVMs. The *<vmredirect>* tag is proprietary and used to

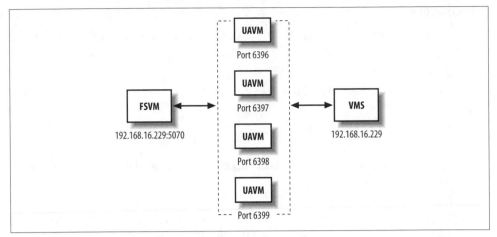

*Figure 14-1. Voice Mail application architecture example*

perform the tasks specific to FSVM. See the upcoming "Call Processing" section for more information. Here is the CPL script used by the FSVM:

```
<?xml version='1.0' ?>
<!DOCTYPE cpl SYSTEM 'cpl.dtd'>
<cpl>
    <incoming>
          <vmredirect/>
      </incoming>
</cpl>
```

The FSVM uses many of the classes associated with the Feature server, including *FsFeature*, *FsConfigFile*, and *FsUsers*. See Chapter 13 for more information about those classes.

## Initialization

Initializing the FSVM involves provisioning the server and instantiating the *VmBuilder* and *VmBasicProxy* classes.

### Provisioning

The FSVM configuration, stored in the file *Features/<IP of FSVM>:<Port>.xml*, normally looks like the following:

```
<featureServer host="192.168.16.229" port="5070"><type value="Voice Mail"></type>
<group value="Voice mailGroup"></group>
<uaVMServers firstPort="6396" host="192.168.16.229" lastPort="6399"></uaVMServers</
featureServer>
```

## Data structures

The FSVM uses the class structures found in the Feature server code including the following:

*VmBasicProxy*
> Derived from the base class *BasicProxy*.

*VmBuilder*
> Derived from the base *Builder* class.

*VmPortContainer*
> Contains a queue of UAVMs (IP and port), which are available (i.e., not busy) to receive calls.

*UAVMContainer*
> Contains a list of UAVM ports that are up and active. The UAVM ports send heartbeat multicast messages to the Voice Mail Feature server. There is no requirement for the Heartbeat server.

*VmCallInfo*
> Derived from *CallInfo*. It adds the UAVM IP/port information to the base class call processing structure.

### VmBuilder instantiation

*VmBuilder* is instantiated in the file *FSVM.cxx* through two steps:

1. *VmProvisioning::Instance()->initialize()*

    a. Gets the configuration file from provisioning and updates the *VmPort-Container* with all the servers and ports contained in that file.

    b. *CreateVmCplScript()* creates the *vm.cpl* script (hardcoded), then calls *createFeatureFromCPL(vm.cpl)* to parse the script and create the components of the Feature server.

2. Adds the feature to the Builder feature container with *addFeature(VmProvisioning::Instance()->getVmFeature())*.

### VmBasicProxy instantiation

During instantiation of the *VmBasicProxy*, from *FSVM.cxx*, it overwrites the *Call-Container* reference pointer so that a *VmBasicProxy* object contains a *VmCall-Container* instead of a regular *CallContainer* (*HeartLessProxy* parent class). Also, it creates the Heartbeat threads that will listen to the *uavmheartbeat* messages sent on the multicast address (226.2.2.5:6000).

# Heartbeat

The FSVM heartbeats and listens for heartbeats from the UAVMs, to find out which UAVMs are up and which are down.

### FSVM receiving UAVM heartbeat messages

Each UAVM sends heartbeat information to the FSVM, which keeps track of active and inactive UAVMs through the *UAVMContainer.cxx* class. In the FSVM *main()*, *fsVm.startUAVMHeartbeatThread()* creates and starts the new Heartbeat thread that listens to these UAVM heartbeat messages (using the multicast address 226.2.2.5: 6000).

### FSVM sending heartbeat messages

The FSVM heartbeats by sending messages to the multicast address configured in the system configuration data via provisioning (default: 224.0.0.100:9000). *fsVm. runHeartbeatThread()* and *FSVM.joinHeartbeatThread()* start the Heartbeat thread.

## Call Processing

Figure 14-2 shows an operator/state diagram.

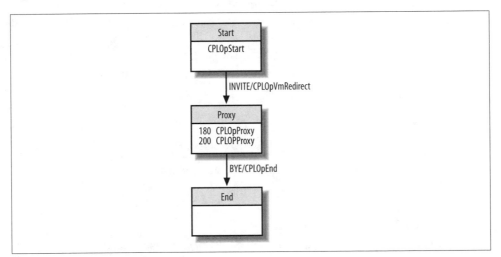

*Figure 14-2. FSVM operator/state diagram*

The states and operators used by the FSVM are the same as those used by the Feature server, except for the *CPLOpVmRedirect* operator, which is the main operator for the FSVM performing the call-processing-specific tasks. Its main function is to redirect calls to available UAVMs.

When the FSVM receives an INVITE message, it first checks for availability of at least one UAVM. If all UAVMs are down, the server replies with a 503 Service Unavailable message. If some UAVMs are active, but all active UAVMs are busy, the server replies with a 486 Busy Here message. If at least one UAVM port is available, the server proxies the call to that UAVM.

When the UAVM receives the INVITE, it contacts the Provisioning server to find the mailbox number associated with the called user. Having the UAVM perform this check enables the Voice Mail administrator to configure just one mailbox per user. For example, Voice Mail messages intended for the phone numbers 4000, 555-4000, or 408-555-4000 go to the same mailbox. Programmatically speaking, if the master user number is 4000, and the originally called number includes the exchange or the exchange plus the area code, *getMailboxFromProvisioning()* will return 4000.

There are two methods for finding the originally called user in the received INVITE message:

- If the CC-Diversion SIP header is present, the originally called number is the first user in the CC-Diversion URL list.

- If there is no CC-Diversion header, the originally called number is the user in the To field.

The CC-Diversion is described in the expired Internet draft *draft-levy-sip-diversion-01.txt*. The FSVM modifies the request line so that the user is the master user of the originally called number and the host and port are that of the available UAVM. If you look at the text of the messages coming from the FSVM, you will notice that the "Vovida: NoAnswer" comment is added in the Via to keep track of the redirection reason. This behavior has been deprecated in the most recent drafts of the SIP standard.

## FS Code Hooks

As new CPL tags are introduced, some hooks must be added in the Feature server's existing code to support them. These hooks include the following:

- *CPLInterpreter::nodeStart()* has been added:

```
case CPLNodeVmRedirect:
returnValue = processVmRedirect(ptr)
```

- *CPLInterpreter::processVmRedirect( xmlNodePtr ptr)* has been added.

- The *cpl.dtd* has been changed and the *vmredirect* element added (*create-CPLDTD in FsConfigFile.cxx*):

```
<! ELEMENT vmredirect EMPTY>
```

# Voice Mail User Agent

The Voice Mail User Agent (UAVM) is a user agent that uses a *vmcpDevice* instead of a Quicknet or sound card device. The UAVM sends events to the VMS through a TCP socket using the VMCP protocol. The UAVM has SIP and RTP threads, and plays the recording of the RTP packets into files, using *Recorder.h*, and plays recorded files, such as user greetings, using *PlayQueue.h*.

---

The UAVM shares the regular UA code, but uses the *-m* option, upon startup, which activates a *vmcpDevice*.

## VMCP Protocol

The VMCP class handles messaging between the UAVM and the VMS and can be seen as a protocol stack. It uses TCP sockets and abstracts the VMCP messages for sending/receiving VMCP messages and is therefore used by the UAVM and the VMS.

VMCP is not an approved protocol, nor is it a standard, and it may not even be a good idea. It is our own proprietary, hacked-together Voice Mail control protocol. Ideally, we would like to replace VMCP with our Real-Time Streaming Protocol (RTSP) stack and media server.

The VMCP text messages are listed here. Upon reception, each message is acknowledged.

| | |
|---|---|
| CLOSE | UAVM client → Voice Mail server |
| SESSIONINFO | UAVM client → Voice Mail server |
| REQSESSIONINFO | Voice Mail server → UAVM client |
| DTMF | UAVM client → Voice Mail server |
| PLAYFILE | Voice Mail server → UAVM client |
| STARTPLAY | Voice Mail server → UAVM client |
| STOPPLAY | Voice Mail server → UAVM client |
| PLAYSTOPPED | UAVM client → Voice Mail server |
| RECORDFILE | Voice Mail server → UAVM client |
| STARTRECORD | Voice Mail server → UAVM client |
| STOPRECORD | Voice Mail server → UAVM client |
| RECORDSTOPPED | UAVM client → Voice Mail server |
| ACK | Both ways |
| NACK | Both ways |
| INVALID COMMAND | Both ways |

## vmcpDevice

The *vmcpDevice* is an object that controls the VMS. It derives from the *ResGwDevice* (residential gateway device) class and the thread where it executes, the *DeviceThread*, receives events from the Worker thread and the VMS sockets.

The main loop is in *VmcpDevice::hardwareMain*, which expects one of two types of events to happen: either a VMCP message arrives or a SIP message arrives.

## VMCP message handling

If a VMCP message arrives, it is handled by *VmcpDevice::process()*. The expected messages are:

*VMCP::Close*
  Closes the VMCP connection

*VMCP::PlayFile*
  Puts a file, whose location is specified in the VMCP PLAYFILE message, into the player queue

*VMCP::StartPlay*
  Issues the command to start playing the file

*VMCP::StopPlay*
  Issues the command to stop playing the file

*VMCP::RecordFile*
  Opens a file, whose location is specified in the VMCP RECORDFILE message for recording

*VMCP::StartRecord*
  Issues the command to start recording the file

*VMCP::StopRecord*
  Writes to the file

Figure 14-3 shows the VMCP messaging sessions.

When a call is received by a UAVM, it connects to the VMS. Then the UAVM sends a SESSIONINFO VMCP message to the VMS. An *eventVmcpEventSessionInfo = 103* event is generated by the *lineVmcp* object and handled by the *State::processLoop()* object, which starts the startup timer. The default is 1 second and is defined in *StateSetup::Process*.

After the startup timer finishes, Voice Mail goes into the *StateSetup* state. *StateSetup:: ProcessTimer* gets the information about the call. Note that in this version of the Voice Mail code the *interactive* flag is never set.

If no mailbox is defined for the called user, *InvalidUser.wav* is played; otherwise, Voice Mail moves to the *StatePlayOwnerGreeting* state if the user has a prerecorded greeting. If not, it will go into *StatePlayGreeting* state and the default greeting, *greeting.wav*, will be played.

A PLAYFILE VMCP message, including the location of the *.wav* file to be played, is sent to the UAVM to put the file in the player queue, as well as a STARTPLAY to start playing it. When a PLAYSTOPPED message is received, the Voice Mail goes into the *StateRecordMessage* state.

A RECORDFILE VMCP message, including the location of the *.wav* file to be recorded, is sent to the UAVM to open this file for recording, as well as a

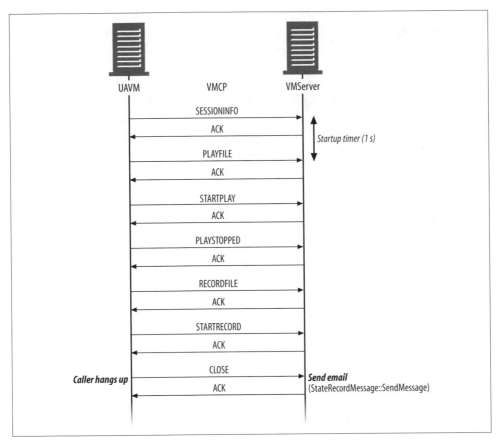

*Figure 14-3. Full VMCP standard messaging session*

STARTRECORD to actually write in the file. When the caller hangs up, a CLOSE message is received, then the recorded file is sent using the *sendMessage* script as configured in *voicemail.config*. *StateReturn* is returned to exit the *State::ProcessLoop* loop and end the session.

If the user presses a DTMF digit instead of hanging up, a message should be sent and the session terminated as well.

### SIP message handling

When the Worker thread receives SIP events and wants to communicate with the *vmcpDevice*, it sends an event to the *vmcpDevice* thread, which will process it through the base class method *ResGwDevice::processSessionMsg()*. The signal type determines which method in *vmcpDevice* is called:

*Vmc provideCallInfo*
    Initializes *vmcpDevice* class members.

*provideRingStart*

> In *OpStartRinging* a *DeviceSignalRingStart* signal is generated and sent to the device thread queue. When received, *provideRingStart* first connects the UAVM to the VMS through a TCP socket: it goes through the list provided in the *ua.cfg* file (see the later "VmServer Redundancy" section) until a *connect()* succeeds. Then it sends a VMCP message to the VMS to initialize the session variables (*vm.setSessionInfo(...)*). Finally it sends a *DeviceEventHookUp* event to the Worker thread to simulate an off-hook and to connect the call (*OpAnswerCall*) on the SIP side.

*audioStart*

> *OpStartAudioDuplex* generates an *audioStart* signal sent to the device queue. Upon receiving this event, *audioStart* performs different tasks, such as creating a new RTP session.

*audioStop*

> *OpStopAudio* sends an *audioStop* signal to the device thread. *audioStop* then sends a *DeviceEventHookDown* event to the Worker thread, stops the player, and closes the socket to the VMS.

*provideDtmf*

> See the later "DTMF Digits."

## Play/Record

This is an overview of the file structures that enable the play and record functions.

### Play

All files are played using the *libsndfile* library (see */vocal-1.3.0/sip/ua/PlayQueue.h*). Right now, the files needed by the Voice Mail system are stored in *VMS/Voice*. These files are in the *.wav* format, but the library is capable of reading other sound file formats.

### Record

These files are saved in *.wav*, μlaw, 8 kbps sample rate. To save the sound file, we use the *libsndfile* library (*vocal-1.3.0/contrib/libsndfile-0.0.22*) with the following options (see */vocal-1.3.0/sip/ua/Recorder.cxx*):

```
SF_INFO info;

memset(&info, 0, sizeof(info));
info.samplerate = 8000;
info.channels = 1;
info.pcmbitwidth = 8;
info.format = SF_FORMAT_WAV | SF_FORMAT_ULAW;
m_iFd = sf_open_write(fileName.c_str(), &info);
```

If you want to change the format in which we record, just change this code with the new format at *http://www.zip.com.au/~erikd/libsndfile/api.html#open*.

## DTMF Digits

The UA currently supports only RTP DTMF (RFC 2833) and the Cisco proprietary variant to transport DTMF. The code is in the RTP stack (see *vocal/rtp/RtpEvent.hxx* and *vocal/rtp/DTMFInterface.hxx* for reception and *vocal/rtp/RtpSession.hxx* for emission).

DTMF digits are handled by Voice Mail for future developments but are not used in the current Voice Mail application. Presently, if a user presses a digit while connected to the Voice Mail system, the digit is transported to the UAVM via RTP. The RTP stack then calls a callback function (*recvRTPDTMF*, defined in the *vmcpDevice* class), which generates an event (*DeviceEventDtmf0....*, *DeviceEventDtmfHash*). When the UAVM receives this event, the *vmcpDevice* thread sends a VMCP message to the VMS for further processing (*vmcpDevice::provideDtmf*).

Currently, the VMS does nothing when it receives a digit, but there is some code in place to handle special digit combinations (see the upcoming "Interactive VMS Mode" section).

## VmServer Redundancy

Whenever UAVM receives the INVITE, it goes to the *uavm.cfg* to look for the *VmServer* IP address in order to connect to it. If none is available, no *DeviceEventHookUp* event will be sent, and the call won't connect.

*uavm.cfg* file is the same as *ua.cfg* except that it has a list of *VmServers* with their IP addresses, for redundancy.

```
VmServer    string  192.170.123.200
.
.
VmServer    string  192.170.123.201
```

# Voice Mail Server

The VMS contains the intelligence for the Voice Mail application and controls the UAVM through the VMCP protocol.

## VMCP Messages Received by VMS

The VMCP messages received by the VMS are:

- *VMCP::SessionInfo*
- *VMCP::Dtmf*

- *VMCP::PlayStopped*
- *VMCP::Close*

The *LineVMCP* class contains a VMCP object and listens to the open sockets for VMCP messages (*LineVMCP::Process*). When a message is received, an event is generated and sent to the state machine for further processing. Events are listed in *vocal-1.3.0/vm/vme/EventList.h*.

## VMS Initialization

When the VMS is started, the following objects are created:

*VMCP*
Needed to communicate with the UAVM.

*LineVMCP*
Receives VMCP messages from the VMCP and puts corresponding events to the event queue. It contains a VMCP object and the socket descriptors for the UAVM connections.

*VmSession (derives from Session)*
Contains all the data associated with the current session, in particular the VMS configuration and a *LineVmcp* object.

The VMS then waits on port 8024 and forks a new process for every connection request received from a UAVM. Then events are generated upon reception of VMCP messages as described later, and the VMS waits for these events in the *State:: ProcessLoop()*.

## IMAP Server

The interface for the Internet Message Access Protocol (IMAP, RFC 2060) library client (*vocal-1.3.0/contrib/imap/c-*) is *vocal-1.3.0/vm/mail/mail.cxx*. This file is used to communicate with the IMAP server that contains the Voice Mail messages. This IMAP client is used when the VMS is used in the interactive mode as explained next.

## Interactive VMS Mode

*Interactive mode* means that the user can interact through DTMF with the Voice Mail application. For example, he can record a greeting, listen to his mailbox on the phone, and skip and delete messages. The code is already in place but has not been tested. The new states are *StateLogin*, *StateMain*, *StateRecordGreeting*, and *MenuState*. Also, the prompts are stored in a table (as many more prompts are needed, and it is easier to keep track of their location this way).

If you use the *-m* option when starting the VMS, which sets the *VmSession::m_xInteractive* flag, then in *VmSession::Start*, uncomment the code that loads the prompt table.

In *StateSetup::ProcessTimer*, the *m_xInteractiveflag* is checked, and if set, the state is changed to *StateMain* and the new code is executed. Code examples as well as a *PromptTable* are given as examples of how to use these new states in the folder */vocal-1.3.0/vm/vme/test*.

In order to use the VMS in interactive mode, using an IMAP server and a VMS IMAP client, do the following:

1. Log in to a user mailbox and enter a password: see *StateLogin.cxx*.
2. Record a greeting on the phone: see *StateRecordGreeting.cxx*.
3. Navigate through the user mailbox with IMAP client commands: see *StateMain.cxx*. To change the options available to the user (add/modify DTMF combinations), change the code here.

One of the *vxml* interpreter project goals was to have a *vxml*-based Voice Mail, which would make it easier and more standard for customers to customize their applications. One problem: we have never had a completely satisfactory method for securely storing the users' IMAP passwords.

The UAVM currently supports DTMF tones via the INFO message. This needs to be replaced with AVT tone support, and we're looking for help from the community to implement this.

# Setting Up a Voice Mail System

Voice Mail Server configuration requires the following steps:

1. Configuring the server
2. Adding a new user
3. Recording the user greeting
4. Running the server
5. Running the client

## Configuring the Server

Voice Mail reads configuration parameters from the file *voicemail.config*, which includes the following default settings:

- State the path for the common Voice Mail files including the greeting and other sound files:

```
voiceRoot     string     /usr/local/vocal/bin/voicemail/Voice/
```

- State the path to the Voice Mail users' home directories, not to be confused with the Unix home directories:

  ```
  homeRoot     string    /usr/local/vocal/bin/voicemail/vmhome/
  ```

- State the path where the temporary files are created:

  ```
  tempRoot     string    /tmp/
  ```

- State the maximum time, in milliseconds, to wait for user input before any action, such as cutting off the call, occurs:

  ```
  InputTimeout     string    10000
  ```

- This parameter is reserved for future use and should be left at 0:

  ```
  RecordOnSelfCall     int     0
  ```

- State the maximum permissible message size in milliseconds:

  ```
  MaxRecordTime     int     60000
  ```

- State the logging level:

  ```
  LogLevel     int     7
  ```

- State the name of the log file:

  ```
  LogFileName     string    /usr/local/vocal/log/vmserver1.log
  ```

- State whether the log should be displayed on screen (0=No, 1=Yes)

  ```
  LogConsole     int     0
  ```

- State the path for the *sendMsg* script:

  ```
  SendMsgScript     string    ./sendMessage
  ```

- State the path for the prompt table:

  ```
  PromptTable     string    /usr/local/head/vocal/vm/wav/PromptTable
  ```

- This parameter is reserved for future use and should be left at 0:

  ```
  AutoPlay     int     0
  ```

## Adding a New User

To add a user, create a directory with the master username/number (4000, for example) in the *homeRoot* directory. For example, if the *homeRoot* is set to */vmhome/* you and the user address is sdv4000, you should create a directory */vmhome/sdv4000*.

In this directory you should place a config file and prerecorded greetings.

The config file has the following records:

```
Name          string <User Name>
SendTo        string <email@user.domain>
Greeting      string <filename.wav>
```

The VM server uses the value typed into the Name field when it sends out voice mail messages to the user. This value is used to populate the From field in these messages. The SendTo field is the email address of the Voice Mail recipient. The Greeting is the name of the *.wav* file used for the prerecorded user greeting. For example, "You have reached my Voice Mail box. Please leave a message after the beep."

Users can set the Call Forward Not Answered or Busy numbers to a number in the dial plan that points to an FSVM server in the user administration GUI.

## Recording the User Greeting

The greeting files should be 8-kHz coded *.wav* files. *Vmclient* does not do any check on the format, so if the file is in an incorrect format, a calling party will hear some noise instead of a greeting.

Record a greeting using the *rec* command:

```
rec -c 1 -b -U -r 8000 greeting.wav
```

If your sound card driver does not support this format, you can make your greeting in two steps:

1. Record a greeting with any format supported by your sound card; for example: 44100-kHz, 16-bit mono.

    ```
    rec -c 1 -r 44100 -w temp.wav
    ```

2. Convert it to 8kHz mμ-Law:

    ```
    sox temp.wav -c 1 -b -U -r 8000 greeting.wav
    ```

This *.wav* file is called each time a calling party is forwarded to the Voice Mail system.

## Running the Server

The server must be built by running:

```
make vmserver
```

Once it has been built, to run the server, type:

**vmserver [OPTIONS]**

The choice of options is as follows:

```
Options: -d  run in daemon mode
         -f <file>  Configuration file with path
         -m  Interactive mode ( This is for future release)
         -v LOG_LEVEL
            LOG_EMERG
            LOG_ALERT
            LOG_CRIT
            LOG_ERR
            LOG_WARNING
            LOG_NOTICE
            LOG_INFO
            LOG_DEBUG
```

```
LOG_DEBUG_STACK
LOG_DEBUG_OPER
LOG_DEBUG_HB
```

For example, type the following:

```
vmserver -f ./voicemail.config -vLOG_DEBUG
```

## Running the Client

Go to the *vocal-1.3.0/sip/ua* directory. The Voice Mail User Agent executable can be built in one of two ways:

```
cd ../vocal-1.3.0/
make ua
```

or by running make from the *ua* directory:

```
cd ../vocal-1.3.0/sip/ua
make
```

To run the UA, type **ua [*options*]**. For example:

```
./ua -s -f /usr/local/vocal/etc/ua1000.cfg
```

For a choice of options, see "Using the UA" in Chapter 2.

To run the UA as a UAVM, you need to specify the *-m* option (interactive mode), for example:

```
ua -m -f uavm.cfg -vLOG_DEBUG
```

The *uavm.cfg* file is same as the User Agent configuration file, *ua.cfg*, except that the LOCAL_SIP_PORT is the port for UAVM (configured in the provisioning GUI), and it should have a list of Voice Mail servers with their IP addresses, for redundancy.

For example:

```
VmServer      string    192.170.123.200
    .
    .
VmServer      string    192.170.123.201
```

Whenever UAVM receives the INVITE, it goes to the *uavm.cfg* to look for the Voice Mail server location in order to connect to it.

This concludes our look at the Voice Mail Feature server. The next two chapters cover the MGCP and H.323 translators.

# CHAPTER 15

# MGCP Translator

Media Gateway Control Protocol (MGCP, RFC 2705) is a master/slave protocol in which the endpoints, known as *gateways*, are slaves of the gateway controllers, the call agents (CAs). The CAs maintain statefulness with all registered gateways at all times, telling them to give dial tone to analog phone sets, interpreting the dual-tone multifrequency (DTMF) tones sent through the gateways by the phones, and enabling all features and functionality.

One of our original designs of VOCAL was based on MGCP, and during its implementation, we had to write our own stack because there was no open source MGCP stack available anywhere else. When we realized that SIP served our needs better than MGCP did, for reasons spelled out in Chapter 1, we took our MGCP Call Agent design and converted it into an MGCP/SIP translator. Our work on the MGCP stack has continued, and the mailing list is highly active, with most of the questions (and answers!) coming from members of the community outside of our core development group.

## Media Gateway Control Protocol

The MGCP standard, derived from a combination of the two draft protocols, Simple Gateway Control Protocol (SGCP) and IP Device Control (IPDC), stipulates a call control architecture that puts network devices, called *call agents*, in control of the endpoints, called *gateways*. Like SIP and H.323, there are dedicated MGCP phones. There are also MGCP gateways that are dedicated hardware units that support large numbers of ordinary analog phones and are often called Internet access devices (IADs). If the Call Agent instructs the gateway to ring a phone, the gateway sends the proper electrical voltage that makes the phone ring. From the phone's point of view and, in many cases, the user's point of view, the phone might as well be connected to a PSTN class 5 switch at a central office.

The call agents (CAs) contain all signaling, feature, and functionality intelligence. The gateways are dumb endpoints, and, in an MGCP network, there is an expectation that all local CAs can communicate a consistent set of instructions to the gateways so that all local gateways act in a similar or consistent manner. This type of architecture permits the development of huge gateway controllers that can process enormous numbers of calls very cheaply and can connect to a Signaling System 7 (SS7)–controlled network.

MGCP is a text-based protocol that runs over the User Datagram Protocol (UDP). To counteract packet loss, the MGCP CAs and gateways work with a positive subscribe and notify mechanism that not only maintains state but also ensures that all transmitted packets are accounted for and retransmitted, if necessary. A newer protocol called MEGACO (H.248) is similar to MGCP and is receiving wide attention in the VoIP development community. There is also a variant of MGCP from CableLab's PacketCable initiative (*http://www.packetcable.com*) called *network-based call signaling* (NCS).

## Comparing MGCP to SIP and H.323

It is difficult to compare performance between systems that use MGCP and those that use SIP. The issue has to with the amount of call processing performed by the gateways. As MGCP is a device control protocol, the gateway can be oblivious to the call: all it needs to do is set up the media path. The call signaling is the responsibility of the call agent and the softswitch.

MGCP softswitches can connect to the public switched telephone network (PSTN) through Signaling System 7 (SS7) links to signal the call setup. The softswitch is responsible for call routing, feature activation, and instructing the gateway about setting up the call between the PSTN and the packet network. The gateway's responsibilities are simple and enable manufacturers to build units that can process an impressive number of calls per second. A problem with this model is that a big, expensive softswitch can become a big scaling bottleneck. Some of the larger gateways claim to be able to support 6000 DS0s and a rate of about 250 cps.

With SIP-based or H.323-based networks, the gateway is responsible for processing signals between the PSTN and the IP network. This added responsibility reduces the ability of the gateways to scale, but it also eliminates the need for a big, expensive softswitch and enables distributing the call signaling intelligence more equally throughout the network.

## Messages

MGCP has a set of unique messages. The most common of them are the following:

*Create Connection (CRCX)*
> The CA sends a create connection message to the gateway to set up and get an RTP port, which can be either one- or two-way.

*Modify Connection (MDCX)*

The CA sends a modify connection message to the gateway after the streaming audio path has been set up between the calling and called endpoints. This can be used to send updated Session-Description Protocol (SDP) data and for media negotiation.

*Delete Connection (DLCX)*

The CA sends a delete connection message to the gateway as the call is being torn down.

*Request Notification (RQNT)*

Request notification is the subscribe/notify message that the CA uses to get events from the gateway. Besides off-hook notification, RQNT can contain requests for notification for phone numbers that match specific dialing patterns and for events such as the gateway hanging up before the dialing is finished. It also signals ringing and dial tone.

*Notify (NTFY)*

The gateway sends a notify message to the CA when an event takes place, such as going off-hook and dialing a number.

*Restart in Progress (RSIP)*

Restart in progress is the registration message that the gateway uses to connect to the CA. The CA responds with RQNT, which includes a request for notification when the gateway goes off-hook.

*200*

This is an OK message, similar to the message used in SIP.

*250*

This is an OK message that requests call statistics. We have not fully implemented statistics gathering in our translator, but we support this message to enable the translator to work with third-party gateways that use this message.

## Two-level implementation

The Vovida MGCP stack is implemented as a two-level stack: the upper layer is the externally callable Application Programming Interface (API), while the low-level code builds, parses, and encodes the messages. The stack has a class for each MGCP command (for example, *RequestNotification* and *DeleteConnection*). The message sender creates an object and calls methods of that object to set the different possible parameters.

When the far end receives the message, an object is generated and passed to the receiving process, which can use methods to examine the contents of the message. The stack also has API support for both blocking and nonblocking sends: the blocking sends block until the other side successfully responds or until a time-out has occurred. The MGCP stack code is in the *vocal/mgcp* directory (not to be confused with the *vocal/translators/mgcp* directory). More information is available in *libmgcp* and *lowlevelmgcp*, which is the part of the library that actually sends the messages back and forth and encodes and decodes them.

### Architecture

MGCP commands are implemented as classes (one class per command). The parameters are all classes, as well. The general strategy for sending and receiving looks like this:

1. Create an object for the MGCP command.
2. Create objects for each parameter in the command, then *insert()* these parameters into the command object.
3. *send()* the command object via an *MgcpTransmitter* to the other side (agent/gateway).
4. The other side will notice the incoming data from a *UdpReceiver* object and pass the data to *parseMessage()*. *parseMessage()* returns an *MgcpCommand* object, which contains all of the parameters in the message, reencoded into objects.
5. The receiver examines the objects that have been built to decode the message. When it is done, it builds a reply using the same mechanism.

## Vovida MGCP API

The current version of this MGCP library conforms to MGCP Version 1.0 as described in RFC 2705.

### Architecture

The MGCP stack has a class for each MGCP command (for example, *RequestNotification* and *DeleteConnection*). The message sender creates an object and calls methods of that object to set the various possible parameters. When the other side receives the message, an object is generated and passed to the receiving process. The process can use methods to examine the contents of the message. The stack has API support for both blocking and nonblocking sends—the blocking sends block until the other side successfully responds or until a time-out has occurred.

### Memory management

The application may instantiate a command either on the stack or on the heap. The stack will make a copy of the object, so it's the application's responsibility to manage the memory of the original object. The application sets up the message using methods of the message object and then passes it to the transport object. On the receiving side, the stack will create a new message on the heap and pass back a pointer to that message. The application does not need to explicitly free the memory for that object but will need to delete all references to that object when done. Freeing the memory is accomplished through the use of smart pointers, which are described in Chapter 8.

## Error handling

At the moment, error handling is very preliminary. The intent will be that error handling will occur by exceptions (in the C++ code) or error code returning (in the C code), but the exact mechanism is not yet defined (one known factor is that some C++ compilers and operating system combinations work poorly with exceptions). We have identified a need to change the API to more clearly define how the error handling is accomplished.

## Interoperability

Some degree of interoperability testing has been performed with this version of the code with products from Audiocodes (*http://www.audiocodes.com*), Cisco (*http://www.cisco.com*), and Telogy (*http://www.telogy.com*). We intend to perform more interoperability testing of the code as opportunities become available.

## Issues

The following are better-known issues with the MGCP implementation:

*Completeness*
> While the majority of the API is ready and working, a number of things are missing: some package items that take arguments (including ADSI) and much of the script. The error-handling routines need to be more robust and complete. Refer to the *todo* file in the upper level for a list of items that are not completed.

*Simplicity*
> A number of interfaces, which do not need to be called by users, are exposed to users. This may well go away.

*Memory*
> We saw memory corruption related to the string library in a multithreaded environment during high volumes of traffic through the stack, mainly in the *parseMessage()* function. This is a known bug in the string class of the compiler, which is not thread-safe.

*Efficiency*
> There are too many gratuitous copies of strings: numerous extra copies in the code, as well as too much dynamic resizing of templates and the like. At the very least, the *encode()* functions should pass back *char* if possible, and the results can be merely concatenated together as late as possible before the *send()*.

> We believe this will be reasonably safe in most instances because the buffers will not be written by more than one function. Implementing a simpler *int* to *char* would also be good. Parsing the buffer in place can save significant time. As it is, the parsing code makes several copies of each subportion of an incoming

message as it tokenizes each piece of the message. A significant savings in copying time would be to scan the buffer linearly and not copy. The messages should be able to be broken down into tokens in a single-pass scan. Even more efficient would be to scan and tokenize in a single pass, but that might add complexity to the code.

However, there is also the "premature optimization" argument—this should probably be done at a later time when we can actually measure the code for its performance.

*Comments*

The code is not as well commented as it should be. This will not improve unless community members contribute changes to the code base.

*Build*

Inline functions are not cleanly separated out from the actual headers. There are a number of warnings in the code. Namespaces aren't used, but if we use them, they may break older versions of *gcc*. There are not enough uses of *assert( )*.

# MGCP Translator

The MGCP translator acts as a SIP User Agent that talks SIP to a SIP proxy on one side and acts as an MGCP Call Agent (CA) that speaks MGCP to control MGCP gateways on the other side. This allows MGCP gateways to be controlled by the MGCP translator's CA, but servers within the SIP world control the routing decision. A call can be made from an MGCP gateway to a SIP entity or from a SIP entity to an MGCP gateway. A call may also be made from an MGCP gateway back to the same MGCP translator via a network of SIP servers, which in turn will attempt to contact another or the same MGCP gateway.

Although two MGCP connections can be made through the same CA and gateway, the translator sends all messages through to the VOCAL network. There is no hairpinning permitted through the translator (where the signal receives responses from the translator without the participation of the VOCAL servers).

The translator sends SIP registration messages on behalf of all the endpoints it controls. It can make outgoing SIP calls and terminate incoming SIP calls on the MGCP gateways. Besides basic calls, it supports transfer, conference, and call waiting. It has limited redundancy support: the backup MGCP translator is able to support connected calls from a failover. For most of these features, detailed call flow diagrams later in this chapter illustrate their operation.

*Startup and registering*

The MGCP translator sends SIP REGISTER messages on behalf of all the MGCP endpoints that it controls. This can be based on an MGCP gateway/endpoint restarting with a Restart In Progress (RSIP) message or when an MGCP translator comes up and finds a new MGCP endpoint.

*Basic call*

The MGCP translator allows MGCP endpoints to make basic outgoing SIP calls and also terminates basic incoming SIP calls onto MGCP endpoints.

*Transfer*

MGCP endpoints can initiate and/or be a part of a blind or consultation transfer scenario. However, the SIP message currently being used is the obsolete TRANSFER message.

 At the time of this writing, a stable and widely accepted replacement for the TRANSFER message called REFER was still under discussion by the SIP Working Group of the IETF. REFER is similar to TRANSFER, and changing the code to use REFER will not be difficult once the message becomes part of the RFC.

*Conference*

MGCP endpoints can initiate and/or be a part of a conferencing scenario. This also uses the obsolete TRANSFER message. The MGCP translator does not do the Real-time Transport Protocol (RTP) mixing of the conference. The conference bridge performs that task.

*Call Waiting*

Call Waiting is supported, as is Call Waiting Block (activated by dialing *70).

*Early and late RTP*

RTP being set up before ringing is heard allows the far end to generate its distinctive ring-back tone. Late RTP lets the near-end endpoint/gateway generate its own local ring-back tone.

*Handling of 302 messages*

The translator allows SIP-controlled features such as Call Forward All Calls, Call Forward No Answer, and Call Forward Busy, for example, to work with MGCP gateways and other endpoints, by redirecting calls that are replied to with a SIP 302 Moved Temporarily message.

*Redundancy*

There is limited redundancy support: the backup MGCP translator is able to support only basic connected calls from a failover.

*MGCP Call Agent*

To enable the MGCP endpoints to talk directly to one another without involving entities from a SIP-based network, the translator can also be configured in a loopback configuration so that the SIP messages are simply looped back to it, allowing the MGCP translator to act as a simple MGCP call agent.

# Key Data Structure

The following is a summary of the objects found within the MGCP translator:

*CallAgent*
> The only structure-type object that is global and contains all global configuration data read from configuration files and the Provisioning server.

*Endpoint*
> Another structure object; carries all information about a gateway including its active and held-call information. *Endpoint* also contains SIP message pointers related to its call status.

*EndpointMap*
> Used to find an *Endpoint* object, given its name in string, which is used by the *sipProxy* module when getting SIP messages.

*QueueEventData*
> A wrapper for all MGCP and SIP events to be inserted and retrieved from the message queue (*iqs[0]*).

*callDataList and qEventDataList*
> Two preallocated global data pools during startup used for call processing and message passing to save time from requesting and returning heap memory.

There is only one *SipStack* object for receiving and sending SIP messages, but there can be several MGCP stack objects for different gateways.

Limited redundancy is achieved by having a backup MGCP translator listening for heartbeats from the primary translator. The backup activates itself if it misses a certain number of continuous heartbeats within a configured number of seconds from the primary translator. Once it comes up, it will send a Request Notification (RQNT) message for off-hook to all endpoints, and if an endpoint returns a 401 Already Off-hook message, it then sends an RQNT for on-hook, assuming that the endpoint is on an active call. The RQNT for on-hook doesn't mean that the user is hanging up; it is a request in advance for notification when the event happens. This mechanism will guarantee that once an endpoint on-hooks, the CA is notified and clears all call data for the endpoint. All connected calls remain active, but billing is incomplete because the backup CA does not have call data to send a SIP BYE message through the SIP network. Future work should have the primary CA synchronize call data with the backup CA when the call is connected, and also support Marshal failover.

# High-Level Flow: Startup

The flow of messages through the translator and VOCAL is best described in the call flows that we have provided later in this chapter. For those of you who are brave enough to delve deeply into the workings of our translator, we have provided a code-level description of how the translator starts up and begins working. Figure 15-1 illustrates the inner workings of the translator from a high-level flow point of view.

---

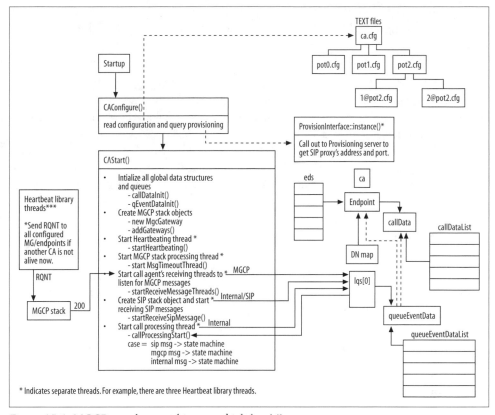

*Figure 15-1. MGCP translator architecture, high-level flow*

Upon startup (in file *callAgent.cxx*), the function *caConfigure( )* is called to read from some text files. This function automatically looks for *ca.cfg*, which contains generic MGC (CA) information as well as the names of the text files that it needs to read to get MG (gateway)-specific information (currently named *pot0.cfg*, *pot1.cfg*, etc.). Within each of these text files are names of more text files that it needs to read to get endpoint information (for example, *1@pot2.cfg* and *2@pot2.cfg*).

To hold all of this information, there is a single global *CallAgent* (*ca*) data structure that stores CA-generic information, an array of gateway structures, and an array of endpoint structures, each of which carries all information about an endpoint, including active call, held call, and SIP message pointers along with other data structures. There's also a *dnMap* to hold directory number information, and an endpoint map for faster access to specific endpoint names, strings, and other less important data structures.

## Provisioning

Once config has read the text files, we then query the Provisioning server via instantiating *ProvisioningInterface::Instance( )* that runs in a separate thread. This callout can add to or overwrite the information retrieved from the preceding text files. Currently, it gets only the SIP Proxy (or VOCAL's Marshal) address and port, but it can be extended to get much more information.

### Starting the Call Agent

The process *caStart()* is then called (in file *callAgent.cxx*), which starts up many things. First of all, it initializes two major pools of global data structures called *callData* and *queueEventData* via the processes *callDataInit()* and *qEventDataInit()*, which you will see being used later. Basically, they are used for call processing and message passing, to reduce the number of requests and returns to and from the heap memory. The function *addgateways()* then instantiates MGCP stack objects per the configuration and Provisioning information retrieved before.

### Starting heartbeating

The *startHeartbeating()* thread is spawned. Actually there are several threads spawned for heartbeating, but the main thread of interest basically sits and listens for heartbeats from any other call agents that may already be up and running. If one is found, it goes into standby mode until it no longer hears any other call agent's heartbeats, at which point a redundancy scheme is initiated. If it is the first one up and does not hear any other call agent's heartbeat, it continues on, sending RQNT messages to each endpoint that it knows about.

### Starting the MGCP stack threads

In the meantime, *startMsgTimeoutThread()* is called to start up the MGCP stack object instantiated before, creating a new thread for each of these stacks. The function *startReceiveMessageThreads()* is called to spawn new threads to listen for MGCP messages. These are the threads that will receive any 200 OK message responses from the RQNT sent out earlier from the Heartbeating thread, as well as any other subsequent MGCP messages from the MGCP endpoints. These messages are encapsulated in a *queueEventData* and put into a generic queue *iqs[0]*.

### Starting to receive SIP messages

The function *startReceiveSipMessage()* is called to instantiate a single SIP stack object, which starts receiving (and is capable of sending) SIP messages. Any SIP messages received will be encapsulated in a *queueEventData* and put into the generic queue *iqs[0]*.

### Starting MGCP state machine

Finally, *callProcessingStart()* is called to spawn a final thread to pull out all of the messages from *iqs[0]*. Based on the type of message in *queueEventData*, it will call either the MGC state machine (if it's a *StackEvent* from the MGCP stack or if it's an internal MGCP event triggered from the SIP side) or the SIP state machine (if it's a message from the SIP stack). What's missing is the internal event triggered from the MGCP state machine that will call the SIP state machine. It was implemented as a direct call to the appropriate SIP-side function instead of going through this queue, due to development time restraints. This is something that we've been meaning to fix but have not yet. Anyone interested in helping here?

There's also a case for *TimeoutEvents* (for example, Call Forward No Answer), which is no longer being used since the SIP feature servers now take care of this.

## Call Flow Sequence

The *sipProxy* module is responsible for sending and receiving SIP messages, and *potStateEngine* and *callProcessing* are the main modules that drive the MGCP Call Agent's state machine. The MGCP state machine is the main controller between the two sides.

# Test Tools

We have a Perl script that hangs in between the MGCP translator and the MGCP gateway/endpoint from a control messaging point of view. It is useful when needing to quickly tweak a message slightly—for instance, during interoperability events.

# Future Development

Here is a list of tasks that we have identified but not yet begun implementing:

- Enable full redundancy with call state/data sharing/syncing, recognizing multiple SIP Proxies in case of a SIP Proxy failure.
- Provision through the new Mascarpone system discussed in Chapter 19.
- Enhance Simple Network Management Protocol (SNMP) support.
- Provide trunk-side support.
- Port to other platforms.
- Equalize IP/hostname comparisons.
- Provide regular expression support for registration using the Restart In Progress (RSIP) message.
- Enable endpoint auditing.
- Add digest and Challenge-Handshake Authentication Protocol (CHAP, RFC 1994) authentication.
- Make the translator fully compliant with the MGCP and SIP stacks.
- Improve the general stability, robustness, and performance of the translator.
- Write a Megaco (RFC 3015) version of the translator.

# Detailed Message Flows

The illustrations in the following sections show ideal message flows. Currently, the translator is not necessarily working this way.

## Registration

Figure 15-2 shows the registration process.

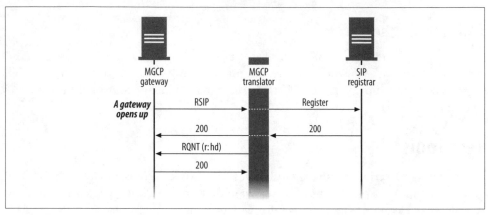

*Figure 15-2. Registering MGCP endpoints via RSIP*

During registration, the gateway starts up and sends an RSIP message to the translator.

 RSIP coincidentally contains the letters *SIP*. This has no correspondence to the fact that the translator is talking to a SIP-based system. RSIP simply stands for *restart in progress*.

The SIP side of the translator, acting as a SIP endpoint, registers with VOCAL, as discussed in Chapter 8, and returns a 200 OK message to the gateway. The translator also sends a Request Notify (RQNT) message requesting notification when the gateway goes off-hook. As long as the gateway is registered, the translator, acting as a gateway controller, maintains stateful control over the gateway. The gateway returns a 200 message in reply to this request.

## Basic Call

Figure 15-3 shows the call flow for a basic call from the calling party's point of view.

### Off-hook

The gateway notifies the CA that it has gone off-hook (*o: hd*), and the CA responds that it received the message (200) and requests notification for when a matching

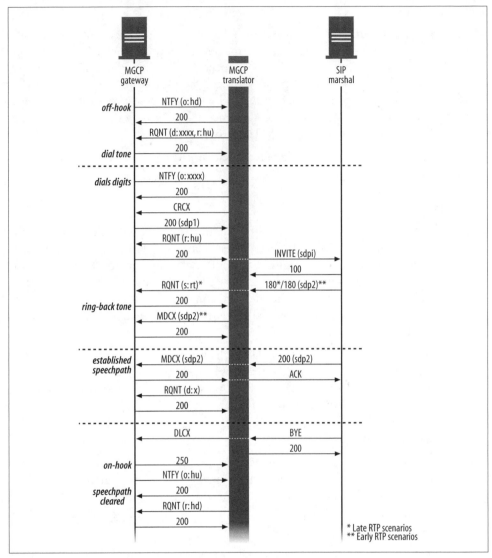

Figure 15-3. Basic call from the caller point of view

pattern of digits is dialed (*d: xxxx*). The CA also asks for notification if the user hangs up before finishing her dialing (*r: hu*).

RQNT (*r: hu*) means requesting notification if the terminal hangs up. RQNT (*r: hd*) means requesting notification if the terminal goes off-hook.

## Dial tone

The RQNT message has enabled dial tone, and the gateway responds (200) that it has received the request from the CA and signals dial tone.

## Dials digits

The gateway notifies the CA that it has dialed a complete phone number. The gateway matches the dialed number to the formats sent by the earlier RQNT message. The CA instructs the gateway to create a connection (CRCX), which leads the gateway to construct a Session Description Protocol message (*sdp1*). The CA requests notification of a hang-up (*r: hu*) before the call has completed processing. The gateway responds positively (200), and the CA sends the SDP data through the translator where it is converted into a SIP INVITE message.

You can follow the path of this INVITE message into the top of Figure 15-4. These figures have been separated for clarity.

## Ring-back tone

Depending on the configuration of the SIP system, either a 180 Ringing message that contains either late or early Real-time Transfer Protocol (RTP) media is returned to the translator. You can see the origin of this SIP message in Figure 15-3. If the 180 message contains early media, the calling party can hear the distinct ringing pattern transmitted by the called party. If the message contains late RTP media, the calling party can't hear the ringing from the called party, and the gateway has to create a substitute sound for the listener.

The translator turns the SIP 180 messages that do not contain early media into MGCP RQNT messages that provide a ringing tone signal (*s: rt*). If a SIP 180 message containing early media is received, the CA sets up a media connection with the gateway that includes the SDP data from the far end (*sdp2*).

## Established speechpath

The far end answers, as you can see in Figure 15-4, sending a SIP 200 message that may include SDP data (*sdp2*). The translator turns this 200 message into an MDCX message that includes the SDP data (*sdp2*). The gateway responds positively and the translator generates a SIP ACK message. At this point, the calling and called parties can speak to each other.

The CA also requests notification if the gateway sends any more DTMF tones (*d: x*) using the REFERS header in a SIP message. This would help the user respond to an interactive voice response (IVR) system that requests digital menu navigation or if the user is required to enter a password for voice mail or another type of service. If the endpoint sends a digit through the RTP stream, the gateways and CA have no concern for its content.

## On-hook

The far end hangs up, as you can see from Figure 15-4, and the translator turns the BYE message into an MGCP delete connection (DLCX) message. The gateway responds positively and notifies the CA that it has hung up (*o: hu*).

## Speechpath cleared

The CA requests notification from the gateway for the next time that it goes off-hook (*r: hd*). The gateway responds positively as it returns to an idle state.

Figure 15-4 shows the same call scenario from the far end, the called party's point of view.

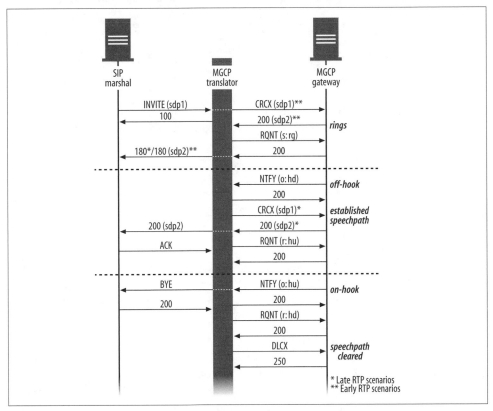

*Figure 15-4. Basic call from the called party's point of view*

## Rings

The translator receives the SIP INVITE message sent by the calling party and turns it into a create connection (CRCX) message with early RTP and/or with the SIP 200 message for late RTP. The gateway responds with a 200 message that includes its SDP data (*spd2*). The CA sends the gateway a ringing signal (*s: rg*) and this signal is eventually sent to the SIP network as a SIP 180/183 Ringing message.

## Off-hook

The user answers the phone, which makes the gateway notify the CA that it is off-hook (*o: hd*). The CA instructs the gateway to create a connection (CRCX) using the

SDP data that it received from the SIP INVITE message (*sdp1*). The (*sdp2*) data is created if there is late RTP in the 200 response to the SIP INVITE message.

### Established speechpath

The SIP 200 ACK messages are sent and received, permitting the speechpath to be established. The CA requests a notification from the gateway if it hangs up (*r: hu*).

### On-hook

The gateway hangs up, notifying the CA, and a SIP BYE message is transmitted.

### Speechpath cleared

The CA requests that the gateway notify it the next time it goes off-hook and deletes the call connection (DLCX). The speechpath is now cleared as the gateway moves to its idle state.

# State Diagram

We've had a number of different engineers working on VOCAL and the protocol stacks. Some have been up-to-date with cutting-edge (and sometimes bleeding-edge) tools and technology, while others have come to us with a more traditional approach to programming. The MGCP translator was developed using a traditional programming approach, and it differs from other VOCAL components both in its C code base and the fact that it has been written with functional rather than object-oriented code.

In line with standard engineering projects, a judgment call was made to strike a balance between designing an enormous state machine that was impossible to manage and a small, cryptic machine that was impossible to understand. While most of the processes pertain to single state/event combinations, some have been reused for similar situations. The state diagram shown in Figure 15-5, rather than being a picture full of boxes and arrows, is a table that describes the behavior of each state during different events. The key to this table follows the diagram, which is followed by descriptions of two call flows as they work their way through the state machine. These descriptions are illustrated in Figures 15-6 and 15-7.

## Key to the State Diagram

The symbols and abbreviations used in the state diagram are described in the following sections.

| | | S0 | S1 | S2 | S3 | S4 | S5 | S6 | S7 | S8 | S9 |
|---|---|---|---|---|---|---|---|---|---|---|---|
| | | Init | Idle | Dialing | Digits collected | Call initiated | Far end alerted | Received call | Alerting | Connected | Discon-necting |
| E0 | Restart | P0 | P1,S0 | P2,S0 | P2,S0 | P3,S0 | P3,S0 | P4,S0 | P4,S0 | P5,S0/– | P6,S0 |
| E1 | Response | P10,S1 | I | I | P13,S4 | I | I | P16,S7 | P21,S8 | I | P31,S0 |
| E2 | NewCall | X | P11,S2 | X | X | X | X | glare | P17 | X | I |
| E3 | DigitsCollected | X | P12,S3 | P12,S3 | X | X | X | X | X | P19 | X |
| E4* | ReceiveCall | reject | P14,S6 | glare | glare | glare | glare | glare | glare | glare | glare |
| E5* | FarEndAlerting | I | I | X | X | P15,S5 | I | X | X | X | I |
| E6* | Answer | I | I | X | X | X | P18,S8 | X | X | P52 | I |
| E20 | ReleaseCall | X | I | P2,S0 | P7,S0 | P30,S9 | P30,S9 | X | X | P30,S9 | P33,S0 |
| E21* | DeleteCall | I | I | P32,S9 | P32,S9 | P32,S9 | P32,S9 | P32,S9 | P32,S9 | P32,S9 | I |
| E30 | Feature | I | I | P43 | I | P45 | P46 | I | I | P49 | I |
| E30 | MDCX | X | X | X | X | X | P51 | X | I | P51 | X |
| ELast | Unknown | X | X | X | X | X | X | X | X | X | X |

*Figure 15-5. MGCP translator state machine diagram*

## General

The following characters and abbreviations are used to describe errors and nonprocessing events:

*X: callEventError*
Invalid transition, report error.

*I: callEventIgnore*
Null transition, ignore event.

*glare: glare (when two messages cross each other on the wire)*
For POTS, resolve the glare by dropping the incoming call and allowing the local call to continue. For trunks, resolve the glare by letting the network side be the "master."

*reject: callReject*
Generate *clearCall* event for the far end.

## State symbols

The following symbols describe the states available within the state machine:

*S0*
Init: the initializing state used when the endpoint registers with the network.

*S1*
Idle: the state of rest for registered endpoints between calls.

*S2*
Dialing: when the endpoint is making a call and the user has dialed some digits but not enough to match any of the dialing patterns anticipated by the call agent.

*S3*

Digits Collected: when the user has dialed an anticipated pattern of digits and the pattern has been sent to the call agent for processing.

*S4*

Call Initiated: the call agent has translated the dialing pattern into a SIP INVITE message and sent it to the User Agent Marshal server.

*S5*

Far End Alerted: the call agent has received a SIP 180 Ringing message back from the far end.

*S6*

Received Call: an INVITE message has been received, translated into a CRCX message, and sent to the endpoint.

*S7*

Alerting: the endpoint has responded to the CRCX message with an MGCP 200 message, which the translator sends out as a SIP 180 Ringing message.

*S8*

Connected: the endpoint is in a call.

*S9*

Disconnected: either the calling party or called party hangs up. The call is torn down, taking the state machine back to Init where the endpoint reregisters with the call agent, taking the state machine to Idle where it rests until the next call.

## Event symbols

The following symbols describe the states available within the event machine:

*E0*

*Restart*: the registration event.

*E1*

*Response*: an MGCP generic response event, which can be a response to any MGCP message; for example, CRCX or RQNT. This could have been split up into a separate event for every message type, but that would have made the machine much larger than necessary.

*E2*

*NewCall*: an off-hook event.

*E3*

*DigitsCollected*: when dialing, the digits are being collected on the endpoint.

*E4*

*ReceiveCall*: a SIP INVITE message comes in to the translator.

*E5*

*FarEndAlerting*: either a SIP 180 or 183 Ringing message comes into the translator.

E6

    *Answer*: a SIP 200 message comes into the translator.

E20

    *ReleaseCall*: an on-hook event when the endpoint hangs up first.

E21

    *DeleteCall*: a SIP BYE message comes into the state machine.

E30

    *Feature*: these are reserved for feature development.

Elast

    *Unknown*: a collection event for errors and illegal messages.

## Process symbols

The following symbols describe processes found in the file *potStateEngine.cxx*:

P0

    *initRestart( )*: the endpoint is on-hook and the translator is in the Init state. The endpoint registers with the translator, taking the state machine to the Idle state.

P1

    *idleRestart( )*: the endpoint is on-hook, the translator is in the Idle state, and the gateway fails. When the service resumes, the gateway sends a new RSIP, and this process takes the state machine back to the Init state.

P2

    *restartTransient( )*: the endpoint is off-hook, the translator is in either the Dialing or Digits Collected state, and the gateway fails. When the service resumes, the gateway sends a new RSIP, and this process takes the state machine back to the Init state.

P3

    *restartOutCall( )*: the endpoint is off-hook, the translator is in either the Call Initiated or Far End Alerted state, and the gateway fails. When the service resumes, the gateway sends a new RSIP, and this process takes the state machine back to the Init state.

P4

    *restartInCall( )*: the endpoint is off-hook, the translator is in either the Received Call or Alerting state, and the gateway fails. When the service resumes, the gateway sends a new RSIP, and this process takes the state machine back to the Init state.

P5

    *restartConnected( )*: the endpoint is off-hook, the translator is in the Connected state, and the gateway fails. When the service resumes, the gateway sends a new RSIP, and this process takes the state machine back to the Init state.

*P6*

> *restartDisconnect()*: the endpoint is on-hook, the translator is in the Disconnecting state, and the gateway fails. When the service resumes, the gateway sends a new RSIP, and this process takes the state machine back to the Init state.

*P7*

> *restartHalfCall()*: the endpoint is on-hook, and the translator is in the Digits Collected state. The user hangs up after dialing but before the call can be initiated, taking the state machine back to the Init state.

*P10*

> *initResponse()*: the endpoint is on-hook, and the translator is in the Init state. The translator sends an RQNT to the endpoint, requesting notification about going off-hook. The endpoint responds with a 200, taking the state machine to Idle where it will wait for the next call.

*P11*

> *newCall()*: the endpoint is on-hook, and the translator is in the Idle state. The user picks up the phone, taking the state machine to the Dialing state.

*P12*

> *dialingComplete()*: the endpoint is off-hook, and the translator is in the Dialing or Idle state. When the user dials a number that matches a pattern in the digit map, an NTFY message is sent from the endpoint to the translator taking the translator to the Digits Collected state.

*P13*

> *connectionReady()*: the endpoint is off-hook, and the translator is in the Digits Collected state. The user has finished dialing, and the translator is constructing a SIP INVITE message. When the translator is ready to send the INVITE, this process takes the state machine to the Call Initiated state.

*P14*

> *receiveCall()*: the endpoint is on-hook, the translator is in the Idle state, and a SIP INVITE message comes in to the translator, taking the state machine to the Received Call state.

*P15*

> *ringBack()*: the endpoint is off-hook, and the translator is in the Call Initiated state. The translator has sent an INVITE message and receives either a 180 or 183 message, taking the state machine to the Far End Alerted state.

*P16*

> *ringing()*: the endpoint is on-hook, the translator is in the Received Call state, and the endpoint starts ringing, taking the state machine to the Alerting state.

*P17*

> *answer()*: the endpoint is on-hook and ringing, the translator is in the Alerting state, and the user picks up the phone to answer the call. The translator remains in the Alerting state.

*P18*

    *connected( )*: the endpoint is off-hook, the translator is in the FarEnd Alerted state, and the far end picks up, taking the state machine to the Connected state.

*P19*

    *dtmfDigit( )*: the endpoint is off-hook and engaged in a call, the translator is in the Connected state, and the user dials a digit, perhaps to work with an IVR system. The translator remains in the Connected state.

*P21*

    *connectionReady2( )*: the endpoint is off-hook and no longer ringing, the translator is in the Alerting state, and the SIP proxy needs to send new SDP data attached to a 200 message. The translator is taken to the Connected state.

*P30*

    *disconnect( )*: the endpoint is off-hook, engaged in a call, and the translator is in either the Call Initiated, Far End Alerted, or Connected state. The near endpoint hangs up, taking the translator to the Disconnecting state.

*P31*

    *disconnectResponse( )*: the endpoint is on-hook, and the translator is in the Disconnecting state. The translator has sent a DLCX message to the endpoint and is clearing the speechpath. When the endpoint responds to the DLCX message, the translator goes back to the Init state.

*P32*

    *clearCall( )*: the endpoint is off-hook, and the translator is in either the Dialing, Digits Collected, Call Initiated, Far End Alerted, Received Call, Alerting, or Connected state. The call is either canceled by the translator before being initiated or torn down by the far end after being connected. The translator is taken to the Disconnecting state.

*P33*

    *disconnectRelease( )*: the endpoint is off-hook, and the translator is in the Disconnecting state. The far end has already hung up, and the near end hangs up, taking the state machine back to the Init state.

Processes P1 through P6 look similar. What makes them distinct are the different data structures that they clean up based on what state the translator is in when the gateway fails.

The following processes are reserved for enabling features, and we are not discussing MGCP-based features in this book:

*P43*

    *featureDialing( )*

*P45*

    *featureCallInitiated( )*

*P46*
    *featureFarEndAlerted( )*
*P49*
    *featureConnected( )*
*P51*
    *transferGetting( )*
*P52*
    *modifyConnect( )*

## Walking Through Basic Calls

Let's look at how making and receiving a call changes the state machine to get a better understanding of its design.

The descriptions following the diagrams have been made from the MGCP translator's point of view. In these descriptions, the near endpoint is always an MGCP gateway. The far endpoint could be any endpoint supported by VOCAL.

### Making a call

Figure 15-6 shows the most likely path through the state machine for calls initiated at the endpoint. At the right of Figures 15-6 and 15-7, the dashed arrows represent two possibilities for the call ending: either the calling party hangs up first, or the called party hangs up first. This distinction has been made in the descriptions following the diagrams.

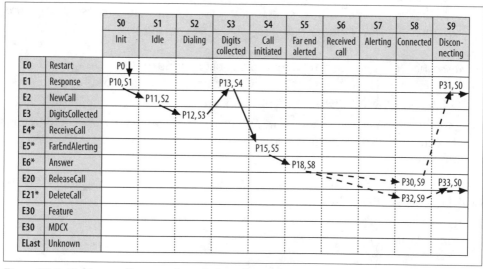

| | | S0 | S1 | S2 | S3 | S4 | S5 | S6 | S7 | S8 | S9 |
|---|---|---|---|---|---|---|---|---|---|---|---|
| | | Init | Idle | Dialing | Digits collected | Call initiated | Far end alerted | Received call | Alerting | Connected | Disconnecting |
| E0 | Restart | P0 | | | | | | | | | |
| E1 | Response | P10,S1 | | | P13,S4 | | | | | | P31,S0 |
| E2 | NewCall | | P11,S2 | | | | | | | | |
| E3 | DigitsCollected | | | P12,S3 | | | | | | | |
| E4* | ReceiveCall | | | | | | | | | | |
| E5* | FarEndAlerting | | | | | P15,S5 | | | | | |
| E6* | Answer | | | | | | P18,S8 | | | | |
| E20 | ReleaseCall | | | | | | | | | P30,S9 | P33,S0 |
| E21* | DeleteCall | | | | | | | | | P32,S9 | |
| E30 | Feature | | | | | | | | | | |
| E30 | MDCX | | | | | | | | | | |
| ELast | Unknown | | | | | | | | | | |

*Figure 15-6. Making a call as seen through the state machine*

When a CA comes up, its view of the endpoint is in Init state. It sends a Restart in Progress (RSIP) message indicating that it is active and ready to communicate. The translator receives the message, remains in the Init state, and sends a Request Notification (RQNT) asking the endpoint to tell the translator when it goes off-hook. This response is represented in Figure 15-6 as a shift from cell S0, E0 to cell S0, E1, and this shift is indicated by the first arrow in the top left.

The endpoint responds to the RQNT message with a 200 OK message, taking the translator from the Init state to the Idle state. Now, from the Idle state, the endpoint can go off-hook or receive a call. The user picks up the phone to make a call, taking the endpoint off-hook. This off-hook event is sent to the translator as a Notify (NTFY) message that takes the state machine to the Dialing state.

In the Dialing state, the translator sends signal events that give the end user dial tone and a digit map that enables it to dial a phone number. The digit map expresses an anticipated dialing pattern, such as xxxx for local calls within a corporate phone system. The translator doesn't want to know what has been dialed until an anticipated pattern has been matched. So if one of the patterns is xxxx and the user dials 1-2-3, no signals are sent from the endpoint to the translator until the fourth digit has been dialed.

Earlier in Figure 15-3, the MGCP message, NTFY (o: xxxx), shows the endpoint notifying the translator that a dialing pattern of four digits had been matched. This is a simplified example. As you can imagine, in real life, the digit map can contain many possible patterns based on configured aliases, local extensions, and E.164 phone numbers.

After a pattern within the digit map has been matched, the endpoint sends an NTFY, which includes the dialed digits taking the state machine to the Digits Collected state. Create Connection (CRCX) is sent to the endpoint so that it can set up a Real-Time Transport Protocol (RTP) port for setting up a speechpath. This information is sent back to the CA via SDP.

A valid response from the CRCX message takes the state machine to the Call Initiated state. At this point, the translator has received all of the digits and set up an RTP port and a gateway is ready to initiate the call. Now the SIP stack on the other side of the translator is triggered to create and send an INVITE message that contains all the collected call information.

Looking at this from a higher level, before the translator can send an INVITE message, it has to go through a long series of information gathering. This gathering is almost serial in nature: each process depends on a previous process completing before it is activated. The RTP port, for example, would not be set up if an NTFY message had not been received with a dialing pattern that matched the digit map. If one process fails, the INVITE message is not sent; instead, error messages are exchanged between the endpoint and translator. The fact that the translator is able to construct an INVITE message means that all preceding processes were successful.

Now we have an outstanding INVITE message that the User Agent Marshal server has responded to with a 100 to prevent retransmission. Eventually, a 180 Ringing message arrives at the SIP Proxy within the translator, taking the translator's state machine to the Far End Alerted state. This is the first time in this call flow that the far side has triggered a change in the translator's state machine.

As it has been shown in the call flows, in Figure 15-1 through Figure 15-8, if the far end sends a 183 message that contains SDP data, the ringing tone can be provided through early media. Otherwise, if the far end sends a 180 message, the translator can re-create the ring tone for the endpoint. If the far end sends a 183, the translator sends a Modify Connection (MDCX) message to the endpoint that includes information sent from the far end within the SDP part of the 183 message. If the far end sends a 180, the connection is not modified, but the translator sends a signal to tell the endpoint to give itself local ring-back tone.

For the Far End Alerted state, the far end picks up the phone and a SIP 200 message comes to the translator, taking it to the Connected state, and the call is established. The process for tearing down a call is the same whether the endpoint originated a call or received a call. It is dependent upon who hangs up first. Let's look at how the state machine behaves when it receives a call; then we'll look at how the call is torn down.

### Receiving a call

Figure 15-7 shows the most likely path through the state machine for calls received by the endpoint.

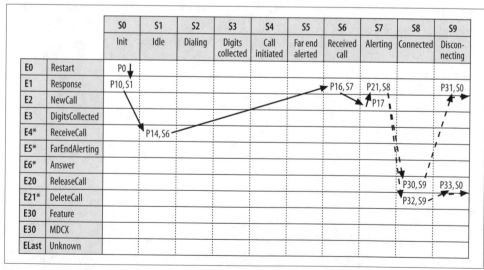

|  |  | S0 | S1 | S2 | S3 | S4 | S5 | S6 | S7 | S8 | S9 |
|---|---|---|---|---|---|---|---|---|---|---|---|
|  |  | Init | Idle | Dialing | Digits collected | Call initiated | Far end alerted | Received call | Alerting | Connected | Disconnecting |
| E0 | Restart | P0 |  |  |  |  |  |  |  |  |  |
| E1 | Response | P10,S1 |  |  |  |  |  | P16,S7 | P21,S8 |  | P31,S0 |
| E2 | NewCall |  |  |  |  |  |  |  | P17 |  |  |
| E3 | DigitsCollected |  |  |  |  |  |  |  |  |  |  |
| E4* | ReceiveCall |  | P14,S6 |  |  |  |  |  |  |  |  |
| E5* | FarEndAlerting |  |  |  |  |  |  |  |  |  |  |
| E6* | Answer |  |  |  |  |  |  |  |  |  |  |
| E20 | ReleaseCall |  |  |  |  |  |  |  | P30,S9 | P33,S0 |  |
| E21* | DeleteCall |  |  |  |  |  |  |  |  | P32,S9 |  |
| E30 | Feature |  |  |  |  |  |  |  |  |  |  |
| E30 | MDCX |  |  |  |  |  |  |  |  |  |  |
| ELast | Unknown |  |  |  |  |  |  |  |  |  |  |

*Figure 15-7. Receiving a call as seen through the state machine*

This flow begins in the Init state when the endpoint sends an RSIP message to the translator, which responds with an RQNT taking the translator to the Idle state. From the Idle state, an INVITE message comes into the translator, which initiates a *ReceivedCall* event, taking the state machine to the Received Call state. The phone is sent an RQNT to ring, and once a response comes back, stating that it is ringing, this takes the state machine to the Alerting state.

The Alerting state means that the endpoint is ringing. When the user answers, the state machine goes to the Connected state. Let's look at how the call is torn down.

### Tearing down the call

This discussion refers to the state machine's transition from the Connected state to the Disconnected state and back to the Init state as shown by the dashed arrows in Figures 15-6 and 15-7.

Either side can hang up, which triggers a *ReleaseCall* or *DeleteCall* event. If the near end hangs up, it triggers a *ReleaseCall*, which takes the state machine to the Disconnecting state. Once it is in this state, a response comes back to a Delete Connection (DLCX) request that takes it back to Init. From Init, the endpoint exchanges messages with the translator to take the translator back to Idle. The endpoint doesn't necessarily register again with the translator, unless its previous registration has expired, but it does return to the Init state before going to the Idle state.

If the far end hangs up first, a BYE message comes to the translator, which triggers a *DeleteCall* event, taking the state machine to the Disconnecting state. The far end is still on-hook, and as soon as the user hangs up, the translator's state machine is taken back to the Init state.

## Graphic State Diagram

Figure 15-8 shows a graphic rendering of the same processes that are described in the earlier call flow diagrams.

Figure 15-8 illustrates the path taken by call events as they make their way through the state machine. Two important points to note in this diagram are:

- The 200 messages are MGCP 200 messages unless they are specifically called SIP 200s.
- *ntfy(hf)* means the endpoint is notifying the translator of a hook-flash event. This is used for transferring calls or connecting to a conference bridge, if one is available.

Having looked at the MGCP translator, let's have a look at the other VoIP protocol translator that we built to work with Microsoft NetMeeting, the H.323 translator.

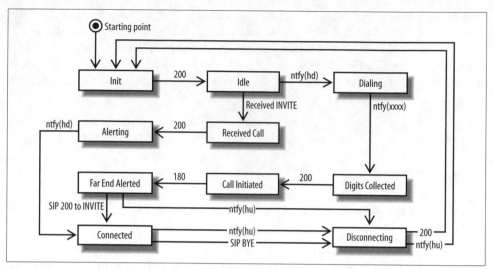

*Figure 15-8. Process diagram*

# CHAPTER 16

# H.323 Translator

At the outset of the VOCAL project, when we committed ourselves to building a SIP-based network, most of the VoIP world was using H.323. We suspected that SIP was going to catch on and become an important basis for VoIP development, but we had to deal with the world the way it was, not with the way we wanted it to be. In 1999, the only softphone with a significant user base was the H.323-based Microsoft Net-Meeting, and there was also a huge install base of H.323/PSTN gateways operating within enterprise phone systems. Rather than waiting (hoping) for VoIP users to come around to SIP quickly, we realized that it would be better for us to provide support for H.323 terminals through a SIP/H.323 translator.

Before we could build an open source translator, we required an open source stack. Fortunately, a group in Australia, known as the OpenH323 Project (*www.openh323. org*), had made such a stack available, which relieved us from having to build our own. Our original plan was to use the OpenH323 library for the H.323 side of the translator and the VOCAL SIP proxy for the SIP side. After evaluating the OpenH323 code, which was mostly geared toward endpoints and much different from the code that we required to build our server, we realized that only the OpenH323 message parsers and builders suited our needs. For example, we used their Q.931 (ISDN) classes to parse and build the Q.931 messages.

## H.323 Background

H.323 is an umbrella protocol that includes several different protocols used for directory-type services (H.225), call signaling (Q.931), capability negotiation and media management (H.245), and other specific tasks. The Q.931 protocol is also used for Integrated Services Digital Network (ISDN) signaling. Some engineers have suggested that H.323 is basically ISDN shoehorned into a VoIP protocol.

The framers of this protocol hoped that reusing ISDN code would help bring the entire H.323 environment to market faster than if it contained unique, original code. This is because many network and telephony service providers already have ISDN code in their systems and would be open to changing a field here and a field there to support VoIP signaling. For those service providers who have several million dollars invested into an ISDN system, changing a few fields to enable VoIP is an appealing proposition.

If you are not familiar with H.323, one important thing worth noting is that, unlike SIP and MGCP, H.323 does not trade ASCII text messages, and the messages do not carry MIME attachments. H.323 messages are Abstract Syntax Notation One (ASN. 1)–encoded binary data. Each of the different protocols within the H.323 umbrella has its own set of requests and responses. Let's have a look at a few of those messages before we look at some call flows.

## Messages

The signaling and call flows in this chapter are described from the point of view of an H.323 terminal trading translated messages with VOCAL. For example, the H.323 terminal, which could be a NetMeeting softphone or an H.323-based phone, doesn't register with a component of the translator; it registers through the translator with the SIP registrar, which, in VOCAL, is incorporated into the Redirect server. This distinction between calling the phone a *terminal* rather than an *endpoint* is important. From the point of view of VOCAL, the terminal combined with the translator becomes a SIP endpoint that acts like a User Agent and talks to a User Agent Marshal server.

The SIP signaling should already be familiar to you; if not, refer to Chapter 7. As for the H.323 signaling, there have been four versions of H.323 released by the International Telecommunication Union (ITU). Our implementation primarily uses H.323 v2 since this is what OpenH323 used. A list of a few of the message types used on the H.323 side of the translator follows. Figures 16-1 through 16-5 illustrate how these messages appear in the call flows.

### Registration, admission, and status (RAS)

The registration, admission, and status (RAS) channel carries messages for gatekeeper discovery and endpoint registration and includes the following message types:

*Gatekeeper Request (GRQ)*
    Sent by the terminal looking for a gatekeeper.

*Gatekeeper Confirm (GCF)*
    Sent by the gatekeeper in response to a GRQ.

*Register Request (RRQ)*
    Sent by the terminal to register the system's registrar.

*Register Confirm (RCF)*
    Returned to the terminal as a translated confirmation of the registration.

*Register Reject (RRJ)*
    Returned to the terminal as a translated rejection of the registration.

*Admit request (ARQ)*
    Sent by a registered terminal when it goes off-hook to make a call.

*Admit Confirm (ACF)*
    Sent by the gatekeeper confirming the terminal's admission.

## H.225/Q.931

H.225 follows the Q.931 (ISDN) structure and is used for managing media and control information within VoIP and includes the following message types:

*Setup*
    Used to initiate calls.

*Call Proceeding*
    Analogous to the SIP 100 message. Signals the endpoint that its message was received and to stop retransmissions.

*Alerting*
    Analogous to the SIP 18x message. Signals that the endpoint is ringing.

*Connect*
    Analogous to the SIP 200 message. Signals that a call session has been established.

*Release Complete*
    Tears down the call. If it is initiated on the H.323 side, it results in a SIP BYE or CANCEL message. A BYE or CANCEL coming in from the SIP side leads to a Release Complete message on the H.323 terminal side.

*Close*
    Analogous in SIP to a 200 after a BYE or a 487 Request Terminated after CANCEL. Completes the call teardown.

## H.245

H.245 is used to advertise the media capabilities on the endpoint and includes the following message types:

*Terminal Capability Set (TCS)*
    A set of messages used to determine the capabilities of each side of the call. The translator uses this information to put together the SDP headers on a SIP message.

*MasterSlaveDetermination*
    A set of messages used for setting up conferencing and determining which endpoint is in charge of the conference call, should such a call be set up. There is no analogous message in SIP.

*Open Logical Channel (OLC)*
    The final message sent after the TCS and master/slave messages have been processed. This leads to the translator sending a SIP ACK or 200 message to the far end, depending on which side initiated the call.

This is only a partial list of possible H.323 messages. There are many more messages used in different applications that are beyond the scope of this chapter.

## Translator Components

As you can see in the call flow diagrams, the translator is illustrated as a combination of the following three components:

*Gatekeeper*
    Keeps track of available endpoints, cataloging them by their IP address and phone number.

*Router/translator*
    In the simplest terms, this is the point of translation where incoming messages from one protocol are translated into the other. The router is combined with a translating device that is not shown in the illustrations.

*SIP Proxy*
    The interface between the translator and the VOCAL Marshal server.

In Version 1.3.0 of VOCAL, the gatekeeper within the H.323 translator can be run as a standalone H.323 gatekeeper. The other components are interdependent but could be turned into standalone units with a moderate amount of development effort.

## Registration and Admission

In H.323, as in SIP, a distinction is made between call-related and non-call-related signaling. In the SIP world, the REGISTER message establishes the relationship between the UA and the registrar. In H.323, there are separate messages for discovery, registration, admission, and potentially location and bandwidth control. We did not implement location and bandwidth control in our translator.

The gatekeeper is the server: the intermediary brokering the calls. The gatekeeper's clients are callable H.323 endpoints, each with a unique IP address that acts as their unique identifier. Most users are not interested in dialing IP addresses: they would rather dial an alias, like a phone number or name, and have it translated into an IP address. In the same way that DNS servers perform this translation on the World Wide Web, the gatekeeper performs this service for the H.323 network.

When a terminal comes online, it can send out a multicast message asking, "Are there any gatekeepers out there that will provide me with telephony service?" Hopefully, there is a gatekeeper that responds, "Yes." If there are no gatekeepers, the terminals are forced into a point-to-point situation in which the users have to dial each other by IP address or use some other out-of-band mechanism for their addressing. The address of a gatekeeper can also be manually configured on H.323 endpoints.

A gatekeeper tells you that there are terminals out there that can accept calls. If you don't use a gatekeeper, you can use something else, like DNS, which can do the alias-to-IP-address translation but can't tell you if the IP address belongs to an entity that's online and waiting for a call. The SIP Redirect server is analogous to the gatekeeper; it is almost identical in its function. Naturally, due to the differences between the protocols, there are differences in how these servers are implemented. For example, compare the SIP registration process, as it is described in Chapter 7, with the H. 323 registration described next in this chapter.

The gatekeeper handles discovery, registration, and admission. Every terminal that registers with a gatekeeper is considered to be in that gatekeeper's zone, and the gatekeeper is in charge of that zone. The gatekeeper is also responsible for giving the OK for terminals to make calls. If there are many, redundant gatekeepers, the terminal can figure this out and independently work with the network as configured.

## Registration

When a terminal comes on, the endpoint sends a discovery (GRQ) message that says, "Is there a gatekeeper out there?" The gatekeeper responds with a Gatekeeper Confirm (GCF) message.

Figure 16-1 shows a NetMeeting terminal registering with VOCAL through the translator.

 At the time this code was initially being developed, the intended and tested terminals were NetMeeting softphones.

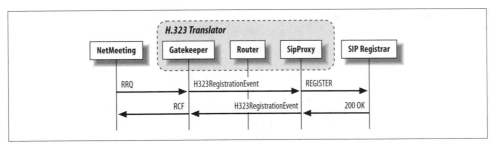

Figure 16-1. NetMeeting successfully registering with VOCAL

In the VOCAL system, once the terminal has found a gatekeeper, it registers (RRQ), sending an internal event, which is translated into a SIP REGISTER message to enable the terminal to register with the Redirect server (RS). If the registration is accepted, the RS sends back a 200 OK message, which is translated into an H.323 registration success (RCF) message and forwarded to the terminal. If the registration is rejected by the SIP system, a SIP 4xx or 5xx response is returned, which is translated into an H.323 registration failed (RRJ) message.

## Admission

In VOCAL, the SIP/H.323 translator does not perform authentication or authorization; the Marshal servers perform those functions as described in Chapter 11. A registered terminal's admission to the translator is guaranteed. However, once the admission is translated into an INVITE message and sent to the Marshal, the user is authenticated based on the parameters set in provisioning, just as with any other SIP endpoint.

## Placing a Call

Figure 16-2 shows the terminal initiating a call.

A registered NetMeeting terminal wants to make a call and sends an admit request (ARQ) message, which is confirmed (ACF) by the gatekeeper. The terminal then establishes a Q.931 TCP connection to the router and sends it a Setup message.

Setup is achieved with an H.225-Q.931 Setup message that is similar to the message used by ISDN to set up phone calls. The Setup message is translated into a SIP INVITE and sent to the SIP network. The network returns a SIP 100 Trying message back to the translator where it is translated into a Call Proceeding message and forwarded to the terminal. The Call Proceeding message is analogous to the SIP 100, which signals the endpoint that its message was received, allowing it to stop retransmissions. Unlike most SIP messages in VOCAL, the H.323 message is sent over the Transmission Control Protocol (TCP).

On the surface, this situation could be considered amusing. With TCP, the connection is guaranteed—dropped packets are not an issue—and sending a message confirming that no packets were lost in transmission seems unnecessary. However, the 100 message is required by the rules of the SIP standard as an indication of a successful transmission between the translator and the Marshal and as a signal to the translator to stop retransmissions.

There is no SDP data in the INVITE message because, with the H.245 Version 1 signaling we are using, all media negotiation is performed after the call-signaling channel has been set up. The difference between the H.323 signaling that we're using and SIP is that SIP does SDP concurrently with the call signaling, while in H.323 Version 1, it is done afterward. To accommodate this difference, we're forced to provide the SDP data with the SIP ACK message. While the first version of SIP did not allow this, later versions made it explicitly clear that providing SDP with the ACK was allowed.

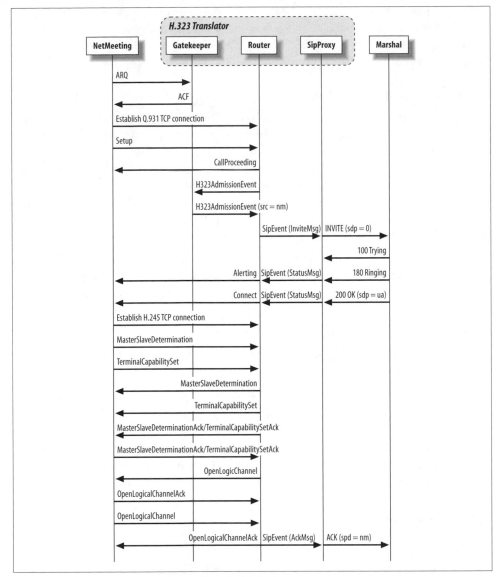

*Figure 16-2. NetMeeting initiating a call with VOCAL*

The signaling for the Setup, Call Proceeding, and Connect messages is done over an H.225/Q.931 socket. When those processes have completed, an H.245 socket is established for sending the terminal capabilities exchange (TCS), which is an advertisement of each side's terminal capability set. Both sides use the TCS message to negotiate commonly acceptable media.

H.245 provides a master/slave relationship that allows conferencing at the endpoint with a master that's in charge of the conference. There is no analogous arrangement in SIP signaling. Setting up the H.245 connection requires exchanging eight messages just to express capabilities and how each side is going to talk to the other. Finally, the message "open logical channel (OLC)" is sent, which is followed by an OLC ACK and then another OLC in parallel. The translator uses its state machine to keep track of the many preconditions that need to be satisfied on the H.323 side before the data required for the ACK can be constructed into a SIP message.

At a macro level, there are two state machines. The H.323 router state machine provides Q.931 and H.245 signaling, and the translator machine provides H.323-to-SIP message translation. There is also a capabilities translation function that takes the terminal capabilities set and translates it into SDP and also translates the SDP from an incoming SIP message into an H.323 capabilities set. The SIP proxy does very little in this schema. The translator tells the proxy to send a SIP INVITE message, and the proxy constructs the message and forwards it to the SIP network. The translator and the H.323 router do the real work.

After this message stream has completed, an RTP channel can be opened between the H.323 terminal and the far end of the call. Since the translator doesn't have to translate the actual media stream, its resources are freed up to process more calls.

## Receiving a Call from VOCAL

Figure 16-3 shows a call coming from the SIP User Agent to NetMeeting through the translator.

As you can see in Figure 16-3, the INVITE message is translated into an Admission event with the gatekeeper, which sets up a Q.931 connection to the NetMeeting terminal. An Alerting message from the terminal is translated into a SIP 180 Ringing message and returned to VOCAL. Between the time that the NetMeeting terminal starts ringing and the user answers, the terminal capabilities set is determined and then passed to the SIP network as SDP data within the 200 message.

## Tearing the Call Down

Figure 16-4 shows the NetMeeting client ending a call with a SIP endpoint.

To achieve an H.323 drop, we send an H.245 end session, which sends an end session back and responds with a close H.245. The Q.931 Release Complete results in a BYE, a CANCEL, or both, depending on the state and direction of the call. A 200 is returned, which is translated into a Q.931 Close.

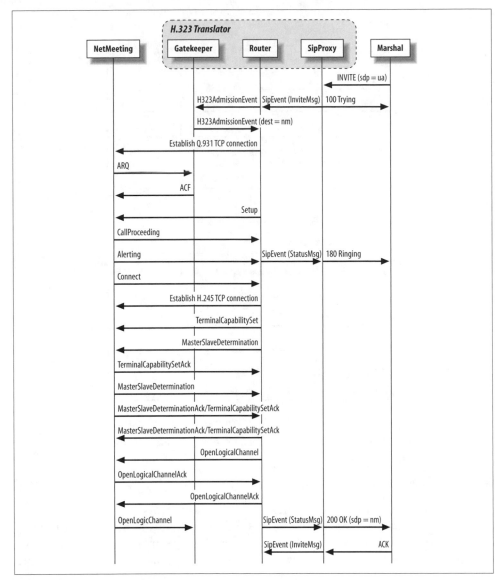

*Figure 16-3. SIP User Agent to NetMeeting*

Figure 16-5 shows a SIP endpoint ending a call with a NetMeeting client.

If the BYE or CANCEL comes from the SIP side, it is translated into an End Session (ES) and Release Complete. There are two sockets per call, which are both closed. And finally, a 200 is sent out to the SIP network.

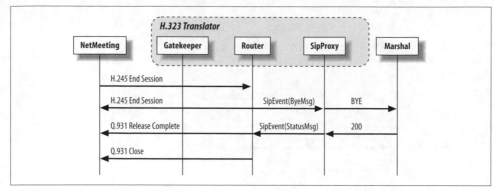

*Figure 16-4. NetMeeting to SIP network: NetMeeting ending the call*

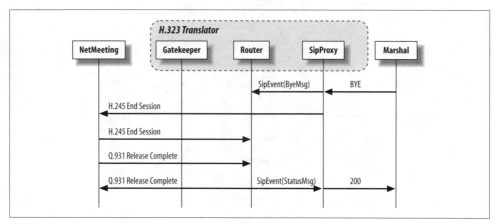

*Figure 16-5. NetMeeting to SIP network: SIP endpoint ending the call*

## Source Code

The code is large, among the largest of all the components within VOCAL. It is approximately 10 times larger than the MGCP translator, including the MGCP stack. A lot of the size has to do with having a large number of templates, each with its own debugging information. The translator starts as a 15MB process that grows as more calls are processed.

The translator is one of the more complicated components of VOCAL because there are two stacks, a large number of H.323 state machines, a translation state machine, an H.323 translator, and the SIP stack. We tried to write the translator and the gatekeeper so that it could be reused and understandable, and it appears that we were successful. The gatekeeper and router are modular and could be reused for other H.323 projects. The translator and proxy are specific implementations and cannot be reused.

# Project Components

As we started planning our project, it occurred to us that the translator could be divided into several subprojects.

*Transport*
> We decided to use a certain type of transport, asynchronous I/O, with the potential to poll a large number of sockets. That approach was completely different from any of the code that was in VOCAL's base. We essentially had to write a different type of socket infrastructure.

*SIP Proxy*
> This piece uses the VOCAL base proxy code.

*Gatekeeper*
> This portion of the code does the registration translation.

*H.323 router state machines*
> The call-signaling framework includes all of the H.323 state machines, and there are many, plus the Q.931 state machine and all the different H.245 state machines.

> H.245 is analogous to the Session Description Protocol (SDP): the mechanism by which you advertise capabilities and open a media channel. H.245 is a separate signaling channel. Once you have established a call, this channel defines how the call will continue. Compared to SIP, there is much more negotiation over H.245. Each terminal has a capability set, which it advertises, and then it has an open logical channel exchange to determine a channel that both sides can agree to open.

*H.323 to SIP translators*
> Once you get through the state machines, you have to do the translation, which goes through the H.323/SIP translation state machine. On one side are the pure H.323 state machines and on the other, the translator state machines.

The router/translator state machines do not correspond to the design and function of the SIP state machines. In the translator, the operators are merged into the events, which know how to work their way through the state machine and what to do when the state machine is in a certain state. The state machine doesn't really know anything about actions: the events know what actions to take.

# Call-Signaling Gateway

The *siph323csgw* component of the VOCAL system provides a call-signaling translation function between H.323 terminals and a SIP proxy. The *siph323csgw* is an H.323 gatekeeper that proxies a SIP Redirect server. Similarly, the call signaling is gatekeeper-routed, so the H.323 terminals are represented to the VOCAL system as SIP user agents. This implementation is currently targeted to work only with Microsoft NetMeeting 3.01.

The following features are supported by the *siph323csgw*:

- Gatekeeper discovery
- Endpoint registration
- Endpoint admission
- Gatekeeper-routed calling via the *siph323csgw*

The following endpoints are known to interoperate:

- NetMeeting to VOCAL's SIP UA (audio only), NetMeeting to Cisco 7960 SIP phone (audio only), NetMeeting to Cisco 5300 SIP gateway (audio only), NetMeeting to NetMeeting (audio, video, T.120 data)
- Cisco ATA 186 Analog Telephone Adaptor H.323 to VOCAL's SIP UA, Cisco 7960 SIP, and Cisco 5300 SIP; Cisco ATA 186 Analog Telephone Adaptor H.323 to Komodo H.323
- Cisco 2600 H.323 gateway to VOCAL's SIP UA, Cisco 7960 SIP IP phone, and Cisco 5300 SIP gateway; Cisco 2600 H.323 gateway to Cisco 2600 H.323 gateway
- Cisco Call Manager to VOCAL's SIP UA, Cisco 7960 SIP IP phone, and Cisco 5300 SIP gateway; Cisco Call Manager to Cisco 2600 H.323 gateway

The following features are not supported:

- Faststart
- H.245 tunneling
- Re-INVITE (SIP)/dynamically opening and closing channels
- TCS 0

  This is an empty terminal capability set (TCS). A TCS is used to advertise an endpoint's capabilities. When an endpoint sends an empty capability set, it means that the endpoint doesn't support media at the current time. The endpoint may reissue TCS later. Sometimes endpoints use TCS to shut down the media so that they can put the call on hold or perhaps do a transfer.

- H.450 supplementary services

## Directory Structure

The different translator entities are located in *vocal1.3.0/translators/h323/translators* with the translators named:

- *VOCAL/translators/h323/translators/mgcp*
- *VOCAL/translators/h323/translators/h323*

For translator-specific code, the directory structure looks like the following:

```
VOCAL/translators/h323/translators/h323
  |
   -- doc
  |
   -- gk
  |
   -- router
  |
   -- translator
  |
   -- proxy
  |
   -- util
```

Extensions to the *util* base—for example, *SelectorFifo* and Transport—are in *vocal1.3.0/ transport/util* if they are not portable across all platforms. If they are portable across all platforms, they are located in *vocal1.3.0/util*.

# Getting Started

This section provides some information about building the translator and starting it separately for testing.

## Platform Supported

The *siph323csgw* currently is supported on Linux systems using the *gcc* compiler. The *siph323csgw* will not build on Solaris using Forte's CC compiler, due to the nature of the *openh323* stack. The gateway should build on Solaris using *gcc*, but this has not yet been fully tested.

## Compile Instructions

From the directory where VOCAL was installed, type:

```
[bash]$ make siph323csgw
```

To build the *siph323csgw*, you will need at least 256 MB of memory; it will use it all during linking.

## Using the Software

The command-line options are available via the *--help* switch, for example:

```
[bash]$ NO_DAEMON=1 siph323csgw --help
```

which reads:

```
usage: ./siph323csgw [ options ]
[ -o ] [ --output ]     directory    Output directory.
                                        Defaults to "./".
[ -l ] [ --loglvl ]     level        Log level.
                                        Defaults to "LOG_INFO".
[ -g ] [ --gkid ]       id           Use for gatekeeperIdentifier.
                                        Defaults to hostname.
[ -e ] [ --eptid ]      id           Use for endpointIdentifier.
                                        Defaults to "endpoint_".
[ -p ] [ --sipport ]    port         SIP UDP port.
                                        Defaults to 12061.
[ -a ] [ --sipremote ] ipaddress     IP address of SIP remote entity.
                                        Defaults to 192.168.5.23:8061.
[ -c ] [ --client ]                  Remote SIP entity is a client.
                                      Defaults to disabled, meaning the
                                      remote entity is a server.
[ -m ] [ --mediacontrol ]            Media signaling is controlled by gateway.
                                      Defaults to disabled, meaning the
                                      media control signaling is directly
                                      between H.323 endpoints.
```

We suggest running the translator as standalone with the command line:

```
/usr/local/VOCAL/bin/siph323csgw \
  -o /usr/local/VOCAL/log/siph323csgw \
  -l LOG_INFO \
  -p 22400 \
  -a 172.19.175.166:5060
```

where 172.19.175.166:5060 is the IP address/port of the SIP proxy (Marshal) that it will be communicating with.

Use LOG_INFO, since that provides the H.323 and SIP messages being sent without too much debug information. You may also want to use LOG_ERR or LOG_WARN.

You can also do a *killall -TERM siph323csgw* or *kill -TERM <process ID>*. This will bring down the translator safely, killing all outstanding listening sockets and unregistering registered endpoints.

CHAPTER 17

# System Monitoring

The Network Management System is based on the Simple Network Management Protocol (SNMP) project that originated at Carnegie Mellon University (CMU) and the University of California at Davis (UCD) and is now available from the Net-SNMP pages on *SourceForge.Net* (*http://net-snmp.sourceforge.net*). Starting from this code, which includes a stack and a daemon, we built a system that provides network statistics; a GUI interface for monitoring, stopping, and starting servers; and a heartbeating mechanism that allows VOCAL servers to be aware of which servers are up and which are down at any given time.

SNMP works through a dedicated port, which creates a problem if you are running a multiserver system like VOCAL on one host. Our solution was to build an agent to interface with each process and send data back to the Network Manager through the SNMP port to a GUI or another collection point. As you will see in our illustrations and discussion, the agent works with a simple, proprietary protocol called Agent API that provides a conduit between the different processes and the SNMP tools.

## SNMP Support

VOCAL supports Version 2 of the Simple Network Management Protocol (SNMP, RFC 1157). If you are new to this protocol, there is good documentation and a set of tutorials on the Net-SNMP site. Before looking at our implementation, let's review a few SNMP terms and commands as reference points for this discussion.

### Elements

SNMP works with these two elements:

*Management Information Base (MIB)*
> A collection of SNMP objects that contain information about their state, but without any other functionality except for the ability to read and write the values that correspond to their state. See the later section "SIP MIBs" for more information.

*Object Identifier (OID)*

An integer or pointer that uniquely identifies each object within the MIBs. In SNMP terms, the OID represents a category of data and its source. For example, the objects *startprocessname* and *stopprocessname* have different OIDs. For more information, see the later section "Implemented MIBs."

## System Components

The system contains these components:

*Agent*

Receives incoming messages and keeps track of which servers are active or inactive.

*Agent API*

A simple User Datagram Protocol (UDP, RFC 768)–based protocol that enables the agents to communicate with one another and with the NetMgnt. See the later section "Agent API" for a code example and a brief discussion of the messaging.

*Multi-Router Traffic Grapher (MRTG)*

A third-party open source tool used to monitor traffic on networks. MRTG generates HTML pages containing graphic representations of network traffic. See the later section "MRTG-Generated Statistics" for more information and an example of the HTML page.

*Network management station (NetMgnt)*

Responsible for monitoring the status of the processes within the VOCAL system.

*SNMP daemon (SNMPd)*

An agent, located on every host within the system, that provides local network management to any SNMP network management station.

*SNMP GUI*

Requests updates from the *NetMgnt* and displays traps and status information. Third-party products, such as Hewlett-Packard's OpenView, can be used to monitor the network.

## Messages

Here are some examples of messages found within SNMP transactions. You can find a complete list with code examples on the Net-SNMP site:

*get*

Retrieves data from a remote host. The *get* command contains the host name, authentication information, and an OID.

*set*

> Modifies data on a remote host. The *set* command contains the OID that is being updated and the variable being set, along with the data type and the value to which the variable is being set.

*trap*

> An unsolicited message that usually reports a change in monitored activity such as a server going down. See the description of Host 3 under the later "Flow Diagram" section for more information about the interaction of traps. The GUI displays trap messages as they occur.

## Process Interaction

Before looking at how the SNMP tools are arranged in the VOCAL network, let's look at a simplified view of how the messages move back and forth between the Network Manager and the different servers and stacks. Figure 17-1 shows some of these components and their interactions.

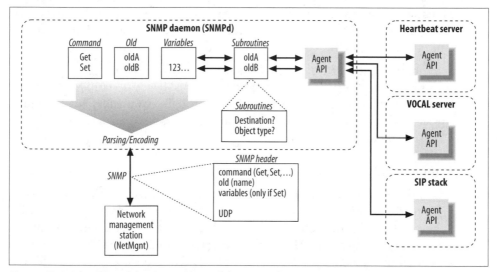

*Figure 17-1. The roles of the Agent API and the SNMPd*

> In Figure 17-1, the label *VOCAL server* represents a call control server such as the Marshal, Redirect, Feature, or Call Detail Record server. The Heartbeat server is included with VOCAL, but its function has more to do with network management than call control.

The *SNMPd* uses SNMP to talk to the *NetMgnt* and Agent API to talk to the other servers and stacks. If the *NetMgnt* sends a *get* command to the *SNMPd*, it parses the command and sends it to the correct server. When the response comes back, the

*SNMPd* encodes the response and returns it as a well-formed message back to the *NetMgnt*.

In Figure 17-1, the SNMP transaction is shown as a set of headers that are transported over UDP. The get message contains an OID representing a type of data and the identity of the server that is being queried. The identity of the servers is provided by a registration process that occurs at compile time, which assigns different ranges of OIDs to different servers. When the *SNMPd* parses the incoming message, it forwards it to a subroutine that interprets the object type and where that object is supposed to be sent. This activity is illustrated in Figure 17-1 as Destination? and Object Type?

Assume that the *NetMgnt* wants to know how many Heartbeat servers are active. It can't send an SNMP message directly to the server because SNMP uses only one port per host, and other servers are sharing this host. Instead, the *NetMgnt* sends a get command to the *SNMPd* with OID A. Assume that OID A represents "number of active servers" and is registered with the Heartbeat server. Therefore, "get OID A" is the same as asking, "How many Heartbeat servers are active?" Once the object type and destination have been determined, the request is sent to the Heartbeat server via Agent API, and the response is sent back from the server to the *SNMPd* where it is parsed and then encoded into SNMP and returned to the *NetMgnt*.

For a *set* command, the SNMP headers need to include a variable and the value to which it is being set. The path is the same from the *NetMgnt*, into the *SNMPd*, to the subroutines, and then out to the intended server through Agent API. This communication process sounds like a convoluted workaround, but it works, and this is not an unusual implementation of SNMP.

### Creating the executable

When you download the SNMP package from Net-SNMP, you get the SNMP stack, a skeleton of the *SNMPd*, and several utilities. You then need to run your MIBs through the MIB-to-C script routine, which is discussed later. The output from this routine gives you a skeleton *SNMPd* with empty subroutines. The subroutines must then be written to enable the OIDs to do their jobs. Once this code is complete, you link together the stack, the *SNMPd*, the subroutines, and the OIDs and their registry and compile them into a single executable.

We wrote Agent API to allow the *SNMPd* to query the servers for information, as it is required. For example, the *SNMPd* does not know how many Heartbeat servers are active; it forwards that request to Agent API.

### Altering the SNMPTRAP daemon

The *SNMPTRAPd*, which is represented in Figure 17-2, is 99% UCD code. We added a few lines of code to enable the *SNMPTRAPd* to write the traps to the GUI.

---

Our code is a call to the Agent API to send the traps to the *NetMgnt*. This Agent API also has to be linked to the *SNMPd* code at compile time.

## VOCAL's Network Management System

Now that we have looked at how the elements work with one another, this section goes into more detail about how the SNMP elements are hosted within VOCAL.

## Flow Diagram

The system-monitoring process is illustrated in Figure 17-2.

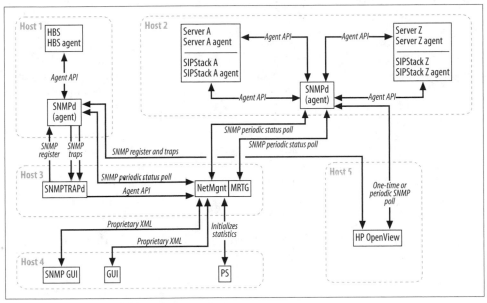

*Figure 17-2. System-monitoring message flow*

Here is a guide to the flow diagram:

- Dotted boxes represent different hosts.
- Solid boxes represent different processes.
- Arrows indicate interfaces.

The diagram shows views of five hosts, each running a different process. These hosts are separated to help clarify this discussion. In an *allinone* VOCAL system, all of these servers are running on the same host.

### Host 1

The Heartbeat server (HBS) and its colocated agent keep track of which servers are up and which are down. The HBS and its agent communicate with the SNMP daemon (*SNMPd*) through Agent API. Every host has an *SNMPd*, which is used to monitor each server.

## Host 2

Server A and server Z represent a series of different VOCAL server processes such as the Redirect, Marshal, and Feature servers. The *SNMPd* communicates with agents that represent both these processes and the SIP stack. This communication path uses the proprietary Agent API, which is defined later in this chapter.

## Host 3

The Network Manager (*NetMgnt*) can reside on one host or on as many hosts as are necessary to provide the required performance.

The SNMP trap daemon (*SNMPTRAPd*) waits for SNMP traps to appear. When it receives a trap, the daemon forwards it to the *NetMgnt*, which updates its database. The *NetMgnt* also tells the *SNMPd*, which is colocated with the HBS, about registrations so that the *SNMPd* knows where to send its traps. This information can be displayed as a Multi-Router Traffic Graph (MRTG); see the later "Agent API" section.

The HBS works with the *NetMgnt*. Every 5 minutes, the *NetMgnt* polls the *SNMPd* for complete system status. If any server has gone down, the HBS notifies the *SNMPd*, which sends a trap to the *SNMPTRAPd*, which then updates the *NetMgnt*. The HBS receives a list of what servers should be in the system from the Provisioning server (PS) and determines what servers are up or down by multicasting heartbeat pulses.

The MRTG application works with the *NetMgnt*. When the *NetMgnt* comes up, it kicks off the MRTG process to obtain the location of every SIP stack. Then, the MRTG queries every *SNMPd*, which is colocated with a SIP stack, on a regular, configurable interval to give it statistics. An example of a statistic is the number of SIP INVITE messages that a given *SNMPd* has received. The *SNMPd* uses SNMP to communicate with the network management stations and Agent API to communicate with all other entities.

## Host 4

The Provisioning server (PS) initializes the *NetMgnt* with information about the location of the HBS.

The SNMP GUI requests updates from the *NetMgnt* and displays traps and status information. The SNMP GUI can be accessed from any computer on the network that is running a Java-enabled browser. Multiple GUIs can run simultaneously on the same network. See "SNMP GUI," later in this chapter, for an illustration and further discussion.

### Host 5

Hewlett-Packard's OpenView can be used to monitor the network. Any network management station can talk to the daemons using SNMP, providing that it registers with the *SNMPd* colocated with the HBS depicted as Host 1 in Figure 17-2.

# MIBs

In a TCP/IP-based network, each device maintains a set of variables describing its state. In SNMP, these variables are known as *objects*, but these objects do not hold the same meaning as those within an object-oriented programming architecture. SNMP objects contain information about their state without any methods, other than the ability to read and write their values. A collection of SNMP objects is known as a Management Information Base (MIB).

As networks are built, they normally use several MIBs to describe the various structures within the overall architecture. It is more common for developers to use public MIBs for the nonproprietary components of their system and to build their own for the proprietary components.

The VOCAL system supports the following public MIBs:

- MIB II (refer to RFC 1213)
- Network Services Monitoring MIB
- SIP MIBs
- VOCAL Enterprise MIB

## Network Services Monitoring MIB

VOCAL supports a subset of the Network Services Monitoring MIBs defined in RFC 2788. This includes the minimum identifiers of applications for use in other MIBs as indexes.

For more information, refer to *vocal-1.3.0/proxies/NetworkManager/NETWORK-SERVICES-MIB.txt*.

## SIP MIBs

VOCAL supports an initial version of the SIP MIB dated July 2000 (*draft-ieft-sip.mib-01.txt*). The MIB is temporarily located under *enterprise.vovida.vovidaTemporary* until the MIB is assigned a permanent location by the Internet Assigned Numbers Authority (IANA, *http://www.iana.org*). For more information, refer to the *vocal-1.3.0/proxies/NetworkManager* directory for these SIP MIBs:

*SIP-COMMON-MIB.txt*
   Initial version of the SIP common MIB module; defines objects, which may be common to all SIP entities

*SIP-MIB-SMI.txt*

> Initial version of the SIP MIB module that defines base OIDs for all other SIP-related MIB modules

*SIP-REGISTRAR-MIB.txt*

> Initial version of the SIP registrar MIB module

*SIP-SERVER-MIB.txt*

> Initial version of the SIP server MIB module

*SIP-TC.txt*

> Initial version of the SIP MIB textual conventions module used by other SIP-related MIB modules

*SIP-UA-MIB.txt*

> Initial version of the SIP User Agent (UA) MIB module

## UCD Enterprise MIB

VOCAL supports and implements the University of California at Davis (UCD) Enterprise MIBs (these were originally called the CMU MIBs after Carnegie Mellon University, where they were first developed). For additional information about the UCD MIBs, refer to the Net-SNMP site. For more information about these specific MIBs, refer to the *vocal-1.3.0/proxies/NetworkManager* directory.

*VOVIDA-LOCAL-GRP-MIB.txt*

> Identifies the local configuration, including the enabled trap destinations and local processes running on this machine

*VOVIDA-NOTIFICATIONS-MIB.txt*

> Defines the VOCAL softswitch MIB extensions for *ServerUp* and *ServerDown*

*VOVIDA-SERVERGRP-MIB.txt*

> Provides configuration and statistics on the software servers located within the system

*VOVIDA-SOFTSWITCHSTATS-MIB.txt*

> Provides system-level statistics

*VOVIDA-SUBSCRIBERSTATS-MIB.txt*

> Identifies the system-level subscriber statistics and their application information

# SNMP Daemon

There is a basic definition of the daemon near the beginning of the chapter; this section gives you a more complete definition. The SNMP daemon (*SNMPd*) is an agent that is located on every host within the system, providing local network management to any SNMP network management solution. The *SNMPd* can manage all VOCAL devices and many system processes. It also provides full SNMP v2 support including get, set, trap, and other message types, along with a simple proprietary UDP interface for colocated managed processes and an implementation for two-way registration with colocated managed processes.

## High-Level Flow

On startup, the *SNMPd* sends out a hostwide multicast registration. It then monitors two User Datagram Protocol (UDP) ports for incoming messages and requests. The well-known SNMP port (161) receives requests from any network management station, and a VOCAL proprietary communication port (33604) processes registration and responses from the colocated managed processes.

The REGISTER message includes a UDP port for communication between the agent and the managed entity, the name of the process from the managed entity (RS1, for example), and the type of managed entity (RS, MS, FS, or SIP stack).

There can be multiple entities within a single process; for example, process RS1 has both an *RS_Agent* for Redirect server (RS)–specific information and a *SIP_Agent* for the Session Initiation Protocol (SIP) stack. The *SNMPd* correlates these entities and tracks their statuses.

These requests can be from any SNMP network management station. The *SNMPd* views information requests (*get*, *getnext*), parses the object identifier (OID), and sends an appropriate response to the Network Manager. If the request asks for information that the *SNMPd* cannot provide, the *SNMPd* forwards the request to the proper local managed entity and waits for a reply. If the request is for an action, such as set, the *SNMPd* parses the OID and either takes appropriate action or forwards an action request to the proper local managed entity.

Local managed entity responses to *SNMPd* requests are processed, and then an SNMP response is sent to the requesting Network Manager. If a trap is received from a local element, the *SNMPd* forwards the trap message to all Network Managers that have registered to receive that specific type of trap.

## Key Data Structures

The *SNMPd* contains the following structures:

*ApplTable*
    Contains information needed to communicate with all of the local managed elements. This can be found in *vocal-1.3.0/proxies/agent/ApplTable.hxx*.

*ServerTable*
    Contains the information about applications that is needed to implement the network application MIB. This can be found in *vocal-1.3.0/proxies/agent/ ServerTable.hxx*.

*TrapAgent*

This class is a C wrapper for the C++ classes that the *SNMPd* uses. This can be found in *vocal-1.3.0/proxies/agent/TrapAgent.hxx*.

*AgentRegister*

The Agent Application Programming Interface (API).

# Network Manager

The Network Manager is responsible for monitoring the status of the processes within the VOCAL system.

Figure 17-3 shows a block diagram of the Network Manager.

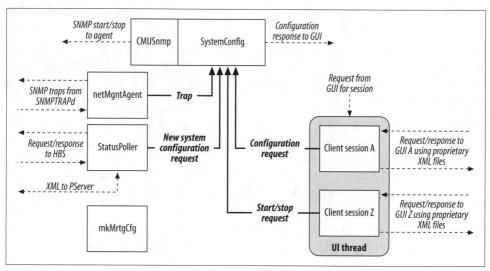

*Figure 17-3. Network Manager block diagram*

# Threads

The Network Manager application has five different tasks that run as separate threads, represented as boxes in Figure 17-3. These threads interact with one another for data and perform different functions. These processes include:

*System Configuration Thread*

The *netMgmtAgent*'s main management is derived from the *agentAPI*, which receives the traps from the *SNMPTRAPd* and forwards them to the System Configuration thread. This thread is a table that contains a database of system configuration, a list of all entities on the network, and the status of all network entities. This thread also sends responses to the GUI on demand.

*netMgntAgent*

This thread processes the incoming trap messages: it receives the interprocess communication (IPC) message and the network address of the sender as the input arguments and processes them. This thread checks whether the message type is a trap. If it is a trap, the thread passes the *hostport*, *servertype*, and status to the system configuration object, which in turn processes the trap. This thread continues running and waits for any trap messages.

*UiThread*

The *UiThread*'s main purpose is to receive connection requests from the graphical user interface (GUI) that have been sent to the *NetMgnt*. This thread waits in an infinite loop, accepting new *clientSession* connections from the GUI.

*StatusPoller*

The *StatusPoller* thread periodically polls the heartbeat server (HBS) for the system status. This thread gets the list of HBSs from the Provisioning server and registers to receive traps. Every polling period thereafter, it clears the trap list and reregisters to receive the *serverTable*. This is used to get the number of servers, which is then compared with the MIB variable indicating the total number of servers from the *applTable*.

*mkMrtgCfg*

This is another thread whose main purpose is to create a configuration file and start the MRTG application to which the config file becomes an input. The MRTG application is a third-party tool, which is used for monitoring and plotting graphs of certain MIB variables. In VOCAL, it is used to plot graphs from some critical SIP variables indicating the performance of the system, like *totalSIPInvites*, *sipInputSummary*, *sipOutputSummary*, and others.

During startup, the Network Manager process accepts the Provisioning server and the redundant Provisioning server's hostname, port number, and SIP port number. In this case, the SIP port number is redundant. The *StatusPoller* is then instantiated with the command-line information about the Provisioning servers in the system.

## Heartbeat Server

The Heartbeat server (HBS) plays an important role in VOCAL's network management process. The Network Manager periodically polls the HBS for the system status. Also, the Network Manager registers to receive traps, and the HBS sends a trap to the trap *agentTrapPort* on the local machine whenever there is a change in the status of the servers. In VOCAL, this machine has been set to port number 33606. For more information about the HBS, see the later "Heartbeat Server" section.

## SNMP++

*SNMP++* is a wrapper around the *SNMPd*'s C language application. The Network Manager process uses the *snmpwalk* and *snmpgetnext* functions to obtain values from the SNMP. These functions are explained with code examples on the Net-SNMP site.

# Agent API

*AgentAPI.\*xx* is the base class for any application agents. It can be found in the *vocal/util* directory. An example of an agent is *vocal-1.3.0/hb/hbs/HbsAgent.\*xx*. If you add a new server to the network, add a SIP stack and make an agent for that server.

Let's have a quick look at some of the code from *vocal-1.3.0/util/snmp/AgentApi.hxx*:

```
typedef enum
{
    Get = 1,
    Set,
    Trap,
    Response,
    Register,
    Register_Req
} actionT;

#define PARM1SIZE 128
#define PARM2SIZE 1024
typedef struct
{
    actionT action;
    AgentApiMibVarT mibVariable;
    unsigned long transactionNumber;
    char parm1[PARM1SIZE];
    char parm2[PARM2SIZE];
}
ipcMessage;
```

This is a very simple protocol: there is only one message type, *ipcMessage*, sent back and forth and a limited number of fields contained within the message. When an *SNMPd* starts up, it broadcasts an *ipcMessage* as a *Register_Req* request. Any servers or SIP stacks that respond also use the same message type with a Register response. When an *SNMPd* requests system status information, it sends out the *ipcMessage* with a *get* command, which is responded to with an *ipcMessage* containing either a set, trap, or response value.

The parameter *parm1[PARM1SIZE]* contains the *get* value. Any received set values are written to the *parm2[PARM2SIZE]* variable.

# Implemented MIBs

The Network Manager application uses the *SNMPd* and the HBS to manage the VOCAL applications. As stated at the beginning of this chapter, the basic SNMP framework used in VOCAL is taken from the UC Davis SNMP project. Proprietary

Vovida MIBs have been implemented, and the agent has been extended to support the VOCAL and SIP stack MIBs.

The Vovida MIBs include:

*applTable*
> The application table that contains information about all of the applications on the specific host where the *SNMPd* is located.

*serverGrp*
> Contains all information related to VOCAL servers. This table is populated by the data obtained from the HBS and is updated only when there is a request made to the agent. Depending on the last time of access, the *serverTable* cache is deleted and a new version is created so that it contains the latest information about the server at all times. The time that has been set is currently 30 seconds, which means that if two queries are received within 30 seconds, the second query is handled with data stored in the cache. If a following query arrives after the 30-second timer has expired, the table is updated with fresh data.

*localGrp*
> Contains a server used to manage each process in the system. There are separate OIDs for *startProcessName* and *stopProcessName*. To start or stop a process, the *startProcess* or *stopProcess* OID is set to an ASN_INTEGER of value 1. Otherwise, by default, it is set to 0. Once the value is set to 1, the *vocalstart* script is used to start, stop, or restart the process. To set the *debugLevel* of a process, a *GetStackPortFromIndex* is done and, using the port, an *ipcRequest* is done to get the old *DebugLevel* or set the new *DebugLevel* of the process. The *DebugLevel* is 0 by default and can be an ASN_INTEGER up to 7. The *DebugLevel* is set to 1 for all processes on a given host.

*SipMib*
> Contains many SIP stack–specific variables including *sipProtocolVersion*, *sipServiceOperStatus*, *sipStatsInviteIns*, and *sipStatsInviteOuts*. Not all of the MIB variables have been implemented, but the SIP statistical specifics have been implemented. These OIDs are updated by the SIP stack agent and are mostly countervalues that provide useful information for tracking the performance of the system. The third-party tool, MRTG, uses some of these OIDs to plot graphs.

## MRTG-Generated Statistics

Multi-Router Traffic Grapher (MRTG) is a third-party open source tool used to monitor traffic on networks. MRTG generates HTML pages containing graphic representations of network traffic.

When the VOCAL system is started, MRTG automatically collects network traffic information and generates graphs from the collected information. MRTG is configured to generate daily and weekly graphs for:

- SIP ACK messages
- SIP CANCEL messages
- SIP INVITE message
- SIP response messages
- SIP request messages
- Failed SIP requests
- SIP success class messages
- Total bytes of SIP messages

The graphs are formatted into an HTML page and can be viewed from a web browser using this URL: *http://<hostnameoftheprovisioningserver>/vocal/NetMgnt/ serverIndex.html*.

You can configure MRTG to collect statistics and generate graphs for other types of messages or network traffic. In addition, you can configure MRTG to collect network traffic data and display it in a daily, weekly, or monthly graph format.

For more information on using MRTG or interpreting MRTG graphs, refer to the MRTG user documentation. A copy of this documentation is provided in the *vocal-1.3.0/proxies/ NetMgnt/mrtg/* directory.

Figure 17-4 illustrates an example MRTG-generated graph. In this example, MRTG collects and displays Ethernet traffic data for the host server *audi.private.vovida.com*. MRTG plots the Ethernet traffic (number of bytes per second) against each day or week.

## Running the SNMP Agent

The file *SNMPd.conf* explains how to set up the SNMP agent for operation in your system and is the default configuration file for the SNMP agent. It should be placed in */usr/share/snmp/SNMPd.conf* or */etc/snmp*. This file is well annotated with many commented-out examples.

The following lines change the access permissions of the agent so that the *community* string provides read-only access to your entire *network* (e.g., 10.10.10.0/24) and only read/write access to the *localhost* (127.0.0.1, not its real IP address).

1. Map the community name (*community*) into a security name (*local* or *mynetwork*, depending on where the request is coming from):

```
#          sec.name       source    community
com2sec    vovidaDevel    default   private
```

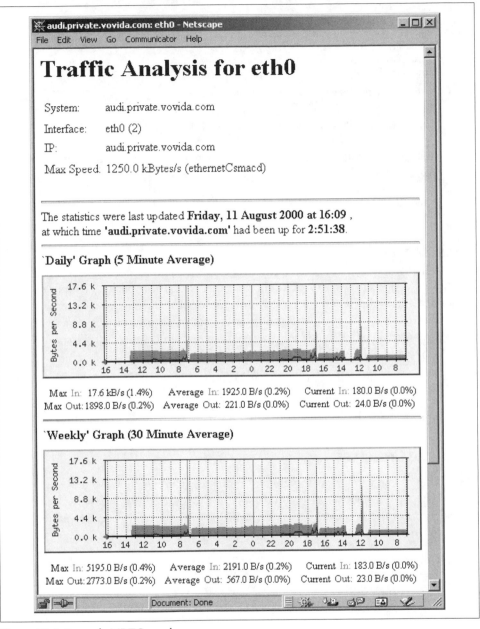

*Figure 17-4. Example MRTG graph*

2. Map the security names into group names:

```
#               sec.model   sec.name
group vovida    v1          vovidaDevel
group vovida    v2c         vovidaDevel
group vovida    usm         vovidaDevel
```

```
group MyRWGroup    v1          local
group MyRWGroup    v2c         local
group MyRWGroup    usm         local
group MyROGroup    v1          mynetwork
group MyROGroup    v2c         mynetwork
group MyROGroup    usm         mynetwork
```

3. Create a view for us to let the groups have rights to:

```
#                  incl/excl   subtree           mask
view   systemview  included    1.3.6.1.2.1.1
view   all         included    .1                80
```

4. Grant the two groups access to the one view with different write permissions:

```
#         context    sec.model sec.level  match read      write notif
access    vovida     ""  any   noauth     exact all        all   all
access    MyROGroup  ""  any   noauth     exact all        none  none
access    MyRWGroup  ""  any   noauth     exact all        all   none
```

# SNMP GUI

Each VOCAL system server sends (via multicast) heartbeat packets to its peers at a predefined interval. The Heartbeat server monitors the exchange of heartbeat packets between VOCAL servers and sends server status trap messages to the network management system. The network management system displays server status on the VOCAL SNMP GUI.

Figure 17-5 illustrates an example of the GUI, which can be accessed by typing this URL: *http://<hostnameoftheprovisioningserver>/vocal/index.html*.

The GUI's screen is divided into four sections. Here is an explanation of each.

## Hosts and Processes

This frame displays the host servers and indicates whether they are active (blue) or inactive (red). If a host server contains several processes, it will display a red ball if one or more of the processes are inactive. Table 17-1 describes the *Expand=Off* and *Expand=On* icons.

*Table 17-1. SNMP GUI screen: icons*

| Icon | Description |
|------|-------------|
| | *Expand=Off:* appears beside contracted folders. Click this icon to expand the folder. |
| | *Expand=On:* appears beside expanded folders. Click this icon to contract the folder. |

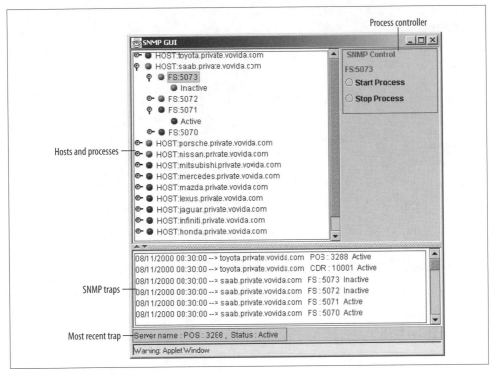

*Figure 17-5. VOCAL SNMP GUI screen*

## SNMP Process Controller

The process controller allows you to start or stop the SNMP control process. Starting or stopping the process controller requires two steps:

1. From the Host view, select a hostname and a process.
2. Select Start Process to begin the SNMP control or Stop Process to stop the SNMP process.

The process or function running on each host server is listed under each hostname. When the process is selected, the process abbreviation and port number appear above the Start Process and Stop Process buttons.

Processes or functions are abbreviated as:

*FS*
    Feature server

*MS*
    Marshal server

*POS*
    Policy server

*RS*
> Redirect server

*CDR*
> Call Detail Record server

*PS*
> Provisioning server

 There is no warning prompt for starting or stopping processes. If you select a process and click either Start Process or Stop Process, the instruction is sent immediately to the host.

## SNMP Traps

The SNMP traps field displays SNMP traps from the system with timestamps.

## Most Recent Trap

This section displays the server name and status of the most recent trap.

# Adding MIBs

The process for adding SNMP device support to the VOCAL system includes working with both the *SNMPd* and agent sides.

## SNMPd Side

Follow these steps on the *SNMPd* side:

1. Create new agent code as described in the upcoming section "Creating New Agent Code."
2. Move new agents to *vocal-1.3.0/snmp/vovidaSNMP/agent/mibgroup*, under the proper subdirectory.

Here are some places to find more information:

- Tables can use code similar to that found in *applTable.c*.
- Traps are similar to *softSwitchNotifications.c*.
- Nontable items are similar to *localGrp.c*.

## Agent Side

Here are some tips about working with the agent side:

- The threads on the agent side are part of the monitored processes for the different VOCAL servers.
- *vocal-1.3.0/util/AgentApi* is the class from which the agents are derived.

---

- *vocal-1.3.0/heartbeat/hbs/HBAgent.cxx* is the current agent for the HBS and can be used as a basis for other agents.
- *vocal-1.3.0/util/AgentApiMibVars.hxx* is a large enumerator that includes all VOCAL-specific MIB variables. New variables should be included here as well.

## MIB Group Directory

The point of having an MIB group directory is to help the developer organize her MIBs in the Concurrent Versioning System (CVS) tree. The directory—*vocal-1.3.0/snmp/vovidaSnmp/agent/mibgroup/*—holds the implementation of individual MIB groups. Currently, this contains the eight MIB II groups, plus a basic template for new MIB groups (*examples/example.c* and *examples/example.h*). See the Net-SNMP manual pages for information about how to use these examples. They are available from *http://www.csc.liv.ac.uk/~daves/Misc/UCD/guide.html*.

Additionally, you may wish to check out the *mib2c* Perl script in the *SRCDIR/* local directory that can convert a written MIB into C template files for use with this package as MIB modules.

Each group consists of two files as shown in the examples in the following section.

# Creating New Agent Code

This section explains how to work with your code.

## Agent Extensibility

The agent that comes with this package is extensible through use of shell scripts and other methods. See the configuration manual pages (like *SNMPd.conf*) and the *EXAMPLE.conf* file for details.

You can also extend the agent by writing C code directly. The agent is extremely modular in nature and you need only create new files, reconfigure, and recompile. No modifications of the distributed source files are necessary. For details on how to go about this, see: *http://www.csc.liv.ac.uk/~daves/Misc/UCD/guide.html*, *agent/mibgroup/examples/example.c*, and *agent/mibgroup/examples/example.h*. Also, see the *local/mib2c* program and its *readme* file for help in turning a textual MIB description into a C code template.

## example.h: Interface Information

The following are the requirements for the *example.h* file:

- Declaration of the initialization function *init_example* (optional).
- Declaration of the function(s) for retrieving variable information (*var_example*).

- Declaration of the function(s) for setting variable information (if appropriate).
- Declaration of the function(s) for parsing the *SNMPd.conf* file (optional).
- A call to *config_parse_dot_conf()* to explain to the *.conf* parser which tokens we want to parse (optional).
- Definitions of the MIB group magic numbers declarations of the public functions.
- A list of the variables in the group, including type information, mapping between magic numbers and OID subidentifiers within the group, accessibility information, and the relevant function for retrieving this variable's value. They *must* be listed in the descending-descending order.
- A call to *config_load_mib*, identifying the location of this MIB group within the general MIB structure.
- A call to *config_require* to identify any other files that are required by the implementation of this group (optional).

These last three are visible only within the "glue" file *snmp_vars.c* (if at all) but are declared here for ease of maintenance.

## example.c: Implementation Code

The following are the requirements for the *example.c* file:

- Need not exist if only the *.h* file is needed
- A list of kernel information needed to report on this group
- An initialization function (optional)
- A routine to parse an *SNMPd.conf* line
- A routine to free resources from above and return to default settings
- Header function(s) to map the OID requested to the next appropriate OID (and similar system-independent setup)
- Function(s) (possibly system-specific) to determine and return the value of the variable requested
- Functions used to set values other functions use internally (optional)

## MIB to C Script

This is a Perl program, which, when given an OID reference as an argument, creates some template MIB module files to be used with the *ucd-snmp* agent. It is far from perfect and will not generate working modules, but it significantly shortens development time by outlining the basic structure.

Using the script requires some manipulation of the code both to make it accessible to the Perl script and to repair some of the flaws created in the script's output. These flaws are the result of some unresolved bugs that remain in the script. Look at *example.h* and *example.c* for more information about how to alter the MIB's syntax for processing through the script and how to modify the script's output into good code.

## Using mib2c

To process a MIB through the *mib2c* Perl script, do the following:

1. Create a new MIB.
2. Alter the MIB's syntax to make it work with the script.
3. Run the altered MIB throughout *mib2c*.
4. Hand-modify the source code to fix any problems caused by the Perl script.
5. Check the hand-modified MIB into CVS.
6. Keep your original MIB and discard your other files.

# Heartbeat Server

The Heartbeat server (HBS) is used by the system to keep track of which servers are up and which are down. *Heartbeating* refers to a server transmitting regular pulses to a multicast address as a signal to the other servers that it is up.

## Transmit Thread

The Transmit thread is the multicast group. Any type of server that uses an HBS sends heartbeat pulses to the multicast group. If a server wants to determine the status of other servers on the network, it polls the Multicast thread regularly. The polling frequency is configurable and set to a default value of 250 milliseconds.

## Receiving Heartbeats

The RX thread specifies which servers a particular server listens to for heartbeats. Some servers receive the TX thread without keeping track of which servers are up; other servers listen for heartbeats from specific servers.

Each server has a data container that it uses to match incoming heartbeats. The matching criteria include server type, IP address, and port. Every server receives every multicast heartbeat but throws away those that do not match the list in its data container. If heartbeats are not received for servers listed in the data container, it is assumed that those servers are down.

## Housekeeping

Every 250 milliseconds, a heartbeat is received and marked. Housekeeping checks if any anticipated heartbeats fail to appear. The Housekeeping thread flags the counter for missed heartbeats and increments the counter by 1. If the counter matches the maximum missed heartbeat value set in provisioning, the thread reports to the *SNMPd* agent that the server is down. The *SNMPTRAPd* looks at the list and sees which servers are marked as being down.

## Using the Heartbeat Software

To use the HBS software, do the following:

1. Create a Heartbeat Transmit thread with your application port identifier (i.e., SIP port) as a parameter.

   For example:

   ```
   HeartbeatTxThread heartbeatTxThread(defaultPort)
   ```

2. Create a Heartbeat Receive thread which takes as a parameter the *Server-Container* used for storing heartbeat data.

   For example:

   ```
   HeartbeatRxThread heartbeatRxThread( ServerContainer::instance( ) )
   ```

   The job of the Heartbeat Receive thread is to pull multicast heartbeat messages off the UDP stack and extract the server host and port info from the message and use it as a key to search against the *ServerContainer*. If the key matches an entry in the Server Container, the heartbeat counter for that server is incremented by 1.

3. Create a Housekeeping thread, which also takes the *ServerContainer* as a parameter.

   For example:

   ```
   HouseKeepingThread houseKeepingThread( ServerContainer::instance( ) )
   ```

   The job of the housekeeping thread is to sleep for 250 ms (or however long the heartbeat interval happens to be) and wake up to do a sweep through the *Server-Container*. During the sweep, we check each server entry to see if the maximum number of heartbeats missed has been reached. If the maximum number has been missed, the server is declared dead. If no heartbeats have been missed, the heartbeat counter is reset to 0.

   *ServerContainer* derives from *BaseServerContainer*.

4. Run the threads. For example:

   ```
   heartbeatTxThread.run( );
   heartbeatRxThread.run( );
   houseKeepingThread.run( );
   ```

5. The main thread will be sleeping. For example:

   ```
   while (true)
           {
           sleep ( 0xFFFFFFFF );
           }
   ```

# CHAPTER 18

# Quality of Service and Billing

This chapter is a combined view of billing and Quality of Service issues. VOCAL does not include a billing server, but has a Call Detail Record (CDR) server that can either cache billing information or send it via the Remote Authentication Dial In User Service (RADIUS, RFCs 2138 and 2139) protocol to a third-party billing server. VOCAL also comes with an Open Settlement Protocol (OSP) stack that permits different VoIP systems to share billing information for cross-network calling through the use of a common clearing house. OSP is a product of the Telecommunication And Internet Protocol Harmonization Over Networks (TIPHON) project at the European Telecommunications Standards Institute (ETSI), *http://www.etsi.org*.

QoS is an advanced topic: while the protocols and technology exist to ensure some sort of QoS between IP networks, the practice has not caught up to the theory. Some organizations offer point-to-point IP backbones that are useful for VoIP traffic; however, the adoption of packet telephony will have to expand by several orders of magnitude before the business incentives to improve QoS push a truly ubiquitous, reliable service into the realm of practical and affordable reality. In many cases, service providers simply engineer a network to have plenty of bandwidth and ignore per-call QoS.

Although these topics are normally diverse and would normally deserve separate treatment there is, in VOCAL, a commonality that brings these topics together. Using OSP for billing between autonomous networks and setting up reserved paths to enable QoS both use the Policy Server and the Common Open Policy Service (COPS, RFC 2748) protocol. Using COPS with OSP was not a major breakthrough; it was, rather, a convenient opportunity to reuse some code.

## Quality of Service

Quality of Service (QoS) is an effort to manage transmission and error rates and to minimize latency, packet loss, and jitter during internetwork calls. VOCAL does admission control based on resource availability. If resources cannot be allocated,

VOCAL resorts to a "best effort only" delivery. Calls are still processed, but they may not be of the best quality.

*Policy* is a broadly used and widely interpreted term that describes the business rules of the organization applied to the operation of its telecommunications systems. The term stems from the same source as *corporate policy*, meaning the rules that guide the behavior of those who work for or with a corporation.

With respect to the practical application of QoS, policy is a combination of enforcement and decision making that permits calls to be initiated, established, and torn down between networks over the Internet or between managed IP networks. Enforcement and decision making are explained in detail in this section.

## Policy Server

The Policy server is the key component used to achieve per-call QoS. Service providers typically will offer QoS only if authorizations and payments are guaranteed by a third party. The Policy server administers admission control for QoS requests based on the installed QoS policies. These policies are either static policies translated from service-level agreements between two network providers or dynamic policies installed by the Internetwork Marshal (policy client) based on the call source and final destination. The Policy server uses the QoS policy database to provide network elements with the decisions necessary to enforce the admitted QoS requests.

## COPS

COPS is an Internet Engineering Task Force (IETF) proposed standard for implementing QoS policies as an end-to-end service. It is a simple query-and-response protocol that enables a Policy server to control devices on the network, such as routers and switches, in order to facilitate a consistent policy for traffic flows based on business priorities. In a COPS-based schema, the Policy server is known as the policy decision point (PDP), and its clients are known as policy enforcement points or (PEPs). The protocol employs a client/server model in which the client (PEP) sends requests, updates, and deletes to the remote server (PDP), and the PDP returns decisions back to the PEP. COPS is a companion protocol to Resource Reservation Protocol (RSVP, RFC 2205).

PEPs can be routers, gateways, and other devices that transfer voice channel signals between subscribers and their calling destinations. When the call is initiated, the PEPs query the Policy server for authorization to reserve bandwidth. Regardless of whether the bandwidth is available, VOCAL allows the call to go through, providing that it is a valid call that is being billed. When the call ends, the Policy server sends instructions to the PEPs to release the bandwidth.

The PDP is the Policy server. When the PEPs query the Policy server for authorization, the Policy server makes a policy decision to either accept or reject the request.

We wrote our own COPS stack, which you can download separately from *Vovida.org* (*http://www.vovida.org*). The stack, which uses TCP as its transport layer protocol, is extensible: it can leverage self-identifying objects and support diverse client-specific information without requiring modifications to the protocol itself. Let's have a look at our stack and its components.

### Key features

The stack is compliant with RFC 2748 and implements all of the functionality outlined in the RFC except for the scheme to support IPv6 addressing. In addition, the stack contains an implementation of the COPS-PR extension to support policy provisioning.

The stack can be used for building applications. One example is a policy-based QoS framework in which a policy client can exercise policy-based admission control over the use of RSVP-controlled endpoints. Another example is implementing application-level authorization, authentication, and accounting (AAA) functions in an IP telephony network for interdomain calls in which COPS works with OSP to process these requests through a third-party service, such as a clearinghouse. This process is explained later in this chapter under "OSP."

### Basic protocol design

The PEPs and PDPs within COPS work in a client/server relationship: the PEPs continuously request information from the PDPs and the PDPs respond accordingly. All transmissions between the PDPs and PEPs are conducted over TCP.

COPs maintains statefulness between requests and decisions sent between the PEPs and PDPs. The PDP keeps track of requests made by a PEP until that PEP sends delete commands to erase the requests from the PDP's memory. PDPs can also generate new, asynchronous decisions regarding active requests at any time until the request has been deleted. The PDP can also create a configuration statefulness when it sends new configuration information to the PEP. The PDP can end this statefulness by terminating the configuration transmission stream after the data has been transferred.

### Messages

Here is a summary of the messages used within the COPS protocol:

*Request (REQ) PEP → PDP*
> Used when the PEP sets up a request/decision state with the PDP to manage the bandwidth allocation.

*Decision (DEC) PDP → PEP*
> The PDP's response to the REQ that either acknowledges the request, authorizes the bandwidth allocation, or returns an error if the REQ is poorly constructed.

*Report State (RPT) PEP → PDP*

Used by the PEP as a statement of success or failure in implementing the DEC from the PDP.

*Delete Request State (DRQ) PEP → PDP*

Used by the PEP to tear down the request/decision state.

*Synchronize State Request (SSQ) PDP → PEP*

A request by the PDP for the PEP to resend the state.

*Client-Open (OPN) PEP → PDP*

Contains information about the types of clients that the PEP can support.

*Client-Accept (CAT) PDP → PEP*

When the PDP responds positively to the OPN message with a timer that sets the interval for keep-alive messages.

*Client-Close (CC) PEP → PDP, PDP → PEP*

Notifies the PEP or PDP that the client is no longer being supported.

*Keep-Alive (KA) PEP → PDP, PDP → PEP*

A heartbeat mechanism that indicates whether the other end is alive. The PEP must transmit this message regularly to maintain state with the PDP. The PDP must echo this message upon receiving it.

*Synchronize State Complete (SSC) PEP → PDP*

A response to the SSQ message after synchronization has been completed.

### Classes

These are the two main classes found within our COPS stack:

*COPSTransceiver*

*COPSTransceiver* is the main class to send and receive any COPS message over a socket.

*COPSMsgParser*

After receiving the raw COPS message, it decodes the message and creates a message object based on the type of the message. It also validates the message and throws *COPSBadDataException* if the message is not valid.

## RSVP

RSVP allows a certain bandwidth on the Internet to be reserved so that voice conversations can be transmitted with minimal delays. We obtained an RSVP stack from the Information Sciences Institute at the University of Southern California (*http://www.isi.edu/rsvp/*).

## Quality of Service Enabled

This section illustrates the messages exchanged to reserve bandwidth over the Internet, as well as the normal SIP messages used for call signaling.

## Suggesting a bandwidth path

Figure 18-1 shows a request for bandwidth from User Agent B being processed through the networks. These signals are identified with letters rather than numbers because they are sent over the voice channel at roughly the same time that User Agent B sends a 180 Ringing message.

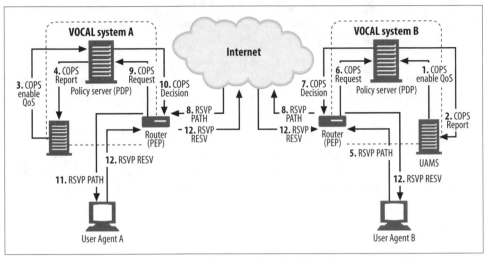

*Figure 18-1. Interactions: suggesting and reserving a bandwidth path*

Table 18-1 describes the messages illustrated in Figure 18-1.

*Table 18-1. Interactions shown in Figure 18-1*

| Interaction | Step | Description |
|---|---|---|
| Enabling QoS | 1–4 | At the time that the UAMS receives either a 180 or 183 message from the called party, it sends a COPS messages to the Policy server (PoS) requesting it to establish QoS and receives a response. |
| Requesting bandwidth | 5 | User Agent B sends an RSVP PATH request to suggest a bandwidth path to the on-network router. |
| Requesting a decision from the PoS | 6 | The router generates a COPS-RSVP request and sends it to the PoS. |
|  | 7 | The PoS responds with a COPS decision, authorizing the request. |
| Sending the request to UA A | 8 | The router sends the RSVP PATH request to the router in the far end system. A. |
| Sending an RSVP RESV | 9 | The router generates a COPS-RSVP request and sends it to the PoS. |
|  | 10 | The PoS responds with a COPS decision, authorizing the request. |
|  | 11 | The router sends the RSVP path to UA A. |
|  | 12 | UA A sends RSVP RESV message to UA B, reserving bandwidth. |

## Managed Networks Versus the Internet

On the Internet, you get whatever is available at the time you make your connection, which is variable in quality. The real issue is seldom bandwidth; well, maybe it is at the lowest end of the connection spectrum (for example, using the G.711 codec over a dial-up connection). For most of us, the problem is jitter. Our practical experience is that VoIP calls can predictably provide land line toll quality for calls between Canada and the U.S. For calls overseas, the signal tends to become much worse; however, it is significantly better than it was only a few years ago.

Correctly designed managed networks work very well. In contrast to the public Internet, managed networks are usually overprovisioned and they control the network to ensure high-quality transmission. These companies can also classify packets and forward them based on their priorities to overcome common QoS issues. Corporations have built managed networks with toll-quality performance all over the world.

Inherently, both the Internet and managed networks use the same equipment. The rule of thumb is that the Internet is random (no one knows his connection is a dog), while managed backbones are manageable and accountable to their customers.

# Billing

At Vovida Networks, our initial trial customers were service providers who needed to invoice their customers for VoIP services. This section discusses how we implemented an interface to a third-party billing system.

## CDR Server

The Call Detail Record (CDR) server is a fairly simple program. Primarily it takes CDRs that represent billing information about a call from the MS and caches them to disk. It can then connect to a billing system using the RADIUS protocol and transfer the records over to the billing system. Because the records are queued on disk, a very large number can be stored, and no billing information is lost if the billing system is down for a day or two.

When the system is configured in a high-reliability configuration, the Marshal dual-writes billing information to a pair of CDR servers. These servers then pass on the data to the billing system whenever it is up. This helps improve performance and reliability. The system can also be configured with just one CDR server, or none if billing is not required.

### The bill record

After receiving notifications from the Marshal servers about the start and end of each call, the CDR server generates a bill record that contains a duration field. This field is the calculated difference between the start and end times of the call. These records are cached on the local disk in record file *billing.dat*. The record file gets rolled over after a configurable size into files with the extension *.unsent* and with a date and time stamp.

The records in files with *.unsent* extension then can be sent from the CDR server to a third-party billing system on a regular schedule using RADIUS messages over UDP. Billing files are normally purged from the CDR server after 72 hours.

### Sending records with RADIUS

The CDR server uses a RADIUS stack to communicate with the billing system. In order to send records to the billing system, you must know which vendor-specific attributes are being used in the billing system's code and modify the CDR server to accept those attributes. The billing data is not sent in real time; it is stored on the CDR server in case the transmission to the billing system fails. You must set up a transmission schedule for the frequency that best suits your needs (for example, for hourly or daily transmission).

RADIUS by itself doesn't have enough information in it to do billing, but you can describe vendor-specific attributes (VSAs), which are custom data formats that are transported between the CDR server and a billing system and are compiled into the RADIUS stack. Our stack originally used a set of Cisco VSAs that were specified by MIND CTI, Ltd. (*http://www.mindcti.com*): the billing system that we used for testing our CDR server. Today, the most current RADIUS stack has been stripped down and optimized for better performance. It is no longer vendor-specific, and if you were to use it with a specific billing system, you would have to add your own VSAs.

### Class descriptions

Here is a rundown of the classes found within the CDR server:

*CdrManager*
> A class that holds *CdrServer* and *CdrCache* and initiates some of the processing.

*CdrServer*
> The main class that listens for Marshal servers that are available for new connections.

*CdrCache*
> Correlates the start and end of calls and writes out the complete records to files.

*CdrFileHandler*
> Writes items to file and manages the files to ensure correct rollovers before the file sizes become too large.

*CdrBilling*
> Runs in its own thread and takes care of deciding when to send the queued records to the billing system, initiating calling the *MindClient* object to build and send them.

*MindClient*

> Takes care of adding VSAs that are specific to the Mind billing system. It inherits from the *RadiusStack* class since it specializes this stack for the Mind billing system. To connect to another specialized billing system, you would need to modify the code to derive a different class from the RADIUS stack that defined its own VSAs.

*RadiusStack*

> Implements a RADIUS stack and has the code to send information to the billing system.

*CdrConfig*

> Loads configuration information when the server starts up.

# Call Detail Records

The CDR server communicates with the Marshal servers over TCP/IP. As has been shown earlier in this book, every call involves both incoming and outgoing Marshal servers. At the time when a call starts and again when it ends, both Marshal servers notify the CDR server. From this notification, the CDR server creates a new billing file, called *billing.dat*, with two Start and two End records, one of each from both Marshal servers.

## What defines the start of a call?

As was shown in Chapter 7, the start of a call happens when the voice channel is established. After the INVITE has been transmitted from the calling party to the called party and the called party starts ringing, the called party picks up and thereby transmits a 200 OK message to the calling party. When the calling party replies with an ACK message, the Marshal servers notify the CDR server to create a Start record.

> You can provision the CDR server to bill for ring time. If you do, the Marshals notify the CDR server to create a start record when it receives the 180 Ringing message from the called party.

## Notifying the CDR server for Start record

Figure 18-2 shows the SIP messages that lead up to the Marshal servers notifying the CDR server to create a Start record. In this scenario, the INVITE has already passed from SIP phone A to SIP phone B.

Table 18-2 describes the message illustrated in Figure 18-2.

Message 1 in Figure 18-2 occurs after the INVITE has been passed from SIP phone A through the system to SIP phone B.

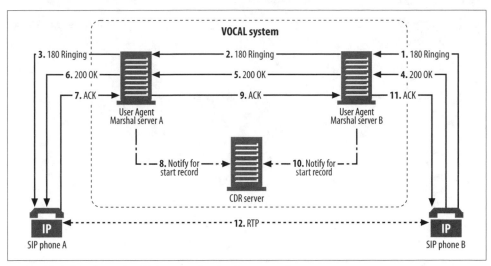

*Figure 18-2. Notifying the CDR server for the Start record*

*Table 18-2. Interactions shown in Figure 18-2*

| Interaction | Step | Description |
| --- | --- | --- |
| Ringing | 1–3 | SIP phone B sends a 180 Ringing response to User Agent Marshal server (UAMS) B, which is forwarded. |
| Pickup | 4–6 | SIP phone B sends a 200 OK response to the UAMS. This means that the phone has been activated and is ready to establish voice channel contact with SIP phone A. |
| Pickup acknowledged | 7 | SIP phone A sends an ACK message to UAMS A confirming that it is ready to connect to a voice channel. |
| CDRS notified | 8 | UAMS A notifies the Call Detail Record server (CDRS) that the call has started. |
| Pickup acknowledged | 9 | UAMS A forwards the ACK message to UAMS B. |
| CDRS notified | 10 | UAMS B notifies the CDRS that the call has started. |
| Pickup acknowledged | 11 | UAMS B forwards the ACK message to SIP phone B. |
| A conversation takes place | 12 | The calling parties talk to each other using Real-time Transfer Protocol (RTP). |

## What defines the end of a call?

The end of a call happens when the first phone hangs up and, thereby, sends a BYE message to the other phone. Upon receiving the BYE message, each Marshal server notifies the CDR server to create an End record. This process is illustrated in Figure 18-3.

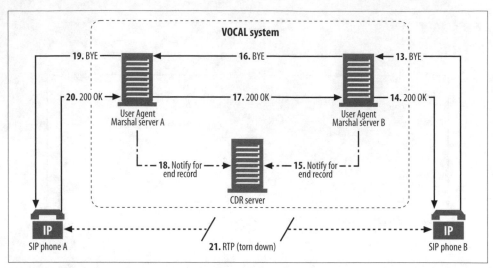

*Figure 18-3. Notifying the CDR server for the End record*

Table 18-3 describes the messages illustrated in Figure 18-3.

*Table 18-3. Interactions shown in Figure 18-3*

| Interaction | Step | Description |
| --- | --- | --- |
| SIP phone B hangs up | 13 | SIP phone B sends a BYE request to User Agent Marshal server (UAMS) B . |
| Notifying the CDRS | 14 | UAMS B notifies the Call Detail Record server (CDRS) that the call has ended. |
| The BYE is forwarded from one UAMS to the other | 15 | UAMS B forwards the BYE message to UAMS A. |
| Notifying the CDRS | 16 | UAMS A notifies the CDRS that the call has ended. |
| The BYE is received by SIP phone A | 17 | UAMS A forwards the BYE message to SIP phone A. |
| SIP phone A sends a 200 and hangs up. | 18–20 | SIP phone A sends a 200 OK message, through the UAMSs, tp SIP phone B. Afterwards, the user hangs up. |
| The voice channel is torn down | 21 | The BYE and 200 messages trigger the voice channel to shut down. |

## RADIUS Stack

The RADIUS stack is fairly simple. It has some calls to connect to the billing server and some calls to records. It is based on the Livingston (*http::/www.livingston.com*) RADIUS code. In general, a message is created and initial headers added using a call such as *createAcctStartCallMsg*. VSAs are added, using the *addAttribute* method, and are then signed with the *addMD5* method The message is then sent using the *transmit* method in the *UdpStack* class from the *util* library in VOCAL. Then the *RadiusStack::handshake* method and *RadiusStack::verifyAcctRecvBuffer* are called to make sure the return value is OK. At this point the data has been successfully transferred to the billing system.

The Marshal servers include a library that contains the *CdrInterface* class. The Marshals write billing information to this interface, and the CdrInterface class takes care of sending it to one or both (if there are two) configured *CdrServers*. A record is written when ringing starts, when the call starts, and when the call ends. Each record contains the To and From of the call as well as the *callID* so that they can be correlated.

### Key features

RADIUS works within a client/server model whereby a client, such as the CDR server, is responsible for passing user information to designated RADIUS servers and returning responses. RADIUS servers receive user connection requests, authenticate users, and return all configuration information necessary to enable the client to deliver service to the user.

### Operation of RADIUS accounting

RADIUS accounting is an extension of the RADIUS protocol that enables billing functionality. Clients configured with RADIUS accounting send an Accounting Start packet that describes the users and the type of service being delivered. The server responds with an acknowledgment to inform the client that billing can proceed. When the service delivery has ended, the client generates an Accounting Stop packet.

### Authentication

Authentication is provided by a shared secret, which is manually entered into both the client and the server without ever being sent over the network. User passwords are hashed with the MD5 algorithm.

### Initialization of service

When a RADIUS client receives authentication information from a user, it may choose to validate this information by creating an Access-Request containing the user's name and password, the client's identification, and the port number that the user is connecting through. The RADIUS server receives the client's request, on behalf of the user, and validates the client based on the shared secret. For validated requests, the server searches for the user's name in a database that stipulates access requirements for the user, which can include the password and the identity of clients and ports that the user is authorized to use. If the user does not meet any of these requirements, the server sends an Access-Reject message back to the client.

Even if all conditions are satisfied, the server will send an Access-Challenge response to which the user must respond. The user's response is a resubmission of the Access-Request with a new request ID and the password replaced by the encrypted response from the server. If the server accepts this challenge response, the server will return a list of configuration values to enable data transmissions as an Access-Accept response.

# OSP

OSP is a proposed European Telecommunications Standards Institute (ETSI) specification for providing interdomain authentication, authorization, pricing information, and accounting standards for IP telephony. We obtained an OSP stack from Trans-Nexus, Inc. (*http://www.transnexus.com*) and wrote C++ wrappers to enable their stack, which was written in C, to work with our Policy server.

While we provide the client side of OSP, on *Vovida.org*, there is an application called OpenOSP that provides the server side of an OSP solution. The server side is geared toward the role that a third party would fill as a trusted broker.

One of the components of the OSP and Policy server architecture is the Secure Socket Layer (SSL) transport layer security toolkit. We downloaded our toolkit from the OpenSSL site (*http://www.openssl.org*).

## Making Internetwork Calls

Figure 18-4 shows User Agent (UA) A initiating a call to User Agent B. In this scenario, the UAs are used together with basic analog phone sets and are attached to different VOCAL systems, and each VOCAL system is known to the others. The call signal routing is carried over the Internet.

The call may be routed through one or more Feature servers before it reaches the Internetwork Marshal (INMS). For the sake of clarity, the Feature servers have been omitted from this scenario.

VOCAL supports multiple Internetwork Marshal servers (INMSs). Each of these servers will accept off-network INVITE messages from one other known SIP-based server. If an INVITE is received from any other off-network entity, it will be rejected regardless of whether it includes a clearinghouse token.

Figures 18-4 and 18-5 show how VOCAL works with OSP between two INMS that are known to each other. In these figures, the 100 Trying messages have been omitted to avoid clutter. Table 18-4 describes the messages illustrated in Figure 18-4.

*Table 18-4. Interactions shown in Figure 18-4*

| Interaction | Step | Description |
| --- | --- | --- |
| SIP phone to INMS | 1–3 | An analog phone attached to User Agent A initiates a call. The User Agent Marshal server (UAMS) authenticates the message and forwards it to the Redirect server (RS). The RS returns a 302 Moved Temporarily message that provides routing information. |
| | 4–5 | The UAMS acknowledges receipt of the 302 message and forwards the INVITE to the Internetwork Marshal server (INMS). |

*Table 18-4. Interactions shown in Figure 18-4 (continued)*

| Interaction | Step | Description |
|---|---|---|
| Requesting and receiving an internetwork token from the clearinghouse | 6 | The INMS generates a COPS authorization request and sends it to the Policy server. |
| | 7 | The Policy server (PoS) composes an Open Settlement Protocol (OSP) authorization request and sends it to an internetwork clearinghouse and receives a response plus a token. |
| | 8 | The clearinghouse verifies the route, by confirming that the dialed digits are correct, and responds with an OSP authorization plus a token. |
| | 9 | The PoS generates a COPS decision, which includes the clearinghouse's token, and sends it to the INMS. |
| INMS forwarding the INVITE message plus the token | 10 | The INMS adds the token to the INVITE message and forwards it to the Internet via the router. |

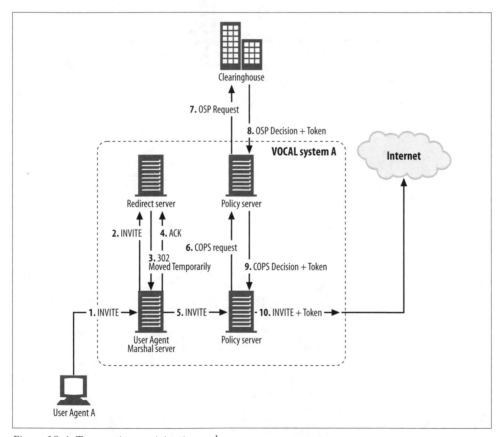

*Figure 18-4. Transactions: originating end*

## The INVITE Message Is Received

Figure 18-5 shows the receiving network processing the INVITE message and the token.

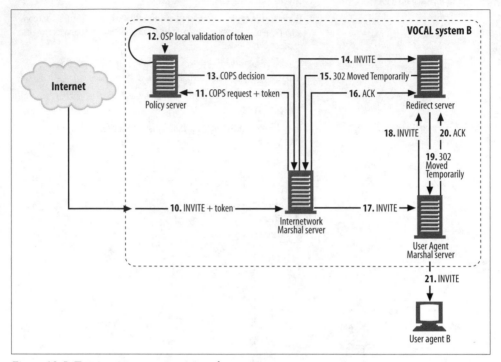

*Figure 18-5. Transactions: terminating end*

Table 18-5 describes the messages illustrated in Figure 18-5.

*Table 18-5. Interactions shown in Figure 18-5*

| Interaction | Step | Description |
|---|---|---|
| INVITE is received from known system | 10 | The INVITE message is received by the Internet Marshal server (INMS). |
| The receiving INMS receives the INVITE and requests verification from the PoS | 11 | The INMS generates a COPS request and sends it, along with the token, to the Policy server for verification. |
| | 12 | The Policy server (PoS) does the local validation of the token received in the INVITE message. |
| | 13 | The PoS generates a COPS decision and sends it to the INMS. |
| The UAMS requests routing information from the RS | 14–16 | The INMS strips the OSP token from the SIP INVITE header and forwards the INVITE message to the RS for routing. The RS returns a 302 Moved Temporarily, and the INMS responds with an ACK message. |
| The INVITE message is sent to SIP phone B | 17–21 | The INMS forwards the INVITE to the UAMS, which forwards it, through the RS, to User Agent B. |

After this one-way reservation has been made, User Agent A makes the same type of reservation, using the same messages, but starting from the reverse direction. We have not added an illustration to show this because it would be the same as Figure 18-4, except the messages would start from the UA Marshal in System A and go in the reverse direction as those shown in Figure 18-4.

## Tearing Internetwork Calls Down

When the calling party tears down the call, it sends a BYE through its system to the far end system and, eventually, to the far end user. As this BYE travels through the servers, it triggers the INMS to notify the Policy server that the call is being torn down. The Policy server exchanges OSP messages with the clearinghouse, but does not send a COPS reply to the INMS. The INMS does not keep track of the changes to Policy after a call has been torn down, and there are no more COPS or OSP messages transmitted as the call is torn down.

# Billing and Toll Fraud

What happens if one user calls another, and then, after the call has been set up, sends a SIP BYE message to stop the billing while continuing to send RTP packets to the other user? If one end of the call is a trusted device, such as a PSTN gateway, it will pay attention to the BYE and tear down the call. If neither user is trusted, the call may stay up, but the PEP in the network will remove the QoS for the call and it will shift to "best effort" quality. Some networks, particularly those running 3G wireless, may be configured to stop packet transport if QoS has been removed.

A new technology being used to fight toll fraud is the Back-to-Back User Agent (B2BUA) that offers the possibility of prepaid calling applications. For more information, see the definition of the B2BUA offered near the end of Chapter 10.

# CHAPTER 19
# Provisioning

In the early days of Vovida, we had a provisioning system that satisfied our needs for testing. As we grew in size and complexity, we recruited a number of talented individuals who expanded a system into the collection of GUI screens, structures, and interactions that are described in Chapters 4, 5, and 6. This system worked well, providing our developers were available to enable access to new servers and configuration setups. The problem was that it would have been very hard for an outside contributor to add her new server to the VOCAL provisioning system.

Late in 2001, our provisioning team was given the opportunity to relieve themselves of their responsibilities for the old system and to design a new system from scratch. This project became known as the Mascarpone project, and this chapter describes how this project was designed. At the time of this writing, Mascarpone has yet to be merged with the rest of VOCAL, but it is included in the CVS tree.

## Old Provisioning System

In the old provisioning system, as shown in Chapters 4, 5, and 6, the Provisioning server is responsible for maintaining and storing all information needed by the VOCAL system and supplying it to the appropriate systems, as well as tracking when important parameters change and relaying that information to interested applications.

### Key Features

The provisioning system provides mechanisms for data retrieval, data storage, and registering interest in data points. It is a high-performance system that is capable of handling many simultaneous transactions from servers. It also provides redundancy: two servers can be connected together. These servers exchange data, synchronize data between themselves, and exchange registration information so that either can serve any requests and update clients, which have registered interest.

## High-Level Flow

The server provides a thread pool, the number of threads being determined by a command-line option. When a new request is received on the listening port, the server dispatches a thread to handle the request. There are several types of requests handled by the Provisioning server, including registrations to providing lists of known servers.

If the request is a registration request, the client provides an IP address and port to which updates should be sent and the data required to complete the registration. Group and Item tags identify all data, and a user may register for a specific item or all items in a group. When the registration is processed, an acknowledgment is sent.

If the request is a list, all items in that particular group are listed and sent to the server. If the request is to retrieve an item, the item is retrieved from the disk (where it is stored as an XML file) and forwarded to the client. There is no processing or validation of the content—the server simply stores the data on the disk.

If the request is to add a new item, the item is written to disk (again, with no awareness of the contents). If any clients have expressed interest in being notified when the information changes, those clients are informed that the item has been added or modified.

If the request is to delete an item, the item is deleted and the listing of registrations is checked to see if any clients are interested in the data, and if so, an update message is sent to the clients that are interested in the data.

Additionally, there are a number of diagnostic messages that return text strings that are used simply to help debug a running server and to ensure that the server is running properly.

## Key Data Structures

Provisioning data is stored as XML files that are indexed by group and item. *PServer* (Provisioning server) stores this data as files indexed by item name within directories indexed by group name. *PServer* is unaware of the file content: it is simply providing a file storage service for the GUI and *PSLib*, which have been designed to interpret the XML tags and use the data.

 In order to support large systems with many users, another level of indirection needed to be added. The Linux filesystem does not support directories with more than 1000 files in an efficient manner; therefore, a hash scheme was used. Some number of bins, labeled AAA, AAB, AAC, for example, are defined, and a directory is created to correspond to each bin. In the default configuration, there are six bins, but the number of bins can be configured from the command line (*/usr/local/vocal/bin/pserver -n <number of bins>*). Within each bin, we repeat the same structure of each group being a directory and each item being a file.

The key classes found in the Provisioning server include the following:

*FileDataStore*

When a file is created, a hash is used to determine which bin to place the data into. At that point, the data is placed in the correct group directory. This increases the total number of users that a system can support. This class is found in the *vocal1.3.0/util* directory.

*RegistrationManager*

A registration map is used internally to map interest in a particular group or group/item pair and to map the interest to a particular client. The information about the client that is registered is stored in the form of the IP address and port where notifications are to be sent. This is to allow for rapid lookup of the clients that need to be notified when a file is modified to ensure that modifications can be completed very quickly.

*RedundancyManager*

A data structure and manager are provided to monitor the redundant server that all information should be sent to and to monitor the state of synchronization between the servers. This class implements a state machine using a few variables and a number of transition functions.

*ConnectPServer*

This is used to store information about connections from existing clients. For each connection, information is stored about whether there is data waiting to be processed by the server and whether a thread has been assigned to process this request. The server uses this to dispatch threads to the appropriate connections to service the requests.

*ServerContainer*

The server container classes are essentially unmodified classes designed to monitor heartbeats. The heartbeats are used only for redundancy in this application. The server does not monitor the heartbeats of any servers except the other Provisioning server and does not export this information to other servers.

## Redundancy

The Provisioning server is designed to be redundant with another server—in other words, it is possible for one server to synchronize with the other. The process is somewhat complicated, so it is perhaps best to explain it in stages.

### No redundancy

A Provisioning server can be run by itself and can function perfectly well—but if the server crashes, all registration information is lost, which causes the system to become unstable because clients expect to be updated when data changes, which is impossible without registration information. This registration is accomplished with provisioning messages and is not to be confused with the SIP method for registering User Agents that was discussed earlier in this book. Running a single server (nonredundant mode) is recommended only for testing and development—not for an actual production system.

## Redundancy

When running in redundant mode, the other server is identified on the command line. When the servers are first started, they synchronize with each other. Both servers set a random timer to decide how long they sleep before trying to synchronize. The server whose timer expires first attempts to contact the second server, and becomes the *master*. The second server that was contacted becomes the *slave*. It is worth noting that this is *only* for the duration of the synchronization—the servers are peers except for the time when the servers are actually in the process of synchronizing. If the servers lose contact with each other, they both reset their timers and resynchronize. This time, the other server may be the master, depending only on the random timer that expires first.

## Synchronization

Once the master/slave relationship has been determined, the server that is the master sends a message to the slave asking for a list of all files stored along with the timestamps of when the files were last modified. It also asks for a list of all registrations. It is the job of the master to compare the list of files and send any files that it has that are newer or do not exist on the slave and to request any newer or nonexistent files from the slave. The master also creates the union of registrations and sends it to the slave, as well as using it itself. After synchronization is complete, the notion of slave and master no longer exists.

## After synchronization

While the servers are synchronizing, and after the synchronization has completed, the servers stay synchronized by reflecting messages to each other—if a write message is sent to server 1, it sends an update to server 2 telling it that there is a new file, as well as a timestamp, to ensure that these servers are synchronized. This is handled similarly for other operations such as delete. Only the server receiving the original message, not the server that is sent the synchronization message, sends an update to any registered clients. As the servers have the same registration map while synchronized, this works fine.

You can send a write to either server. At the end of the transaction, both servers have complete data trees, and only the server receiving the message sends an update to the clients.

### Resynchronization

If the servers become disconnected, they need to resynchronize. If the disconnect was for a short period of time, the cost to synchronize—the cost of exchanging the list of files and the few files that may have been modified on one side or the other—is low.

The heartbeat mechanism waits until some number of beats has been missed before it declares the servers to be out of contact with each other.

## Miscellaneous

Because the *PServer* cannot contact the Provisioning server for the configuration data it needs, as the other servers can, it uses command-line options to set the options needed to operate. For a complete list of these options, type **./pserver - help**.

This list includes the following information:

```
./pserver [Option]...

Where [OPTION] is one of:
  -v <log_level>                   set verbosity
  -r <psRootFileSystem>            set filesystem root
  -f <logfile>                     set log file
  -b <redundant server host:port>  set redundant ("backup")
                                   server hostname and port
  -p <local port>                  set port to listen for
                                   incoming requests
  -d                               disable daemon mode (run in
                                   foreground)
  -u <number of users (1000)>      number of users.
  -n <number of bins>              number of hash bins
                                   (overrides number users)
  -t <number of threads (25)>      number of request process
                                   threads.
  -M <multicast host (224.0.0.100)> multicast host.
  -P <port (9000)>                 multicast port.
```

The option *log_level* can use one of the following options:

```
  -v[<log-level>]    Set the lowest log level
                     Levels:
                       - LOG_EMERG
                       - LOG_ALERT
                       - LOG_CRIT
                       - LOG_ERR
                       - LOG_WARNING (default)
                       - LOG_NOTICE
                       - LOG_INFO
                       - LOG_DEBUG
                       - LOG_DEBUG_STACK
                     Note that there is no ' ' between 'v' and the level.
```

To use redundancy, all of the redundancy options must be provided, and none should be provided if you do not intend to use redundancy at all. Most of the options have default values, as shown.

# PSLib

The *PSLib* is a library that is used to access data from the Provisioning server, making it easier to obtain information without having to do a great deal of coding.

## Classes

At the most fundamental level, the library is made up of *PSReader*, *PSWriter*, and *PSListener* classes, which are responsible for managing read, write, and listen (for update) operations from the server. Additionally, the *ProvisionServer* interface serves as a low-level interface to connect to the servers—it manages the connections to two servers if this is a redundant system, each of which is connected using a *HostInterface* class. The *ProvisionServer* interface has all the commands to send and receive messages from the server and swaps between two *HostInterface* objects to ensure that one server or the other handles the connection.

Built on top of these classes is a library that provides calls with returning information specific to a particular server. Examples include *getCdrServers*, which returns a list of CDR servers, and *getMarshalData*, which returns a *MarshalData* object. These data objects have accessors for each and every item of data within the object. *PServer* maintains a list of all the servers as well as a file containing the data for each server. Additional things, such as maps of Marshal hosts to the group that that marshal is in, are also provided as needed.

## Caching

While it is not used in any server at present, a mechanism is built in to cache the list of each type of server, and objects to store the lists while they are cached are provided.

Because of these functions, retrieving data from the Provisioning server is hidden from the user. While writing is possible using the base *write* command (implemented by *PSWriter*), no applications presently use this—all the writes to the Provisioning server in the current implementation are initiated by the Java GUI, which does not use the *PSLib*.

That is all we would like to say about the old system. The remainder of this chapter is devoted to the new provisioning system, code-named Mascarpone.

# Mascarpone Provisioning System

The Mascarpone Provisioning server is a flexible open source provisioning tool that provides for provisioning subscriber parameters on multiple devices and platforms. It also supports the provisioning of new or custom-developed features, including the end user data required for feature operation, and has the ability to support different databases, back office systems, and large-scale customer deployments.

## Architectural Overview

This design, illustrated in Figure 19-1, provides a generic, extensible provisioning system primarily for provisioning of VoIP solutions, although the generic approach taken to the development makes the program extensible to potentially provision other types of systems.

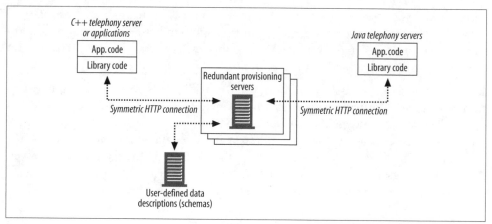

*Figure 19-1. Overview of architecture*

### Top-level components

The system consists of a number of top-level components designed to integrate together to provide an overall provisioning solution that is easily extensible for provisioning new devices, as well as to provide reusable components that allow others to extend or replace components of the system to suit their unique needs.

### Modular Provisioning server

The modular Provisioning server provides the core of the system and was designed with maximum modularity in mind. The first example of the modular architecture is the backend design. Rather than tying the server to a particular database, the backend consists of an abstract base class, which defines the calls for searching, accessing, and modifying the data required by the server. By implementing different concrete classes that access different backend data storage mechanisms, the system can be easily interfaced and adapted to work with whatever type of database might be available at the customer's site. In the initial release, we supplied a SQL backend and provided the ability to add LDAP backends in subsequent releases. Customers in actual deployed environments can use these backends with large numbers of users, when rapid access, searching, and manipulation are required.

Similarly, while the system was designed to use a symmetric pair of HTTP connections as a transport layer and an XML-based command and payload structure, this release is being designed in the form of a module, to ensure that users could eventually replace this with a different protocol if needed.

The system provides a mechanism for registering for callbacks when an attribute of the system is modified. This allows for improved performance of servers using the information, because they do not need to query a database on a per-call basis, but rather can read the information from the Provisioning server once, cache this information, and ask to be notified as the information changes. Applications can safely use the most recent value at all times without the need to contact the server.

New features can be easily added to the system since the information stored in the system is defined in XML description files. The developer of a new application or feature, which he wishes to provision, generates these description files. These files define the fields that the developer would like to have stored in the server, and the server allocates appropriate storage. These fields are then available to the developer just as any other existing field in the database is available.

## Autogenerating reference GUI

The second component of this system is an autogenerating reference GUI implementation. The GUI also uses the XML description files mentioned earlier to generate a simple GUI for editing this information. The GUI can be used to provision a new feature or server with little effort required on the part of the developer. This GUI allows users to add, modify, or remove new servers and users.

The GUI is a reference, in that the scope of what can be provisioned automatically is limited—only servers and user features can be added using the XML description method. Additionally, the GUI displays this data in a usable, navigable format, with the option of embedding a user-provided logo, but it does not provide for layout control, color modification, or other customization that may be desirable to an end user.

One part of the GUI allows a system administrator to view the settings for all the servers in the system, modifying any and all settings for these servers. Viewing all the servers of a particular type as a whole is supported.

Another portion of the GUI displays all users in a system and allows for searching, sorting, modification, and manipulation of this data. As searches, sorts, and paging of the information are provided on the server side, the GUI minimizes the data stored, which increases its speed. The server sends only the information that the GUI is currently displaying, not all of the data that matched the search. This GUI also allows an individual user to be selected, and allows for modification of the features and settings associated with this user.

A user-level GUI, intended for the final end user to access her features, is also provided. This GUI displays the choices for the user and in many ways resembles the screen mentioned earlier for selecting an individual user.

All of the GUIs described use the XML description files to generate the content of the GUI automatically. A new feature or server, which defines the appropriate description files, can be automatically provisioned by the system in this way.

Authentication and access can be controlled at several levels so that different levels of administrators may see more or fewer servers or users. This allows for environments in which a service provider gives administrative access to one client organization's IT manager but wants to restrict this administrator from seeing the information for other client organizations.

### Bulk-provisioning tool and other GUIs

A bulk-provisioning tool, allowing for rapid generation of entries associated with new endpoints (phone devices) and users, is in development. This tool incorporates features to define a user by means of a template. A template defines the settings common to a group of users. A bar code scanner can be used to enter information into the system to speed deployment of systems.

Finally, small hardcoded GUIs have been provided for things such as dial plans. Due to their complexity, dial plans cannot be XM- driven.

### Interface library

An interface library in both C++ and Java enabling applications to store, retrieve, search, sort, and monitor data stored in the Provisioning server is provided. The purpose of this library is to abstract away details of the protocol, transport, and other implementation of the mechanism used for communication between the clients and Provisioning server. The client code simply makes calls to this library, transparently hiding the retrieval, parsing, and caching of the data. Data is retrieved from the server using tag names. These tag names must be known by the application retrieving the data. Queries of the server are also supported.

### Daemon

A daemon allowing a machine at startup to query provisioning for configuration information is under development. This daemon will be able to determine what software (server) should be started and obtain configuration information, if needed, for these applications. The daemon starts the applications on the machine, eliminating the need for startup scripts on these machines. Changes to configuration are handled by modifying the startup parameters stored in the database and by restarting the machine or sending a reset message to the daemon.

## Protocol

The protocol used for communication is a symmetric HTTP-based protocol. Both the Provisioning server and clients communicating with it act as both HTTP servers and HTTP clients. The transport mechanism is standard, unmodified HTTP protocol—only the standard *get*, *put*, *delete*, and *post* methods are used. XML-based information, allowing a richer vocabulary of commands, is imbedded in the content of the messages passed between the servers.

The Provisioning server functions as a server in the sense that incoming messages to search for data, retrieve data, or store data are treated as standard HTTP messages. The request is processed, and a result is returned. Specifics of the command (what one is searching for, what items should be deleted, and the like) are determined from the XML content of the message.

Analogously, a client requesting information from the Provisioning server is functioning as an HTTP client when it makes these requests. The request is sent to the server, and the client interprets the result. For most commands (searches, sorts, requests for data, etc.), this mechanism is used.

The protocol is reversed with the ability to monitor data using callbacks, and one can see why we refer to this as *symmetric HTTP*. The client program posts a request to be informed when a data item changes, and the server replies, acknowledging this request. Some time later, a point of data in which the client has expressed interest is changed, and the server needs to inform the client. At this point, the roles of HTTP server and client are reversed.

When the client requests that it be updated on a data change, it needs to provide a port on which to listen for incoming updates from the server. This port acts as a standard HTTP server. We can optionally have the client respond to a standard *get* request for information, but the primary use of this port is to listen for incoming updates from the server. When the server needs to send an update, it contacts the client at the port the client specified using a standard HTTP *post* command, containing a list of the data items that the server was interested in that have changed.

The actual content of the modified field is not sent. The data is not sent is for efficiency: the data may not be needed immediately and in fact may change again several times before it is needed, so the fact it is changed is all that is sent. The client then knows that the data has been changed and that any local information stored pertaining to it is invalid. The next time the data is needed, it is requested again. If the client program is concerned with the time cost of obtaining the new data at access time, it can always immediately query the data when it receives notice that the data has changed but, with this mechanism, is not required to do so.

With the transport mechanism defined, let us now discuss the commands that are allowed to be sent in the payloads. Searching and sorting of data are accomplished by means of queries sent to the server from the clients. To begin a search, the client first must obtain a token to begin a session. The client uses the token in all subsequent requests, which can be data requests, further searches, or sorts, for example. The token is a large integer.

A few terms used in this document when referring to data are defined in Table 19-1.

*Table 19-1. Provisioning terms defined*

| Term | Definition |
|---|---|
| Domain | The type or category of information. Examples would be users, a particular type of server, etc. This is analogous to the idea of a table in SQL. |
| Record | A particular entry in the database and the data associated with it; for example, a user and all the information about a user. This is analogous to a row in SQL. |
| Field | A particular piece of data, of which one exists for each record. An example would be a phone number—there is one associated with each user. This is analogous to a column in SQL. |
| Value (content) | The contents of a particular field—a particular phone number in the case described previously. |
| Filter | A filter using the grammar defined in this document. |
| Field list | A list of fields, usually specifying which fields one wants to work with. |

With these definitions, we can define the commands the API supports, as shown in Table 19-2.

*Table 19-2. Provisioning APIs*

| Command | Description |
|---|---|
| SessionOpen | *Description:* Returns a token that can be used for subsequent searches/sorts/batched data requests. An optional finite expiration time can be provided, after which a new token must be obtained before one can continue requests. |
| | *Arguments:* Expiration time for the token |
| | *Returns:* Token to be used for later requests, or fail |
| SessionClose | *Description:* Closes the session associated with the token passed. All searches performed and results stored on the server side will be removed, and the token becomes invalid. |
| | *Arguments:* Token defining search |
| | *Returns:* Success/fail |
| SessionSearchDB | *Description:* Given a token, searches database, the result then being associated with the token. Subsequent calls using the same token will search within the result of the previous calls, *unless* the domain is changed, in which case the search is reset. |
| | *Arguments:* Token defining search, domain, search filter, field list, and sort order (ascending or descending), field to sort on |
| | *Returns:* The number of records matching the request, or fail |

*Table 19-2. Provisioning APIs (continued)*

| Command | Description |
| --- | --- |
| SessionGetDBResult | *Description:* Within the search results defined by token, return the fields specified in field list. Index specifies the range or records to return (i.e., to return the first 100, index would be 1–100). |
| | *Arguments:* Token defining search, index of records to return, list of fields to be returned |
| | *Returns:* Field/value pairs, or fail |
| GetRecordsInDomain | *Description:* Within domain, find the items associated with the supplied filter and return the fields requested. |
| | *Arguments:* Domain, filter, field list |
| | *Returns:* The requested fields for the item specified, or fail |
| SessionSetDBResult | *Description:* For every item in the range identified by token, change the item to reflect supplied field/value pair. Since all of these exist (returned by a search), it updates only the column/values specified for each item. |
| | *Arguments:* Token defining search, field/value pairs |
| | *Returns:* Success/fail |
| SetRecordsInDomain | *Description:* Put record (possibly more than one) within domain into the database. The field/value pairs specify the new values for the selected fields. Any fields omitted are left unchanged. |
| | *Arguments:* Domain, search filter, field/value pairs |
| | *Returns:* Success/fail |
| AddRecordInDomain | *Description:* Add a new record to the database in the given domain. All nonspecified fields are left unfilled. If the record would overwrite an existing record, an error is returned. |
| | *Arguments:* Domain, field/value pairs |
| | *Returns:* Success/fail |
| SessionDeleteAll | *Description:* Removes all items in the range identified by token. |
| | *Arguments:* Token defining search |
| | *Returns:* Success/fail |
| DeleteRecordsInDomain | *Description:* Deletes the specified records from the database. All fields associated with this identifier are removed. |
| | *Arguments:* Domain, search filter |
| | *Returns:* Success/fail |
| SessionSortDBResult | *Description:* Sort the results that are defined by the specified token. The field to sort by is provided, along with the order (ascending or descending) to sort. |
| | *Arguments:* Token, sort order (ascending or descending), field to sort on |
| | *Returns:* Success/fail |
| RegisterDomain | *Description:* Registers to receive an update on host/port whenever the contents of a category change. The information sent on change is the identifier and whether that was added, removed, or modified within the category. |
| | *Arguments:* Domain, host/port to send updates |
| | *Returns:* Success/fail |
| RegisterRecord | *Description:* Registers to receive an update when the record defined by domain/identifier is modified. The information sent on change is the domain/name of the item changed. |
| | *Arguments:* Domain, identifier, host/port to send updates |
| | *Returns:* Success/fail |

Table 19-2. Provisioning APIs (continued)

| Command | Description |
|---------|-------------|
| Update | *Description:* This is sent to the clients from the server to indicate that data has changed. The domain, and a list of the relevant items, is sent with the message. |
| | *Arguments:* Domain, list of records that have been updated |
| | *Returns:* NA |
| Trigger | *Description:* This is sent by a third-party server that modifies a record by directly manipulating the database. While this is strongly discouraged, a mechanism is provided to support possible legacy software. This trigger is required to allow updates to be sent and to get XML descriptions. |
| | *Arguments:* Domain, field |
| | *Returns:* NA |

The actual content data is returned in the form of XML. Data is stored in the form of simple value/attribute pairs or value/attribute list pairs. Any specialized data that does not conform to this (tables, special XML structures, and the like) can be retrieved as a string and interpreted by the application as needed, but items or lists of items are as fine as the mechanism is designed to support. It is expected that the library would have overloaded functions to retrieve a number of lists and treat them as though they are a table for ease of access, but this has not changed the fact that the data is simply stored as single elements or lists.

To support redundancy, when one server receives a new data item, a modification, a delete, or a registration, all the servers must be notified. We thought of two possible ways to do this. One was having the server listen on a separate port for synchronization requests (these would have been treated differently than ordinary messages, to ensure that several updates were not sent to clients registered). A second possibility was to have a synchronize equivalent of the previous messages—a *SYNCPUT*, for example. In this case, the server could use the same port for both types of messages. We opted for the second option.

# Provisioning Server

The Provisioning server stores all the information needed by the different components of the system and is responsible for maintaining the information needed to handle callbacks, redundancy, and data manipulation, such as searches, sorts, and batching of information for sending to the GUIs. The server is designed in a highly modular fashion—each module of the server can be removed and replaced with a different component to allow the end user maximum flexibility—and the user can simply define a new module and use that in place of the modules in place. This is accomplished via a compile-time mechanism.

# Architecture

The flow of data within the Provisioning server, as well as the basic structure of the server, is illustrated in Figure 19-2. This diagram shows the way that data in the system flows (not always the same as the call flow).

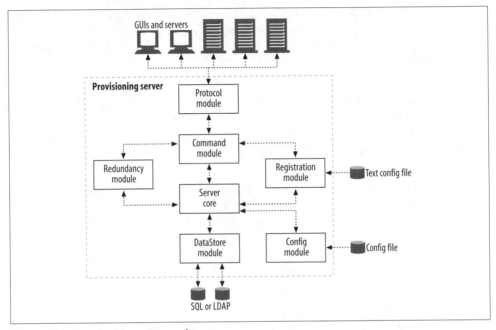

*Figure 19-2. Provisioning server architecture*

The server is composed of a number of pluggable modules, which could be overloaded and defined by the user, as well as a core server engine that drives these components. The component organization is shown in detail in the class diagram following these descriptions:

*Transport/Protocol module*
> Responsible for communication with the outside world. This module is responsible for socket management, as well as transport-level protocol. In our implementation, we have derived an HTTP protocol module from this. The HTTP protocol module is responsible for parsing the HTTP messages, ensuring the message is the proper length, attaching headers to outbound HTTP messages, and so on. This could be changed if the user wished to use some other mechanism, such as SNMP, for sending information.

*Command module*
> Responsible for interpreting the commands sent to the server, after they have been read by the transport/protocol module. This is where the XML commands defined in the architecture section are translated and calls made into the server

core to handle these commands. This could be replaced if the user wished to convert the server to speak some other command dialect.

*DataStore module*

Modular layer to handle different backend databases. Essentially, one can switch between different databases (we intend to offer SQL and LDAP backends) without having to modify the higher-level server components or functions. The top-level API is defined in this document, and each implementation of the module needs to provide implementations (or return errors) for each of the member functions defined in the API.

*Configuration module*

Reads information from a configuration location—in our implementation an XML configuration file that defines basic parameters needed to run the server, such as the location of the data, configuration for the other modules, and so on. At startup, the core queries this module.

*Registration module*

Responsible for keeping a list of who needs to be called back when a given item changes in the database. When the command module informs the core that a file has changed after the database update, this module is called to see what sort of callbacks are needed. If callbacks are needed, the derived modules call the transport module to send an update to the appropriate server or call the local action module to trigger an event. One could imagine redefining this module to call using its own protocol if needed or to trigger some other type of event.

*Redundancy module*

Calls the transport module to send an update to redundant servers to keep them synchronized. This module also is responsible for synchronizing files and communicating with the other server, again using the transport module to actually format and transmit the messages (although this could be changed in a derived implementation). At the time of this writing, this module has not been completed.

The class diagram for the Provisioning server is shown in Figure 19-3.

## Redundancy

The design for the Provisioning server provides for redundancy and fault tolerance through a mirroring mechanism. The system is designed to allow synchronization of two servers. The design is such that the servers are identical—there is no notion of one being a slave or master except during the actual act of synchronizing, and even this role is randomly determined at synchronization time. This design is intended to allow the system to be load balancing as well as redundant. The servers use a heartbeat mechanism to keep track of whether the other servers are up and running.

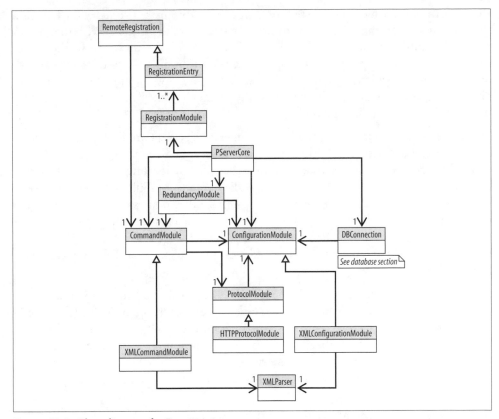

*Figure 19-3. Class diagram for Provisioning server*

When the servers are started or when they lose connection with another server, they use their configuration information to determine what other servers are running and where they are located. When the servers determine another server they are supposed to synchronize with is up, they attempt to synchronize. When a server is aware of another server that it should be synchronizing with, but is not, it sets a random timer. When this timer expires, it tries to synchronize. The other server does the same thing—the situation is completely symmetric. When the timer on one of the servers expires, it attempts to initiate synchronization. This is achieved by sending a message to the second server initiating a synchronize request. The server that initiates this request becomes the *master* and the other server the *slave*—although only for the duration of the synchronization, not for an extended period. If the servers lose connection with each other and synchronize again later, this role could be reversed; it depends strictly on which timer expires first.

After transmitting the synchronization initiate message, the master generates a list of the information stored in its database, along with the time that each item was last modified. The master asks the slave for a similar listing, which is transmitted over the wire to the master. The master compares these two lists. If one server has data the other does not, the data is sent to or requested from the other server (the master makes the calls). Similarly, for shared files, if one server has a newer copy, the newer copy is propagated. Finally, the listing of clients with monitors posted is exchanged, and each server uses the union of these listings as the new monitor list. While synchronizing, any changes to the database can trigger updates to ensure that clients have the most current data possible.

As soon as synchronization starts, and as long as the servers are in contact, the servers reflect messages to each other. If data is added or deleted, the request is forwarded to the second server to ensure that they stay synchronized. Only the server that actually receives the message from the client sends a monitor update, however, to prevent a storm of update messages. If a client sends a request to be updated when data changes, it relays that to the other server, which then adds that monitor request to its list. By having both servers store all data but only the one receiving the change send an update, load balancing is achieved—different clients can communicate with different servers.

If a server in the redundancy group goes down, the other servers can continue to operate but need to synchronize again when they see each other again. Clients can switch freely between servers and not worry about the synchronization: if one fails, then it can simply use the other. The library provided for clients to use handles this transparently for a client. The server stores and transmits a list of servers with which it is synchronizing, or the client can simply use a predefined list of servers to use.

## Provisioning Server Modular Database API

The Provisioning server supports a modular database API. This feature allows the system administrator to decide whether to integrate with the mySQL or LDAP backends.

The customer may choose to set up a new database or integrate with an existing SQL or LDAP database. If the customer chooses to use an existing mySQL or LDAP database, he may do so provided he adheres to the following three requirements:

- When using an existing SQL or LDAP database, the customer must add the database structure provided with the Provisioning server. The database structure can be defined as the LDAP schema or SQL table space.
- The customer must create XML description files for his data.
- Changes made directly to the provisioning database schema must be coupled with a callback to the Provisioning server to make corresponding server-side calls.

The fundamental requirement for the pluggable database APIs is that each API presents the same exact interface up to the PS. This prevents any code changes directly to the PS to accommodate a particular database.

## Design

The PS parses a configuration file at runtime that specifies the following regarding the backend database:

*DB_type*
    mySQL or LDAP

*mySQL_host*
    Host that the mySQL database runs on

*mySQL_port*
    Port for the mySQL database

*mySQL_db*
    Name of the mySQL database

*mySQL_user*
    Username for the mySQL database

*mySQ_pw*
    Password for the mySQL database

The *ConnectionManager* object reads the information in this configuration file and returns the corresponding *DBConnection* object. The *DBConnection* class is an abstract class with various virtual functions for searching and modifying the database. These functions are implemented in any concrete derived class. Currently, the supported derived classes are a *mySQLConnection*, with the possibility of an *LDAPConnection* coming in the near future.

The derived classes from *DBConnection* are hidden from the *DBConnection* clients using the *abstract factory pattern*. This ensures no code changes need to be made to the client to support additional database backends. In relation to the abstract factory pattern, the abstract factory is the *DBConnection* class, and the abstract product is the *DBResultset* class. A *DBResultset* object encapsulates the results of a database search. Accessor methods of the *DBResultset* allow for the client to iterate through the result records, retrieving one or more records at a time. Figure 19-4 shows a class diagram for the modular database API.

## DBConnection class methods

Table 19-3 describes the *DBConnection* methods.

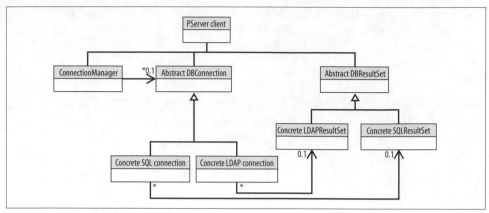

*Figure 19-4. Class diagram for modular database API*

*Table 19-3. Provisioning database API functions in DBConnection class*

| Function | Description |
| --- | --- |
| DBSelect | *Description:* This function is used to search the database. |
| | *Arguments:* |
| | *Result set:* Reference to object that *DBSelect* will create resulting from the database query. |
| | *Domain:* SQL table name or LDAP object class. |
| | *Search filter:* String describing database search. |
| | *Field:* Object describing SQL column name, or LDAP attribute name, and type. |
| | *Order type:* Enumerated type describing order records are returned: *ascending* or *descending*. |
| | *Order field:* String describing field to order by. |
| DBUpdate | *Description:* This function is used to update a field in the database. |
| | *Arguments:* |
| | *Domain:* SQL table name or LDAP object class. |
| | *Search filter:* String describing database search using a pseudo regular expression. |
| | *Result set:* Object created by *DBSelect* that takes the place of *SearchFilter*. |
| | *Field/value set:* Data structure containing map of field/value pairs to update. |
| DBInsert | *Description:* This function is used to insert a record (SQL row, LDAP entry) into an existing SQL table or LDAP object class. |
| | *Arguments:* |
| | *Domain:* SQL table name or LDAP object class. |
| | *Field/value set:* Data structure containing map of field/value pairs to insert into record. |
| DBDelete | *Description:* This function is used to delete one or more records (SQL row, LDAP entry) from an existing SQL table or LDAP object class. |
| | *Arguments:* |
| | *Domain:* SQL table name or LDAP object class. |
| | *Search filter:* String describing database search using a pseudo regular expression. |
| | *Result set:* Object created by *DBSelect* that takes the place of *SearchFilter*. |

| Function | Description |
|---|---|
| DBCreate | *Description:* This function is used to create a new SQL table or LDAP object class. |
| | *Arguments:* |
| | *Domain:* SQL table name or LDAP object class. |
| | *List of fields:* Data structure containing all SQL column names, or LDAP attributes, and their respective types. |
| | *Primary key:* String containing SQL table primary key or LDAP CN (common name). |
| | *Existence check:* Boolean value passed to either ignore the current existence of a table to be created, or to check existence of a table to be created. *CHECKEXISTENCE* is true, *IGNOREEXISTENCE* is false. |
| DBAlter | *Description:* This function is used to alter the identity of a SQL table or LDAP object class. |
| | *Arguments:* |
| | *Domain:* SQL table name or LDAP object class. |
| | *Field:* String containing SQL column name or LDAP attribute name of field to be removed or added. |
| | *New field type:* Data type of new field. This argument is overloaded. If it is passed, *DBAlter* adds a field to the domain. If this field is not passed, *DBAlter* removes the field from the domain. |
| DBDrop | *Description:* This function is used to delete an existing SQL table or LDAP object class. |
| | *Arguments:* |
| | *Domain:* SQL table name or LDAP object class. |
| | *Existence check:* Boolean value passed to either ignore the current existence of a table to be dropped or check the existence of a table to be dropped. *CHECKEXISTENCE* is true, *IGNOREEXISTENCE* is false. |
| DBVerify | *Description:* This function is used to verify whether a SQL table or LDAP object class exists. |
| | *Arguments:* |
| | *Domain:* SQL table name or LDAP object class. |
| | This is the only function in *DBConnection* that returns a value. The value it returns is Boolean true if the domain exists, false otherwise. |
| DBDescribe | *Description:* This function is used to describe information about a given domain or a given database. |
| | *Argument:* |
| | *Domain:* SQL table name or LDAP object class |
| | The return value of this function is not a resultset; it is a table or a record, depending on whether an argument was passed. If a domain is passed as an argument, a three-column table is returned containing *fieldname, fieldtype, indextype*. If zero arguments are passed, a record is returned containing a list of all the domains in the database. |

## DBResultset class methods

Table 19-4 describes the *DBResultset* class methods.

*Table 19-4. Provisioning database API functions in DBResultset class*

| Function | Description |
|---|---|
| GetRecords | *Description:* This function is used to retrieve a group or all of the records in a resultset depending on the arguments passed. |
| | *Arguments:* |
| | *Offset1:* Offset of first record to be retrieved. Offset is an integer much like an array offset. |
| | *Offset2:* Offset of last record to be retrieved. For example, *DBResultset::GetRecords(0,25)* would retrieve the first 25 records from the resultset. *GetRecords( )* also takes 0 arguments, in which case it would return all the records in the resultset. |
| operator[offset] | *Description:* This function is used to retrieve a single record at position [offset]. |
| | *Argument:* |
| | *Offset:* Offset of record to be retrieved. |
| GetFields | *Description:* This function is used to retrieve a list of the fields in a resultset. Normally, the client passes the field list as an argument to *DBSelect( )*, in which case the same client would doubtfully ask[a] for it back. However, the client may pass * as the field list to *DBSelect( )*, in which case the client would use this function to retrieve the list of all the fields in the given domain. |
| | *Arguments:* None. |
| GetRecordCount | *Description:* Return the number of records in the resultset; type is unsigned integer. |
| | *Arguments:* None. |
| GetSearchFilter | *Description:* Return the *SearchFilter* string that was passed to *DBSelect( )* to obtain this resultset. |
| | *Arguments:* None. |

[a] "Doubtfully ask" means that one usually calls *DBSelect* with a list of fields as an argument, and in that case, one would not likely call to ask what fields the *DBResultset* contains, since these fields are specified exactly in the *select* command (recall that *select* returns a *result*). However, if *DBSelect* is called with a wildcard '*', which means return everything, the one who called for this data may want to know what fields were returned, and "doubtfully ask" is the method used to obtain that information.

### Search filters

The *DBSelect*, *DBUpdate*, and *DBDelete* functions accept a search filter as an argument. The search filter is expressed with the follow grammar:

```
<filter> = '(' <filtercomposition> ')'
<filtercomp> = <and> | <or> | <not> | <item>
<and> = '&' <filterlist>
<or> = '|' <filterlist>
<not> = '!' <filter>
<filterlist> = <filter> | <filter> <filterlist>
<item> = <simple> | <substring>
<simple> = <field> <filtertype> `"' <value> `"'
<substring> = <field> '=' `"' <initial> <any> <final> `"'
<filtertype> = '=' | '>=' | <=
<initial> = NULL | <value>
<any> = '*' <starval>
<starval> = NULL | <value> '*' <starval>
<final> = NULL | <value>
```

Here are some examples of search filters (whitespace outside of value strings is ignored):

```
"(&(phonumber="925*")(voicemailserver="vm1"))"
"(&(!(vocalcfb = "1234") ) ( voicemailserver="vm2")(vocalfna=4321))
"(vocalusertimestamp >= "20010509221829").
```

# Provisioning Interface Libraries

Two sets of libraries are available for client applications that require *PServer* support. One set of libraries, *PSLib*, works with C++, while the Comm package is used with the Java platform.

## PSLib

The PSLib is a C++ library, which acts as an interface for client software to communicate with the Provisioning server and vice versa. It is responsible for providing message transport (HTTP-encapsulated XML data content) to accomplish the following:

- Request Provisioning server to perform data retrieval (search/sort).

- Send update (creation/replacement/deletion) requests for data stored in the Provisioning server. An application can insert a new database record, update the values of a set of selected fields within a domain, or update the values of a selected set of fields confined by a database search result.

- Monitor data changes during runtime by providing client registration and notification of data updates. An application can request the Provisioning server to notify it whenever some data in a domain or a specific database record is changed. The application provides a callback function to *PSLib*. When changes occur in the domain or record of interest, the Provisioning server sends a message to *PSLib*, which invokes the registered callback method of the client.

- Parse responses from queries sent to the Provisioning server and cache the XML content in a searchable structure.

- Provide a set of simple, generic APIs for clients to access and manipulate individual pieces of data based on tag/field names.

In future releases, it also provides seamless switching to a redundant Provisioning server in the event that the primary Provisioning server cannot be reached.

This library no longer embeds detailed customized knowledge about individual sets of data. Instead, applications that employ this library need to know the specific tag names/fields. This approach allows the *PSLib* to remain unchanged as the database content evolves.

## PSLib components

Figure 19-5 shows a diagram of the classes found within the *PSLib*.

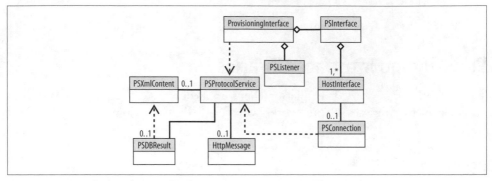

*Figure 19-5. PSLib class diagram*

The classes found within the *PSLib* are as follows:

*ProvisioningInterface*

Singleton object[*] into which clients can manipulate (get, put, delete, search, and sort) information to and from the Provisioning server. Through this object, clients can request *PServer* to monitor and send notification if changes occur on information of interest. *ProvisioningInterface* contains *PSInterface* and *PSListener*. It also maintains a list of callback functions that the clients submit.

*PSInterface*

Contains two *HostInterface* objects for redundancy. It maintains a reference to the primary *HostInterface* that communicates with the primary *PServer*. In the event a disconnection occurs, *PSInterface* swaps the active server to point to one of the secondary *HostInterfaces*.

*HostInterface*

A wrapper for the *PSConnection*.

*PSConnection*

Responsible for sending requests and receiving responses to and from *PServer* via a *TCPClientSocket* object.

*PSListener*

Runs on a separate thread and listens for database notifications from *PServer*.

*PSProtocolService*

A helper class that offers message building and parsing services to *PSTransaction* objects. It hides the implementation details of the message protocol from other components, thus minimizing the scope of changes if *PServer* and *PSLib* adopt a different communication protocol. It drives two other helper classes,

---

[*] *Singleton object* is a C++ design pattern signifying that there is only one copy of this object in the entire system.

*HttpMessage* and *PSXmlContent*, to actually compose and parse the requests and responses between *PSLib* and the Provisioning server. It instantiates a *PSDBResult* object to store the database response.

*HttpMessage*
Responsible for building and decoding HTTP-formatted messages.

*PSXmlContent*
Responsible for composing and parsing the database response in the form of an XML document.

*PSDBResult*
Holds the parsed XML content of a database response from the Provisioning server into an array of records (field/value pairs).

### Client/PSLib call flows

*ProvisioningInterface* offers a single point of contact for client applications to access information from the Provisioning server.

A client invokes one of the API methods in *ProvisioningInterface* to make a database query. *ProvisioningInterface* employs a *PSProtocolService* object to formulate the database query into an HTTP-encapsulated XML content message and sends the request via a *PSConnection* object. *PSProtocolService* is also in charge of deciphering the response sent back from *PServer* and populating the database result as a table of records in a *PSDBResult* object.

When a client requires multitransaction operations like large-scale searches and sorts, it opens a session with the *PServer* via *ProvisioningInterface*. A session ID is returned as a result. The session ID helps identify subsequent queries belonging to that session so that they share the same context. *PSLib* does not maintain a list of outstanding session IDs. Only the client and the *PServer* maintain the session ID. The client can specify a timer (also maintained by *PServer*) to ensure that the session does not become stale.

Figure 19-6 shows the query and response interaction that takes place within the *PSLib*.

## PSLib API

The *PSLib* API is a combination of *ProvisioningInterface* and *PSDBResult* class functions.

### ProvisioningInterface class functions

The client software can invoke the methods in Table 19-5 via a *ProvisioningInterface* object to do single database lookups, updates, and requests for data updates.

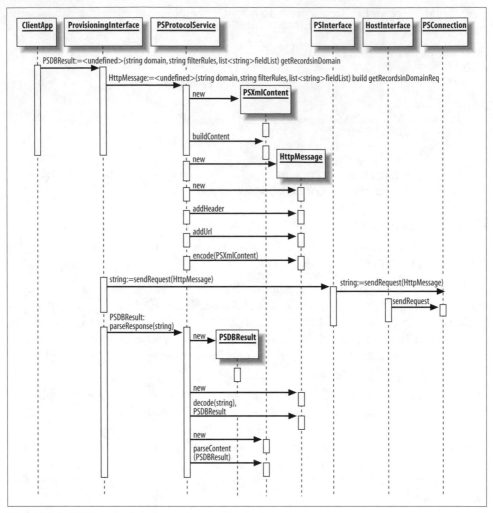

*Figure 19-6. PSLib query/response sequence diagram*

*Table 19-5. Transaction-based functions*

| Function | Description |
| --- | --- |
| getRecordsInDomain | *Input:* String domain, string *filterRules*, list*<string> fieldList* |
| | *Update: PSDBResult* |
| | Object given a field (or list of fields) within a domain, e.g. *Accounts*, and a set of constraints, *filterRules*, send a request to *PServer* to retrieve all the matching records. |
| getFieldsInDomain | *Input:* String domain |
| | *Update: list<string> fieldNames* |
| | Given a domain name, get the field names (corresponding to the column names of, say, a SQL record). |

*Table 19-5. Transaction-based functions (continued)*

| Function | Description |
|---|---|
| setRecordsinDomain | *Input:* String domain, *map<string, string> fieldValuePairs* |
| | Given a list of fields within a record of a domain, for each matching field, replace its value with that in the *fieldValuePairs*. |
| addRecordInDomain | *Input:* string domain, *map<string, string> fieldValuePairs* |
| | Add a new record (using the values to populate the fields in the record) to the given domain. |
| deleteRecordsInDomain | *Input:* String domain, string *filterRules* |
| | Request *PServer* to remove *all* records for a domain given a set of constraints. |
| registerDomainForUpdate | *Input:* String domain, *callBackFunc* |
| | Instruct *PServer* that the client is interested in monitoring any data change within a domain. *ProvisioningInterface* is to listen for update notifications and update the client by invoking the callback function. |
| deRegisterDomainFor-Update | *Input:* String domain, *callBackFunc* |
| | Instruct *PServer* to stop monitoring data changes as requested by *registerDomainForUpdate( )*. |
| registerRecordForUpdate | *Input:* String domain, string *filterRule*, *callBackFunc* |
| | Similar to *RegisterDomainForUpdate*, except the client is interested in monitoring a specific record within a domain. |
| deRegisterRecordFor-Update | *Input:* String domain, string *filterRule*, *callBackFunc* |
| | Instruct *PServer* not to monitor data changes as requested by *registerRecordForUpdate( )*. |
| getCurrentRecordsInDo-main | *Input:* String domain, string *filterRules* |
| | *Update: DBResult* object similar to *getRecordsInDomain( )*, except it caches a snapshot of the data locally for the client and automatically monitors and updates the data. The first invocation of this function sends a request to *PServer* to look up the value(s) associated with the field. Subsequent calls to this function result in accessing the resultant data locally instead of doing a "true" *get* via *PServer*. |

For more complicated database manipulation that may involve multiple query transactions, such as search and sort, *ProvisioningInterface* needs to set up a query "session" containing a set of constraints. Once a session context is established, the client can just refer to the session identifier to do subsequent queries.

Table 19-6 explains the session-based functions.

*Table 19-6. Session-based functions*

| Function | Description |
|---|---|
| sessionOpen | *Input: int timeInSeconds* |
| | *Update:* long *sessionId* |
| | Initiate a query session with Provisioning server that returns a unique identifier, session ID, that subsequent request(s) can refer to. |
| sessionSearchDB | *Input:* Long *sessionId*, string *filterRules*, *list<string> fieldList, sortType*, string *sortField* |
| | *Update:* int *numRecords* |
| | Query *PServer* to look up all records satisfying the supplied constraints and sorting criteria. This method returns the number of matching records. |

*Table 19-6. Session-based functions (continued)*

| Function | Description |
|---|---|
| sessionGetDBResult | *Input:* Long *sessionId*, int *startIdx*, int *endIdx*, list<*string*> *fieldList* |
| | *Update:* PSDBResult object |
| | Query *PServer* to look up a range of records that matches the supplied list of fields based on the result of the previous *sessionSearchDB( )*. |
| sessionSetDBResult | *Input:* Long *sessionId*, map<*string, string*> *fieldValPairs* |
| | Apply the set of field/value pairs to the search result associated with the sessionID, and update the database accordingly. |
| sessionSortDBResult | *Input:* Long *sessionId*, *sortType*, string <*sortField*> |
| | *Update:* PSDBResult object |
| | Query *PServer* to sort the result from *sessionSearchDb( )* by the supplied field. |
| sessionDeleteAll | *Input:* Long *sessionId* |
| | Remove all items found in *searchDB( )*. |
| closeSession | *Input:* Long *sessionId* |
| | Terminate the query session with *Pserver*. |

## PSDBResult class functions

Since information returned by the *PServer* is a string (of XML documents), *PSLib* provides some basic atomic data retrieval and conversion methods via the *PSDBResult* class (Table 19-7).

*Table 19-7. Data entity accessor functions*

| Function | Description |
|---|---|
| getFieldValue | *Input:* String field |
| | *Update:* string value |
| | Get the value as a string object associated with the supplied field. |
| getFieldValueAsBool | *Input:* String field |
| | *Update:* bool *boolVal* |
| | Convert the field's value to a Boolean, true or false. |
| getFieldValueAsInt | *Input:* String field |
| | *Update:* int *intVal* |
| | Convert the field's value to an integer value. |
| getFieldValueAsLong | *Input:* String field |
| | *Update:* Long *longVal* |
| | Convert the field's value to a long integer. |
| getFieldValueAsDouble | *Input:* String field |
| | *Update:* int *doubleVal* |
| | Convert the field's value to a double integer. |

Table 19-7. Data entity accessor functions (continued)

| Function | Description |
|---|---|
| getFieldValueAsList | *Input:* String field |
| | *Update: list<string> listResult* |
| | Convert the content associated with the field name into a list of strings. |

# Java Comm Package

The Comm package is the Java equivalent of *PSLib*: it provides the communication interface between the Provisioning GUI and the Provisioning server. The intent of the Comm package is to shadow the functionalities of *PSLib*. The API is identical between the two packages, but the internal implementations differ slightly, as shown in this section.

## Comm package components

Figure 19-7 shows a diagram of the classes found within the Java Comm package.

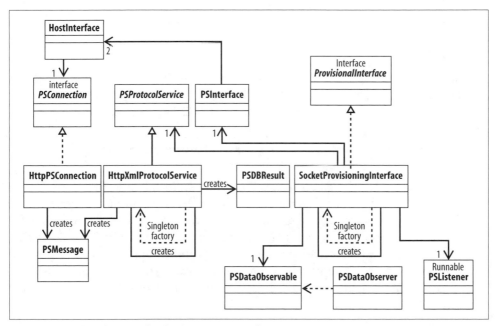

*Figure 19-7. Class diagram for the Java Comm package*

*ProvisioningInterface*

An interface that defines the methods for clients to manipulate (*get, put, delete, search,* and *sort*) information to and from the Provisioning server. Also defines the methods for clients to request *PServer* to monitor and send notification if changes occur on information of interest. Having *ProvisioningInterface* as an interface allows clients to conform to an API that is completely independent of the underlying communication protocol and also makes it easier to plug in a new communication protocol.

*SocketProvisioningInterface*

Singleton implementation of the interface *ProvisioningInterface*. It handles communication if the message is sent via basic socket communication. Contains *PSInterface* and *PSListener*. It also maintains a list of callback functions that the clients submit.

*PSInterface*

Contains two *HostInterface* objects for redundancy. It maintains a reference to the primary *HostInterface*, which communicates with the primary *PServer*. In the event a disconnection occurs, *PSInterface* swaps the active server to point to one of the secondary *HostInterfaces*.

*HostInterface*

A wrapper for *PSConnection*.

*PSConnection*

An interface defining the methods for sending requests and receiving responses to and from *PServer*. It is defined as an interface so specific implementations of the connection object can be made (for example, TCP, UDP).

*HttpPSConnection*

An implementation of *PSConnection* that handles sending and receiving HTTP TCP messages. The handling of HTTP messages is slightly different from normal TCP communication because of persistent HTTP connections.

*PSListener*

Runs on a separate thread and listens for database notifications from *PServer*.

*PSDataObservable*

When the client application requests *ProvisioningInterface* to be informed of a specific database update, it passes a *PSDataObserver* object for registration. *ProvisioningInterface* then instantiates a *PSDataObservable* object to watch for data changes and attaches the *PSDataObserver* object to the *PSDataObservable*. When *PSListener* receives change notification from *PServer*, *ProvisioningInterface* passes it to its list of *PSDataObservable* objects. If the notification applies to a *PSDataObservable*, it invokes all the *PSDataObserver*'s *update()* methods.

*PSDataObserver*

Any client that is interested in receiving database update notification implements an *Observer* interface and requests *ProvisioningInterface* to attach this object to the subscriber list of the *PSDataObservable* object.

*PSProtocolService*

Defines the methods for message building and parsing services. It is meant to be used to hide the implementation details of the message protocol from other components, thus minimizing the scope of changes if *PServer* and *PSLib* adopt a different communication protocol. It returns a *PSDBResult* object to store the database response.

*HttpXmlProtocolService*
An implementation of *PSProtocolService* that provides building and parsing services for HTTP/XML messages.

*PSDBResult*
Holds the parsed XML content of a database response from the Provisioning server into an array of records (field/value pairs).

*PSMessage*
A container for either outgoing or incoming messages. Consists of a header and body.

Figure 19-8 shows the query and response interaction that takes place within the Java Comm package.

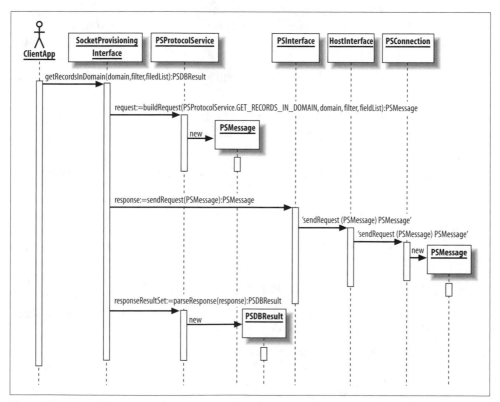

*Figure 19-8. Java Comm query/response sequence diagram*

# Java User Interface

The Java user interface is multifaceted, including a dynamic GUI and methods for maintaining and provisioning data.

## Dynamic GUI from Text Description

Most of the GUI consists of screens that are not hardcoded, but are instead generated from XML files that describe the fields to be presented to the end user. This description includes any labels, tooltips, or other descriptive markers associated with the fields; conditions under which the fields are enabled/disabled; sources of data for the fields; and the rules for validating data entered into these fields. We call these *dynamic GUI screens*. This design was intended to make the GUI very easy to change.

Dynamic GUI screens are used for server and endpoint provisioning because the provisioning needs of servers and endpoints often change. They are also used for feature provisioning of users. One important goal of this product has been to make new features easy to write and provision. Using dynamic GUI screens for features solves the provisioning half of that problem. In order to write the description for a new server or feature screen, a developer begins by writing an XML file similar to the one shown later.

The XML file is parsed by both *PServer*, to build up a database definition, and by the GUI software, to build up a screen. *PServer* looks for the following tags:

*DataDefn*
> Gives a name and optional category (e.g., *FeatureServer*) for this data type.

*Key*
> Indicates a key field for this data type.

*DBEntry*
> Indicates a field for this data type. Attributes of *DBEntry* give the field name and the data type.

All key fields must be repeated as *DBEntry* fields.

All other tags are ignored by *PServer* and used only by the GUI to define the widgets and their relationships to one another on the screen.

Some of the tags that the GUI software looks for include:

*DataDefn*
> Gives a name to the screen.

*DBEntry*
> Understands that a widget is needed to accept or display the data. The type attribute is used to validate data entered into the widget, and the widget

attribute is used to specify the type of widget that must be used. If the widget attribute is not present, the software makes the best possible choice.

*Default*

Provides a default value for the data.

*Label*

Gives the widget a user-friendly label. The name attribute of *DBEntry* is not used for the widget label if this tag is not present.

*Tooltip*

Gives pop-up instructions for filling in the data.

*Table*

Provides the additional instructions for building a data entry table.

*Content*

Provides a source of choices for the widget. These may take the form of a hard-coded list of values or a *PServer* query.

The following is a sample data definition file:

```
<DataDefn name="sample">
 <Key>Host</Key>
 <Key>Port></Key>
 <DBEntry name="Host"/>
 <DBEntry name="Port"/>
 <DBEntry name="NoResponseTime" type="Integer">
  <Label>No Response Time (ms) </Label>
  <Default>1200</Default>
 </DBEntry>
 <DBEntry name="AllowUnbillableCalls" widget="ComboBox">
  <Label>Allow Unbillable Calls</Label>
  <Content>
   <Enumeration>
    <Entry>Yes</Entry>
    <Entry>No</Entry>
   </Enumeration>
  </Content>
 </DBEntry>
</DataDefn>.
```

At the end of this chapter is a Document Type Definition (DTD), which can be plugged into most XML editors to assist in the writing of the file. The developer may then test his file by starting an application that reads the file and builds from it a GUI screen. Once the developers are satisfied that the GUI screen is to their liking, they can then add this new file to a repository and the new screen becomes available in the Provisioning GUI. Figure 19-9 shows the screen that would be generated from the earlier XML description.

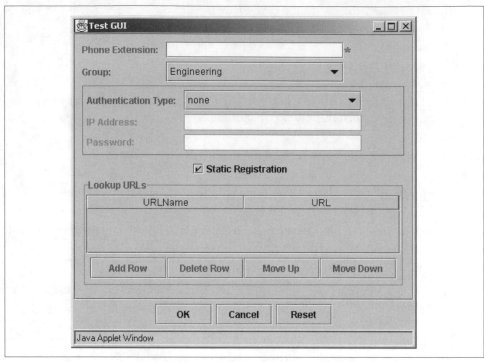

*Figure 19-9. GUI screen generated from data definition*

## Architecture

The main classes used for building dynamic GUIs are *ProvisionableType*, *ProvisionableInstance*, *ContentData*, and *GuiComponent*. Each XML description file is parsed once to create a *ProvisionableType* object, which stores the instructions from the XML description in a memory model. *ProvisionableType* is hashed on the unique name of the XML description and may be retrieved by any object in the GUI. *ProvisionableType* may be queried for the set of data fields it contains and may also be requested to create one or more *ProvisionableInstance*s. A *ProvisionableInstance* object manages the storage, collection, and display of a single instance of the data fields of its *ProvisionableType*. Each data field in the *ProvisionableInstance* has an associated *ContentData* object, which encapsulates the rules for getting, setting, and validating the data in that field. Each *ContentData* object has an associated *GuiComponent* object, which causes the data field to be rendered in a GUI screen.

Figure 19-10 is a diagram of the major classes used for this GUI.

To put this in perspective, *ProvisionableType* is a description of a particular type of server. This server requires certain provisioning data, such as hostname and port number, for example, which is described in the XML description file. There may be many instances of this server process running in the system, each of them with the

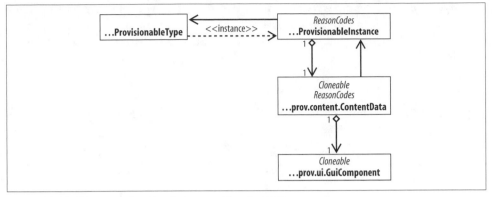

Figure 19-10. Diagram of major classes used for the GUI screen

same data fields but with different data in the fields. The provisioning information for each server instance is gathered by a different *ProvisionableInstance*.

## GuiComponent

Figure 19-11 is a diagram of the classes derived from the *GuiComponent* class.

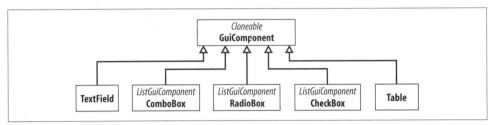

Figure 19-11. Classes derived from GuiComponent

Classes derived from *GuiComponent* control the rendering of a data field and its associated labels and tooltips, for example. The type of class selected for displaying a data field depends on the *widget* attribute of that data field (*DBType*) in the XML description. The default widget type is *TextField*. The code is extensible so that a new type of widget can be added. In addition to writing a new class for the new type of widget, you would have to extend the XML description language to allow the name of this new type to be a value for the *widget* attribute.

The *GuiComponent* holds a reference to the *Swing* class that best represents that type of widget. In cases in which a label would make sense, it also holds a reference to a *Swing JLabel*. *TextField* and *ComboBox*, for example, both hold references to a *JLabel* since these *Swing* components have no internal labels. *CheckBox*, however, includes a label in the *Swing* component, so it does not have a separately rendered *JLabel*.

*GuiComponents* are rendered inside a container that takes care of the layout details. The *SmartLayoutManager* class is an example of a layout manager that takes care of the layout details. *SmartLayoutManager* identifies the type of each widget being added to its container and applies an internal algorithm for placing that widget relative to the others. This allows the XML description file to remain free of layout details.

## ContentData

Figure 19-12 is a diagram of the classes derived from the *ContentData* class.

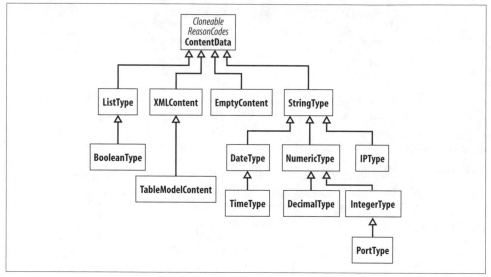

*Figure 19-12. Classes derived from ContentData*

Classes derived from *ContentData* store and validate the data for a particular data field. To some extent, *ContentData*s may be mixed and matched with *GuiComponent*s. For example, a *BooleanType* may be associated with a *CheckBox* that is true if checked, or with a *ComboBox* that has the choices True and False. In many cases, however, there is only one pairing of a *ContentData* with a *GuiComponent* that makes sense. A *ListType* that holds values other than True and False can be visually represented only as a *ComboBox*. The type of *ContentData* associated with a data field is based on the *type* attribute of the *DBEntry*. If no type is specified, *StringType* is the default. The code is extensible to allow new types of *ContentData* to be added. In addition to writing the new class, you would have to extend the XML description language to allow the new *ContentData* type to be a value for the *type* attribute of *DBEntry*.

Whenever data is saved to a *ContentData*, it is first validated by passing it to the validate method of that *ContentData* class. The validation performed by a particular class is a combination of the rules for that class and any additional rules imposed by the XML description. For example, a numeric type must contain only numeric data, but the XML description may additionally specify a valid range for this data. If the data does not pass as valid, a pop-up box alerts the user. Each *ContentData* class contains additional information for the pop-up box, such as a "hint" string that gives the type of allowed input and the range, if any.

## Getting and setting data

The data of a *ProvisionableInstance* can be set either through an entry made to the *GuiComponent* widget or by reading stored values directly from the database. When the data is set through the widget, the updates to the data model in the *ContentData* are handled through the usual Swing mechanism. When the data is set from the database, the data model in the *ContentData* is changed directly through the method *setContent*. Figure 19-13 is a sequence diagram showing the *setContent* method being called on *StringType*.

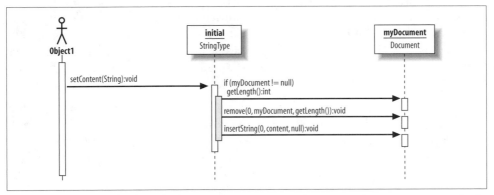

*Figure 19-13. Setting the data of a StringType*

The data model for *StringType* is a document, which is the normal data model for a *JTextField*.

Data coming from the database comes as an XML string containing values for all fields of a particular *ProvisionableInstance* (or GUI screen). The *ProvisionableInstance* class uses an XML parser with Simple API for XML (SAX, *http://sourceforge.net/projects/sax/*) callbacks to parse the XML string and call *setContent* on each *ContentData* as its name is encountered. Here is the XML string that would be used to set the fields of the screen shown in Figure 19-9:

```
<fieldValues>-
<Host>5000</Host>
<Port>1000</Port>
<NoResponseTime>1200</NoResponseTime>
<AllowUnbillableCalls>Yes</AllowUnbillableCalls>
</fieldValues>.
```

Each of the tags nested inside *fieldValues* has the name of a data field that is defined in the original XML description file as a *DBEntry*. As this XML string is parsed, each tag encountered triggers a SAX callback. If the name of the tag matches the name of a *ContentData*, the value of that *ContentData* is set to the content of the tag. For example, the first tag to be encountered would be *Host*. The *ContentData* with that name would have its value set to 5000, which is the content of the *Host* tag.

The same protocol is used for getting data from a *ProvisionableInstance*. When *getData* is called on a *ProvisionableInstance*, each of the *ContentData*s is queried for its content. This data is then printed into a string that looks like the XML string shown before.

## Tabular Display of Data

Several cases in the provisioning user interface call for the display of data in a tabular format. Examples of this include the table of all user accounts seen by the Centrex administrator or tables of various server types seen by the technician configuring the system. To this end, a set of classes has been developed to allow the display of data retrieved from the Provisioning server in a tabular format that may be reused by the various parts of the user interface. These classes also provide a frontend that allows the data displayed in the table to be sorted or searched, relying on the Provisioning server to actually perform the operations.

These classes accommodate large sets of data (up to a million entries) by including paging functionality, which loads data from the server only if it is currently visible on the screen. As the user scrolls or resizes the screen containing the table, new data is retrieved from the server. To facilitate this, a set of threading classes has been designed to perform such tasks in the background.

Figure 19-14 shows the classes involved in displaying a table of data.

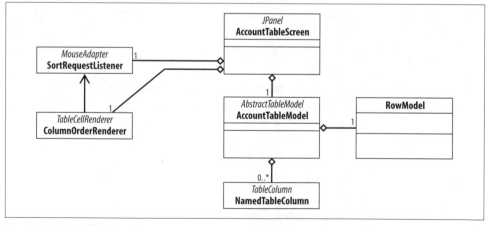

*Figure 19-14. Table class diagram*

The classes used for table functions are as follows:

*AccountTableScreen*

A panel, which contains a scrollable table for displaying data, retrieved from the Provisioning server.

The default way to create a table of data is to pass, as an argument to the constructor, a textual description (the VOCAL XML GUI definition file) of the data it displays. The table then extracts all the Provisioning server fields from the file and creates a column for each field. Alternatively, a list of field names (as recognized by the Provisioning server) may be passed in instead and used to generate columns in the table. Note that the table is not responsible for retrieving its own data from the Provisioning server. Rather, some other class (typically a *Task*; see the next section) retrieves the data and then adds it to the table's model.

*AccountTableModel*

The data model used by the table to get the data it displays. This is the required implementation of the *TableModel* interface. This model does not actually store its own data but, instead, uses a *RowModel*, as described next.

*RowModel*

Maps the order in which the data is displayed in the table to the order in which it is stored. These two may be different because the order in which the data is displayed changes depending on the sort criteria applied to it while the order in which the data is stored does not change after it has been loaded into the model. It is desirable to load data records from the server only once for efficiency.

This class stores a mapping between the row number in the table and an identifier of each data record (the account name) and another mapping between the identifier of the record and the actual data for the record. Whenever the server is requested to perform a sort of the data, the first mapping is invalidated and the newly ordered set of identifiers is retrieved from the server and used to refill the map. These may then be used to index into the second mapping to display the data in whatever the current ordering is without having to reload it from the server.

*SortRequestListener*

An event handler registered on the header of the *JTable*, which displays the data. It detects when a user clicks on a column header and keeps track of which column is currently sorted on as well as the sort order (ascending or descending). It then contacts the Provisioning server to request it to perform the selected sort operation. Finally, it invalidates the current row mapping, loads the new data order from the server, and causes the table to be repainted.

*ColumnOrderRenderer*

A renderer for the header of the table displaying the data. Responsible simply for drawing an icon in the header indicating the sort order.

*NamedTableColumn*
> An extension of the Swing *TableColumn* that stores a column's display name and the identifier of the data, which is displayed in that column.

Figure 19-15 illustrates a table being created.

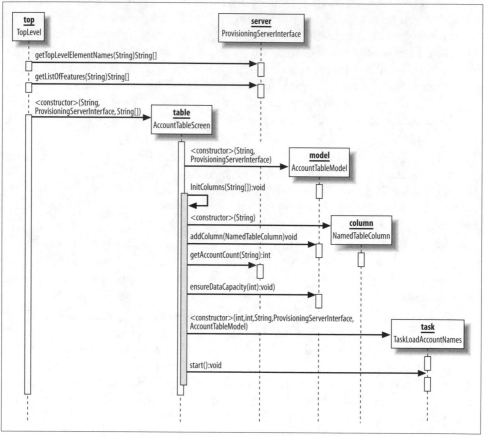

*Figure 19-15. Table being created*

## Tasks

Figure 19-16 shows a class diagram for the tasks.

The task classes are as follows:

*Task/TaskListener*
> A simple implementation of a task, which has its own thread and so may be run in the background. Concrete implementations of the abstract *task* class must override the *doWork()* method to supply the implementation, which is executed by the task.

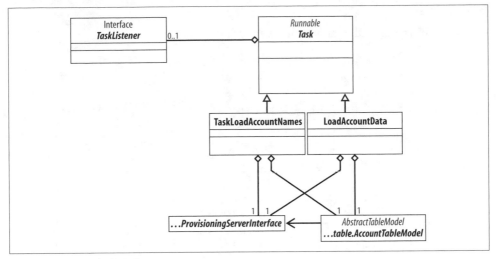

*Figure 19-16. Task class diagram*

An object implementing the *TaskListener* interface may be supplied when the task is created. This object receives a callback whenever the task finishes, either normally or by throwing an exception. It is up to the called object to retrieve the exception (if any) and any return value from *Task*. For simplicity, the return value is of type *Object* and it is up to the called-back object to know the actual type.

*TaskLoadAccountData*

An implementation of *task*, which loads data records from the Provisioning server. This class is used to implement the paging functionality in which only a screenful of data is loaded at a time. This task is typically called from the event handler, which is registered on the data table and listens for scroll or resize events. If it determines that the user scrolled to a portion of the table for which there is no data currently stored in the model, it initiates one of these tasks to load the data.

This task takes a range of rows to load data for as well as the server interface to contact and the table model to place the data in as arguments to its constructor.

*TaskLoadAccountNames*

Another implementation of *task*, which loads only record names from the Provisioning server. This task is used whenever the user has requested to re-sort the data displayed in the table so that the new order of the data may be displayed without having to reload the complete data records from the Provisioning server.

The behavior of this task is analogous to that of *TaskLoadAccountData*.

## Event handlers

Figure 19-17 is a class diagram of the event handlers.

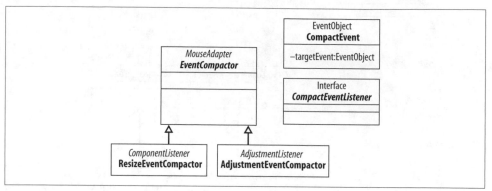

*Figure 19-17. Event-handler class diagram*

The event-handler classes are as follows:

*EventCompactor*

This class compacts events that occur between a *mouse-pressed* and *mouse-released* event when both the mouse events and the other event type are generated by the same class.

To implement the functionality of loading data a screenful at a time, it is necessary to determine when the user has finished scrolling the table of data. However, in the Swing event model, many component-scrolled events are generated while the scrolling is in progress. It is not possible to trigger the loading of data from these events, since we want to load data one time, once the scrolling has finished. Hence *EventCompactor* is used to compact (discard) the multiple events that occur between pressing the mouse down and releasing the button.

This class implements the *MouseListener* interface. Concrete implementations of this class should also implement an event listener interface for some other type of event (for example, an *AdjustmentEvent*). The concrete implementation is then registered as an *xxxEventListener* on some class and also as a *MouseListener* on that same class. When any *xxxEvents* are generated after a mouse-pressed event, they are sent to this class and discarded except for the last one, which occurs before a mouse-released event. The class interested in the compacted event is notified only once, for the last time that event occurred.

*CompactEvent/CompactEventListener*

An extension of the Java event model to include a new *CompactEvent* type.

The event compactor converts the sequence of a mouse-pressed event, numerous *xxxEvents*, and a mouse-released event into a single *CompactEvent*, which it fires at the end of the sequence.

Classes interested in *CompactEvents* should implement the *Compact-EventListener* interface and register with an *EventCompactor*.

*ResizeEventCompactor*

Implements the *ComponentListener* interface and listens for the *Component-Resized* events.

This class compacts (discards) successive *ComponentEvents* generated by resizing a window, which occur between *MousePressed* and *MouseReleased* events.

*AdjustmentEventCompactor*

Implements the *AdjustmentListener* interface and listens for the *Adjustment-ValueChanged* events.

This class compacts (discards) successive *AdjustmentEvents* generated by scrolling a scrollbar, which occur between *MousePressed* and *MouseReleased* events.

Figure 19-18 shows a table of data being scrolled.

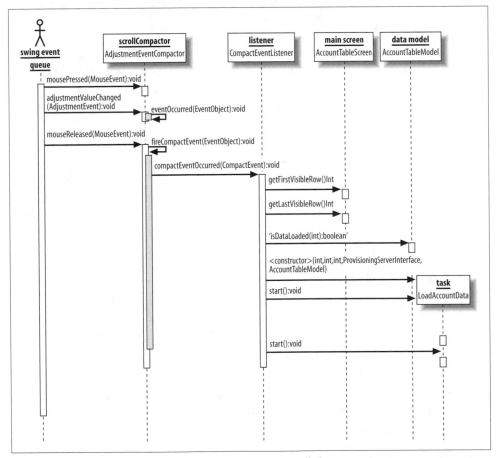

*Figure 19-18. Sequence diagram of a table of data being scrolled*

# Template Builder

The Template Builder is a mechanism to build a set of default values, which then aid in creating database entities. Examples of database entities include users, phones, and servers.

A user of the system has the ability to create templates and then apply a template when adding data to an entity. Basically a template defines default values for the components, or fields, of an entity. Because a template is just a framework, these default values can be manually overridden on a per-entity basis if necessary.

A template consists of the following attributes:

*Name*
> A unique name used to identify this template.

*Entity reference*
> A reference to the database entity this template describes. An example is the table name in the database.

*Focus field*
> A GUI field that receives the first focus of the cursor when adding a new entity. This is to accommodate usability when provisioning many items. For example, if input for this field comes from a bar code scanner that directs its data to *stdin*, the user does not have to position the cursor manually on this field before scanning.

*Confirmation field*
> A GUI field that, when a special control sequence is typed, simulates an OK click. This is to accommodate a bar code scanner that is able to send special unreadable control codes before and/or after reading a bar code. For example, sending the ASCII codes *stx* and *etx* for representing start of text and end of text, respectively.

*Field DataProvider mappings*
> The assignments of a *DataProvider* instance to a field of an entity. *DataProviders* provide the default values for fields in an entity.

## DataProviders

A *DataProvider* is an abstract object that provides access to data. This data could come from several different sources but is accessed in a generic way. Figure 19-19 shows the *DataProvider* classes.

Data from a *DataProvider* is stored as a list of elements that is accessed element by element. A *DataProvider* maintains a cursor that points to the current element.

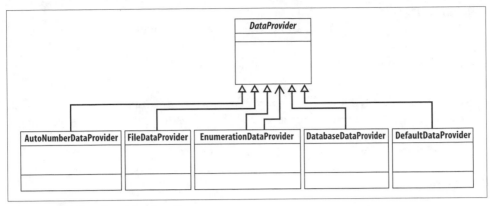

*Figure 19-19. DataProvider class diagram*

Elements can be accessed only sequentially and only in a forward direction. The important methods of *DataProvider* are:

*next( ):boolean*
> Advance the cursor to the next data element. Initially the cursor is positioned before the first element so *next( )*, must be called before the first element in the series can be accessed.

> Returns true if there is at least one more element to provide, false otherwise.

*getData(index:int):Object*
> Get the current data element.

*createUI( ):void*
> Create and return the user interface that is used to input configuration details for the *DataProvider*. For example, when using a database query, the user has to enter the database name, filter, fields, and *orderType*.

> Also, this user interface should have a Test button, which then displays a sample of the data, perhaps just a few elements.

*connect( ):void*
> Connect to the source of the data (acquire resources).

*close( ):void*
> Release the resources acquired by this *DataProvider*.

The implementation of a *DataProvider* can be of several forms:

*AutoNumberDataProvider*
> Consists of sequential numbers between the start and end constraints, inclusive.

*DatabaseDataProvider*
> Consists of the results of a database query. Actual database requests are sent through to the Provisioning server using the communication module of the client. Thus, the Provisioning server makes the actual connection to the database.

*DefaultDataProvider*
> Always returns the same hardcoded value. Can be thought of as an infinite list.

*FileDataProvider*
> Consists of the data read from a newline-delimited file.

*EnumerationDataProvider*
> Returns a list of data for each successive call to *next()* and *getData()*. Can be thought of as an infinite list of a finite list. The list received from *getData()* would most likely be used for a GUI drop-down list. This list can be driven by a set of hardcoded values or another *DataProvider*.

### User interface

Figure 19-20 shows an example of the Template Builder user interface. This user interface allows one to perform many functions:

1. Save the template.
2. Create a new blank template.
3. Create a new template that is a copy of an existing template.
4. Delete the template.
5. Assign a *DataProvider* to a field.

    The user selects a type of *DataProvider* from the drop-down list. Editable parameters pertaining to each type of *DataProvider* are shown below the drop-down list.

6. Assign a focus field.
7. Assign a confirmation field.

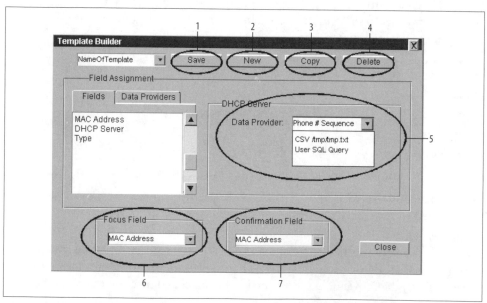

*Figure 19-20. Template Builder interface*

## Dial Plan

Another option the technician may select is to create or edit a dial plan. Dial plans fall into two categories:

*Phone versus IP*
> A phone dial plan is used to process PSTN-like telephone numbers. An IP dial plan is used to process IP addresses (e.g., *johnsmith@vovida.org*).

*Digit manipulation versus finding a gateway*
> A digit manipulation dial plan gives rules for changing digits matching a given pattern. For example, if an address starts with 9, remove the 9 and replace it with $USER. A gateway-finding dial plan gives rules for routing calls to a particular gateway if they match a given pattern.

In order to accommodate both dial plan categories, there are four types of dial plan:

- Digit manipulation phone dial plan
- Digit manipulation IP dial plan
- Gateway-finding phone dial plan
- Gateway-finding IP dial plan

One set of dial plans is defined at the global level, and another set is defined for each Centrex group.

## Security/Authentication/Encryption

To gain access to the Provisioning server via the Java GUI, the user must enter a username and a corresponding password. No other authentication method has been implemented for this release. In the future, the authentication mechanism can be changed to use a more sophisticated algorithm to provide more secure access.

Encryption has not been implemented for this release. However, the Provisioning server and its the clients use HTTP as the communication protocol, and it would be possible to use HTTPS (secure hypertext transfer protocol) as an encryption scheme in the future.

## Server Provisioning

Server provisioning is performed by someone who is responsible for setting up and maintaining servers at an Application Service Provider (ASP). We call this role *technician*. The technician decides how many servers of each type run in the system and assigns each server a host machine, a port number, and other provisioning data as needed. The technician may be responsible for composing the XML description files that describe the provisioning needs of the servers, or those files may be written by the developers who wrote the servers themselves. Another responsibility of the technician is to write dial plans, at both the global level and the Centrex level. The technician is also able to monitor the runtime status of the servers in the system to see which servers are running (heartbeating) and which are not.

When the technician logs into the system, she is presented with options that allow her to view all provisioned servers, grouped by type or by host. The technician has the option of editing an existing server or creating a new one. The technician may also view the runtime status of all servers.

## Account Provisioning

The provisioning system is to maintain and manipulate data for different types of accounts. These include individual subscribers (end users) and multiple levels of administrative-type access (service provider, distributor, Centrex administrator, for example). In each case, a person with an account at a particular access level should be able to have full access to provisioning data for accounts at all levels below his access level but should have no access to accounts at levels above his or at the same level as his.

In each case, the data of interest for an account is the account identifier (account name, such as phone number for end users), a list of groups the account belongs to, a list of groups the account defines, and a list of features the account has access to. In addition to this, the end user accounts also need to store per-user configuration data for each feature the user has access to. An account may also include additional fields. In such a case, a textual description of the data for that account should be available for use by the user interface and Provisioning server to determine the necessary fields.

For all administrative-level accounts, the data for that account (list of subaccounts) is displayed in a tabular format in which the columns consist of the account identifier, a column for each feature available to that account, and additional fields (if any) that are defined for that account.

For an end user account, the data for that individual account is displayed in a screen that contains user interface elements that allow the editing of all the fields associated with each feature available to that user and any additional (non-feature-related fields).

The access level a particular account is associated with is a string identifier of the form:

    isp1.distibutor3.centrex2

which would identify an account belonging to an administrator of `centrex2`, or:

    isp1.distributor1.centrex3.enduser493

which would identify an account belonging to a particular end user.

Consequently, an administrator at some particular level is able to view data for only accounts whose identifiers start with her particular identifier.

---

# GUI Screens

A number of GUI screens are currently under development. Check with the *Vovida.org* web site for updated examples.

# DTD for Data Definition

The following example shows the DTD for the provisioning XML files:

```
<!-- Root element. Gives the name of the data type
(i.e., ForwardAllCalls). This name must be unique. -->
<!ELEMENT DataDefn (Category?, Key*, (DBEntry | Border)*)>
<!ATTLIST DataDefn name CDATA #REQUIRED
>

<!-- Optional set of nested categories provide a hierarchy for
    categorizing this data type. For example, the DataDefn name
    could be ForwardAllCalls and the top-level Category name could be
    Features. Nesting child categories within the top level allows
    a finer granularity of distinction between different types of
    Feature. -->
<!ELEMENT Category (Category*)>
<!ATTLIST Category name CDATA #REQUIRED
>

<!-- Gives one or more DBEntries that this data type is keyed on.
    For example, a server would have a Host Key and a Port Key.
    PServer uses the Key information to build database tables. -->
<!ELEMENT Key (#PCDATA)>
<!-- DBEntry is an entry in the database. PServer uses this to
    build the database table for this data type. The database
    field will be called by the name attribute. The type attribute
    will be either a Schema Type or Xml. If it is Xml, the database
    will store it as a blob. The enabled attribute is set to true
    if the field can be edited. -->
<!ELEMENT DBEntry
(Label?,Tooltip?,Alert?,Default?,Required?,MinInclusive?,MaxInclusive?,Content?,
Column*,Condition*,Border?)>

<!ATTLIST DBEntry name CDATA #REQUIRED
            type ( String | Integer | Boolean | Port | Decimal | Date | Time
            | XML ) "String" enabled ( true | false ) "true"
            widget (TextField | ComboBox | RadioButtons | CheckBox | Table)
            "TextField".
>

<!-- Border is used to visually group DBEntries together for
    association under a common heading. -->
<!ELEMENT Border (Label?, DBEntry*)>

<!-- Used to specify a label for presentation in the GUI -->
<!ELEMENT Label (#PCDATA)>
```

```
<!-- Used to specify a tooltip for presentation in the GUI -->
<!ELEMENT Tooltip (#PCDATA)>

<!-- Used for fields that need a special message to alert the user
     when the field is changed -->
<!ELEMENT Alert (#PCDATA)>

<!-- The default value for this field. If a choice of values is given
     through an Enumeration, then this must be one of those choices. -->
<!ELEMENT Default (#PCDATA)>

<!-- Present if this DBEntry is a required field.
     Note that Key fields are automatically assumed to be required. -->
<!ELEMENT Required EMPTY>

<!-- The minimum value for this field. Ignored unless type is numeric -->
<!ELEMENT MinInclusive (#PCDATA)>

<!-- The maximum value for this field. Ignored unless type is numeric -->
<!ELEMENT MaxInclusive (#PCDATA)>

<!-- Specifies where the DBEntry value comes from. Content could be
     a list of hardcoded strings (Enumerations), the result of a
     PServer query, or a reference to another DBEntry. -->
<!ELEMENT Content (PServer | Enumeration | Reference)>.

<!-- This is the genesis of a PServer query. For example, the query
     could return a list of all hostnames in the system so that the
     end user can choose a hostname from a drop-down list. The syntax
     will be determined when we know more about the PServer API.
     If the userInputOK attribute is true, a value that does not
     appear in the drop-down list can be typed in. -->
<!ELEMENT PServer EMPTY>
<!ATTLIST PServer GET CDATA #REQUIRED
     userInputOK ( true | false ) "false">

<!-- An Enumeration contains one or more Entries that give hardcoded
     values for the end user to choose from. Each Entry must be of the
     type specified for the parent DBEntry.
     If the userInputOK attribute is true, a value that does not
     appear in the drop-down list can be typed in. -->
<!ELEMENT Enumeration (Entry+)>
<!ATTLIST Enumeration userInputOK ( true | false ) "false">

<!ELEMENT Entry (#PCDATA)>

<!-- A Reference takes the name of another DBEntry and puts that value
     into the content of this DBEntry. If the userInputOK attribute of the
     parent is true, the end user may overwrite that value with his
     own. -->
<!ELEMENT Reference (#PCDATA)>

<!-- A table is stored as raw XML and the parent DBEntry must specify
     Xml as its type attribute. A table may have any number of columns.
```

```
      Each column may have a Label, which is the column heading. -->
<!ELEMENT Column (Label?, Alert?, Default?, Content?)>
<!ATTLIST Column name CDATA #REQUIRED
      type CDATA #IMPLIED
      widget (TextField | ComboBox | CheckBox ) "TextField">

<!-- The Condition element specifies a condition to check and an action
      to take when that condition is true. The condition applies
      to the parent DBEntry. The value attribute contains a string which
      will be matched by the parent DBEntry when the condition is true.
      The target attribute refers to the DBEntry that is targeted
      by the Action.
      Its child elements are the possible types of actions that can be
      executed. The Content child follows the
      same rules as a Content child of DBEntry in setting the content of the
      target field. The Enable and Disable child elements cause their target
      field to be enabled or disabled. -->
<!ELEMENT Condition (Action+)>
<!ATTLIST Condition value CDATA #REQUIRED>
<!ELEMENT Action (Content|Enable|Disable)>
<!ATTLIST Action target CDATA #REQUIRED>
<!ELEMENT Enable EMPTY>
<!ELEMENT Disable EMPTY>.
```

# Examples of Protocol Transmissions/Replies

In each case, we indicate what the call would look like if it were a C++ call. This is
not intended to imply that such a call will directly exist in the library—it is simply a
convenient mechanism for illustrating the information passed and the result
expected. In most cases, when a simple 200 OK or 4XX error message is expected,
we list that as a void return type.

For error conditions, we show only some of the possible error codes that could be
returned. In almost all cases, other error codes could possibly be returned, but we do
not list all return codes.

```
void GetRecordsInDomain (string domain, string filter, list <string> fieldList)
```

```
Request:
GET GetRecordsInDomain HTTP/1.1
Content-Length:
<arguments>
  <domain> Accounts </domain>
  <filter> uid = 100* </filter>
  <fieldlist>
    <listitem> uid </listitem>
    <listitem> phone </listitem>
    <listitem> xmlStuff </listitem> <--- Contains xml content
    .

    .

    .
  </fieldlist>
```

```
    </arguments>

Response (Success):

200 OK HTTP/1.1
Content-Length:
<response>
  <domain> Accounts </domain>.
  <resultcount> 5 </resultcount> <-- Maybe? Need to think.
  <match>
    <fieldvalue>
      <uid>1001</uid>
      <phone>4083831001</phone>
      <xmlStuff> <--- Xml Content
      ...
      </xmlStuff>
    </fieldvalue>
  </match>
  <match>
    <fieldvalue>
      <uid>1002</uid>
      <phone>4083839999</phone>
      <xmlStuff>
      ...
      </xmlStuff>
    </fieldvalue>
  </match>
  .
  .
  .
</response>

Response (Success but no matches):

204 No Content HTTP/1.1
Content-Length:0

Response (Fail):

400 FAIL HTTP/1.1
Content-Length:
<error> No response from database </error>.

void SetRecordsInDomain (string domain, string filter, list <string> fieldvalue)

Request:

PUT SetRecordsInDomain HTTP/1.1
Content-Length:
<arguments>
  <domain> Accounts </domain>
  <filter> uid = 100* </filter>
  <fieldvalue>
    <marshalgroup>hostname</marshalgroup>
```

```
      <authenticationtype>none</authenticationtype>
      .
      .
      .
  </fieldvalue>
</arguments>

Response (Success):

200 OK HTTP/1.1
Content-Length:
<response>
   <domain> Accounts </domain>
   <resultcount>8</resultcount>
</response>

Response (Fail):

400 FAIL HTTP/1.1
Content-Length:
<error> No response from database </error>

void AddRecordInDomain (string domain, list <string> fieldvaluelist).
Request:

PUT AddRecordToDomain HTTP/1.1
Content-Length:
<arguments>
   <domain> Accounts </domain>
   <fieldvalue>
     <uid>328</uid>
     <password>6f7d5634f65192a3b9c8479cee3b655</password>
     <ipaddr>192.168.5.8</ipaddr>
     <xmlStuff>
...                                <--- Unparsed xml content
     </xmlStuff>
     .
     .
     .
   </fieldvalue>
</arguments>

Response (Success):

200 OK HTTP/1.1
Content-Length:0

Response (Fail):

400 FAIL HTTP/1.1
Content-Length:
<error> No response from database </error>

void DeleteRecordsInDomain (string domain, string filter)
```

Request:

```
DELETE DeleteRecordsInDomain HTTP/1.1.
Content-Length:
<arguments>
  <domain> Accounts </domain>
  <filter> uid = 100* </filter>
</arguments>
```

Response (Success):

```
200 OK HTTP/1.1
Content-Length:
<response>
  <domain> Accounts </domain>
  <resultcount>8</resultcount>
</response>
```

Response (Fail):

```
400 FAIL HTTP/1.1
Content-Length:
<error> No response from database </error>
```

void SessionOpen (unsigned long expiretime)

Request:

```
GET SessionOpen HTTP/1.1
Content-Length:
<arguments>
  <expiretime> some unsigned long </expiretime>
</arguments>
```

Response (Success):

```
200 OK HTTP/1.1
Content-Length:
<response>
  <token> some unsigned long </token>
</response>
```

Response (Fail):

```
400 FAIL HTTP/1.1
Content-Length:
<error> Server Error </error>
```

void SessionClose (unsigned long token)

Request:

```
POST SessionClose HTTP/1.1
Content-Length:
```

```
<arguments>
  <token> some unsigned long </token>
</arguments>

Response (Success):

200 OK HTTP/1.1
Content-Length:0

Response (Fail):

400 FAIL HTTP/1.1
Content-Length:

<error> No Token Found </error>

400 FAIL HTTP/1.1
Content-Length:
<error> Server Error </error>

void SessionSearchDB (unsigned long token, string filter, list <string> fieldList,
string sortType, string sortField)

Request:

POST SessionSearchDB HTTP/1.1
Content-Length:
<arguments>
  <token> some unsigned long </token>
  <domain> Marshals </domain>
  <filter> host = cktam-lnx </filter>
  <fieldlist>
    <listitem>bridgeNumber</listitem>
    <listitem>gateway</listitem>
    <listitem> ... </listitem>

    .
    .
    .
  </fieldlist>
  <sortType>ascend</sortType>
  <sortField>bridgeNumber</sortField>
</arguments>

Response (Success):
Found 3 results.

200 OK HTTP/1.1
Content-Length:
<response>
  <domain> Marshals </domain>
  <resultcount>3</resultcount>
</response>

Response (Success but no matches):.
```

```
204 No Content HTTP/1.1
Content-Length:0

Response (Fail):

400 FAIL HTTP/1.1
Content-Length:
<error> No response from database </error>

void SessionGetDBResult (unsigned long token, string start, string end, list <string>
fieldList)

Request:

GET SessionGetDBResult HTTP/1.1
Content-Length:
<arguments>
  <token> some unsigned long </token>
  <start>10</start>
  <end>12</end>
  <fieldlist>
    <listitem> uid </listitem>
    <listitem> phone </listitem>
    <listitem> ... </listitem>
    .
    .
    .
  </fieldlist>
</arguments>

Response (Success):

200 OK HTTP/1.1
Content-Length:
<response>.
  <domain> Accounts </domain>
  <match>
    <fieldvalue>
      <uid>1010</uid>
      <phone>4083831001</phone>
      .
    </fieldvalue>
  </match>
  <match>
    <fieldvalue>
      <uid>1011</uid>
      <phone>4083839999</phone>
      .
    </fieldvalue>
  </match>
  <match>
    <fieldvalue>
      <uid>1012</uid>
      <phone>4083839999</phone>
      .
```

```
    </fieldvalue>
  </match>
</response>
```

Response (Fail, bad index (start/end) ):

```
400 FAIL HTTP/1.1
Content-Length:
<error> invalid index </error>
```

Response (Fail):

```
400 FAIL IITTP/1.1
Content-Length:
<error> Token expire </error>
```

void SessionSortDBResult (unsigned long token, string sortType, string sortField)

Request

```
POST SessionSortDBResult HTTP/1.1
Content-Length:
<arguments>
  <token> some unsigned long </token>
  <sortType>descend</sortType>
  <sortField>bridgeNumber</sortField>
</arguments>
```

Response (Success):
Sorted 3 results.

```
200 OK HTTP/1.1
Content-Length:
<response>
  <domain> Marshals </domain>
  <resultcount>3</resultcount>
</response>
```

Response (Fail):

```
400 FAIL HTTP/1.1
Content-Length:
<error> Token Expire </error>
```

void SessionSetDBResult (unsigned long token, list <string> field)

Request

```
PUT SessionSetDBResult HTTP/1.1
Content-Length:.
<arguments>
  <token> some unsigned long </token>
  <domain> Accounts </domain>
  <filter> uid = 100* </filter>
```

```
  <fieldvalue>
    <marshalgroup>hostname</marshalgroup>
    <authenticationtype>none</authenticationtype>
    .
    .
    .
  </fieldvalue>
</arguments>
```

Response (Success):

```
200 OK HTTP/1.1
Content-Length:
<response>
  <domain> Accounts </domain>
  <resultcount>8</resultcount>
</response>
```

Response (Fail):

```
400 FAIL HTTP/1.1
Content-Length:
<error> No response from database </error>
```

void SessionDeleteAll (unsigned long token)

Request

```
DELETE SessionDeleteAll HTTP/1.1
Content-Length:
<arguments>.
  <token> some unsigned long </token>
</arguments>
```

Response (Success):
Deleted 5 results.

```
200 OK HTTP/1.1
Content-Length:
<response>
  <resultcount>5</resultcount>
</response>
```

Response (Fail):

```
400 FAIL HTTP/1.1
Content-Length:
<error> Permission Denied </error>
```

# VOCAL SIP UA Configuration File

This appendix covers the configuration file, *ua.cfg*. This file is stored in */usr/local/vocal/sip/ua*.

This file has many comments throughout its different sections offering you tips and examples for configuring a user agent for testing VOCAL. In this section, we have elaborated on these comments and separated the different tag areas to make the file more understandable. If you use another softphone or IP phone set, use the settings suggested in the *ua.cfg* file to guide you through your setup.

The title of the file appears as:

```
# VOCAL SIP User Agent (ua) Configuration File
# Updated 05/01/2001
```

The configuration for each section follows the same basic line format:

```
tag type value
```

Tags and values are case-sensitive. In Version 1.3.0 of VOCAL, types are not checked.

## General

The first area of the file that you need to change is the general identification tags:

```
Device_Name          string          /dev/phone0
User_Name            string          1000
Display_Name         string          Phone
Pass_Word            string          test
```

*Device_Name*
> Specifies the device to use. If you are using a Quicknet card, use */dev/phone0*. If you are using a sound card, use */dev/dsp*.

*User_Name*
> Specifies the phone number or extension for this User Agent.

*Display_Name*

Specifies the name that appears in the titlebar of the panel used for this User Agent.

*Pass_Word*

Specifies the password that will be used for authentication if you are using a Proxy server. This value is optional and is checked only if you are using digest authentication.

## SIP Port and Transport

These tags help you set up a listening port for the SIP stack and your preferred transport protocol:

```
Local_SIP_Port        string        5060
SIP_Transport         string        UDP
```

*Local_SIP_Port*

Specifies the listening port number of the SIP stack. Normally the default port is 5060.

*SIP_Transport*

Specifies the transport protocol—UDP or TCP. Version 1.3.0 of VOCAL runs over UDP.

## Proxy Server

In the VOCAL system, the SIP proxy server that talks to the user agents is called the User Agent Marshal server. The syntax for this configuration is *Proxy_Server string <host>[:<port>]*:

```
Proxy_Server          string        192.168.22.12
```

*Proxy_Server*

Specifies the IP address and port number of the SIP Proxy server

If you do not have a Proxy server, comment out the line.

## Transfer and Conferencing

Transfer and conferencing are *not supported* in Version 1.3.0 of VOCAL, so it is best to leave these parameters alone:

```
Ua_Xfer_Mode          string        Off
Conference_Server     string        6000@192.168.5.4
```

*Ua_XferMode*

Turns on transfer or ad hoc conferencing. The options are *off*, *transfer*, or *conference*.

*Conference_Server*
> Specifies the URI for the Conference server. The URI consists of the conference bridge number and the IP address of the Proxy server or the conference bridge itself (if no proxy is being used).

# Registration

Use these parameters to enable or disable registration. Enabling the option registers the User Agent with a Registration server.

```
Register_On        bool        False
Register_From      string      192.168.22.12
Register_To        string      192.168.22.12
Register_Expire    int         60000
Register_Action    string      proxy
```

*Register_On*
> To enable (True) or disable (False) registration.

*Register_From*
> Specify the IP address of the Proxy server. Usage:
>
> ```
> Register_From string <host>[:<port>]
> ```

 The Registration server is often the same as the Proxy server.

*Register_To*
> Specify the IP address of the Registration server. Usage:
>
> ```
> Register_To string <host>[:<port>]
> ```

*Register_Expire*
> Specify the time in seconds when the registration will expire.

*Register_Action*
> Specify either *proxy* or *redirect*. This tells the server to act as a Proxy or Redirect server for future requests. Most servers will use *proxy*. This feature has been deprecated in SIP.

# Ringback

This is used for testing purposes only. It is best to leave it as False:

```
Provide_Ringback    bool        False
```

*Provide_Ringback*
> Provide (True) or disable (False) the ringback tone option

# RTP

These tags help you set up how the user agent works with Real-time Transport Protocol (RTP):

```
Network_RTP_Rate      int      20
Min_RTP_Port          int      10000
Max_RTP_Port          int      10099
```

*Network_RTP_Rate*
    Specifies the network RTP packet size in milliseconds. You can specify a range of RTP port numbers here. This is useful when passing RTP media through a firewall.

*Min_RTP_Port*
    Minimum RTP port. Specify the *Min_RTP_Port* as an even number.

*Max_RTP_Port*
    Maximum RTP port. Specify the *Max_RTP_Port* as an odd number.

# Call Waiting

Call waiting is *not supported* in Version 1.3.0 of VOCAL. It is best to leave it as False:

```
Call_Waiting_On       string       False
```

*Call_Waiting_On*
    Turn on (True) or turn off (False) the call waiting feature

# Call Progress Timer

This tag allows the UA to time out a call initiation if the receiving end is unresponsive:

```
Ua_Timeout            string       10
```

*Ua_Timeout*
    Specifies a limit (in seconds) on the amount of time it takes to receive a response (other than 100 Trying) from the UA server at the far end of the network

# Subscribe/Notify

This is used for testing purposes only. It is best to leave it as False:

```
Subscribe_On          string       False
Subscribe_Expires     int          60000
Subscribe_To          string       192.168.22.12
```

*Subscribe_On*
    Enables (True) or disables (False) sending Subscribe messages

---

*Subscribe_Expires*
    Specifies the subscription period in seconds

*Subscribe_To*
    Specifies the IP address of where the Subscribe messages are going to be sent

# Dialing Timers

Specifies dialing timers in milliseconds:

```
Initial_Digit_Timeout    int         16000
Inter_Digit_Timeout      int         8000
```

*Initial_Digit_Timeout*
    Specifies a limit (in milliseconds) on the amount of time it takes to receive the first dialed digit after the user goes off-hook

*Inter_Digit_Timeout*
    Specifies a limit (in milliseconds) on the amount of time it takes to receive the subsequent dialed digits after the first one

# Dial Patterns

Dial patterns enable the UA to send completed phone numbers or URLs to VOCAL. These patterns include a set of variables that match commonly dialed numbers, such as 411 or 9-###-####. Dial patterns can also match dialed IP numbers or URIs. Unlike dial plans, dial patterns do not provide routing information, they simply let the UA know that a completed number has been dialed and can be sent to VOCAL for routing.

There are two timers at work with dial patterns. First, the Initial Digit Timeout clocks the time that transpires between the UA going off-hook and the first digit being entered. If this times out, the UA will need to go back on-hook, then off-hook again before the user can resume dialing. If the user enters a digit before this timer expires, the Initial Digit Timeout stops, and the Inter Digit Timeout begins.

The Inter Digit Timeout measures the time taken between digit entry and the time the dialed number matches a dial pattern. If the dialed numbers match a specific dial pattern, such as 411, this timer stops and the number is sent to VOCAL. If the user stops dialing but the number dialed does not specifically match any dial patterns, this timer times out and the UA sends the dialed number to VOCAL as is. This method of combining timers with dial patterns helps manage calling and permits variable-length dial patterns to be sent to VOCAL, which is useful for international dialing or local dialing in countries that permit variable-length phone numbers at the local central offices.

The dialing patterns use regular expressions. The second data field in the dialing pattern is the method for constructing the SIP URI. Usage is as follows:

```
Dial_Pattern string <type> <pattern>
```

 To make a simple User-Agent-to-User-Agent call you do not need to modify the dial patterns.

```
Dial_Pattern          string    0    ^#[0-9][0-9][0-9]
Dial_Pattern          string    1    [0-9][0-9]*#
Dial_Pattern          string    2    ^911
Dial_Pattern          string    2    ^611
Dial_Pattern          string    2    ^411
Dial_Pattern          string    2    ^\*69
Dial_Pattern          string    2    ^6[0-9][0-9][0-9]
Dial_Pattern          string    2    ^9[2-9][0-9][0-9][0-9][0-9][0-9][0-9]
Dial_Pattern          string    2    ^91[0-9][0-9][0-9][0-9][0-9][0-9][0-9][0-9][0-
9][0-9]
Dial_Pattern          string    3    ^[*][0-9][0-9][0-9]
```

For *SPEED_DIAL(0)* and *NORMAL_DIAL(2)*, the dialed digits will simply be prefixed to the destination IP address to form the SIP URI.

For *INTERNAL_IP_DIAL(3)*, if your User Agents are part of the same internal network, you can use *INTERNAL_IP_DIAL*. You simply dial the last three digits of the IP address, and the SIP URI is constructed by adding these three digits to the IP address of the internal network. The dial pattern is represented by:

```
Dial_Pattern      string    3    ^[*][0-9][0-9][0-9]
```

For example, suppose you have two User Agents with IP addresses 192.168.5.130 and 192.168.5.135. Since these User Agents are on the same subnet, you do not dial the full IP address. You can simply dial *130 to call the User Agent at 192.168.5.130.

See the *readme* file for limitations of *INTERNAL_IP_DIAL*.

For *CALL_RETURN(4)*, whatever string you defined and dialed will be translated into the last incoming call's URL in the From field. This is the same as the *69 functionality in the North America PSTN world.

The dialing types are:

   0 — *SPEED_DIAL*

   1 — *INTERNAL_DIAL*

   2 — *NORMAL_DIAL*

   3 — *INTERNAL_IP_DIAL*

   4 — *CALL_RETURN*

# Speed Dial List

The speed dial list is used to set up a list of dial numbers, phone numbers, or IP addresses of the parties that you wish to call.

Specify each entry as *Speed_Dial string <digits> <destination>* where *<digits>* is the number that you intend to dial on your keypad and *<destination>* is the phone number/IP address of the party you wish to call.

```
Speed_Dial    string    #150         150@10.0.0.2
Speed_Dial    string    #151         151@10.0.0.3
Speed_Dial    string    #398         6398@192.168.5.130
```

Suppose you specified the following:

```
Speed_Dial    string    #123         123@192.168.5.5
```

To call 123@192.168.5.5, you would simply use the keyboard or phone keypad to dial *#123*.

Speed dial calls will bypass the Proxy server if an IP address is provided in the destination.

# RSVP Configuration

This is used to interface with an RSVP daemon and is for testing purposes only. It is best to leave it as False:

```
Rsvp_On               bool      False
Provisioning_Host     string    bass
Provisioning_Port     int       6005
Use_Policy_Server     bool      False
```

# Manual Call ID

These parameters are provided for testing purposes only. It is best to leave it as False:

```
CallId_On             bool      False
CallId                string    1234567890
```

*CallId_On*
   Enables (True) or disables (False) fixing the SIP call ID

*CallId*
   The fixed call ID

# Load Generation

The following parameters are used for load generation. These parameters do not need to be modified if you want to make User-Agent-to-User-Agent calls:

```
LoadGen_On          bool      False
RunMode             string    Calling
StartTime           int       5000
NumKickStarts       int       1
CallUrl             string    sip:7399@mendel
NumEndpoints        int       1
CallDuration        int       1000
CallDelay           int       1000
CallRestartTimer    int       17000
AnswerRestartTimer  int       12000
AnswerDelay         int       100
MonitorMsgOn        bool      False
MonitorMsgInterval  int       10
DialNumber          string    6000
NumOfCalls          int       -1
CJTime              bool      False
RtpGenOn            bool      False
```

*LoadGen_On*
Turns load generator on (True) or off (False).

*RunMode*
Specifies if the UA is in the *Calling* or the *Receiving* mode.

*RSTest*
Specifies *RSTest* for Redirect server testing.

*StartTime*
Specifies an initial delay (in seconds) before making calls.

*NumKickStarts*
Number of parallel calls.

*CallUrl*
Specifies the SIP URL to call if *RunMode* is set to *Calling*.

*NumEndpoints*
*Not implemented.*

*CallDuration*
Specifies the hold time (in ms) for an active call.

*CallDelay*
Specifies the time (in ms) before making a new call.

*CallRestartTimer*
Guards timer at caller side (in ms). Should be *CallDuration* + *CallDelay* + more.

*AnswerRestartTimer*
Guards timer at called party's side (in ms). Should be less than *CallRestartTimer*.

*AnswerDelay*

> *Not implemented.*

*MonitorMsgOn*

> Turns the statistic messages on (True) or off (False).

*MonitorMsgInterval*

> Specifies the interval (in seconds) to print statistics messages.

*DialNumber*

> *Not implemented.*

*NumOfCalls*

> Specifies the number of calls before terminating. Set to -1 for infinite calls.

*CJTime*

> Uses Cullen Jennings' version of *gettimeofday*.

 You need to set the hardcoded CPU clock time in *CJTime.cxx* to get the correct result.

*RtpGenOn*

> Transmits silent audio packets during load generation if it is set to True.

# APPENDIX B
# Testing Tools

Some of the VOCAL test cases call for generating 20,000 or 100,000 users. The scripts included in this appendix provide assistance to this process.

## genPutUsers.pl

This script will generate a large number of users and put them in the correct directories within the provisioning data. It starts at the specified start user # and increments from there. This file's location is */vocal-1.3.0/tools/ldgen/genPutUsers.pl*.

To use this file, type:

```
./genPutUsers.pl <start user #> <number of users>
```

The script relies upon a C++ executable called *PutMultipleFiles*. In order to use this, you will need to do the following:

1. Type:

   ```
   ../vocal-1.3.0/provisioning/conversion/make
   ```

2. Copy the resulting executable from the bin directory into the directory where you want to run *genPutUsers*.

3. Create an XML file called *tplate* that will be used as the template for the generated users. It is best to generate a valid XML file via the GUI and then copy the file over.

You may want to modify *genPutUsers.pl* if your provisioning data is not in */usr/local/vocal/provisioning_data*.

Once you have the users generated, you can begin registering them.

 This creates only the user files, not their associated calling/called feature files.

# Netcat

Use *Netcat* to inject SIP messages into the RS. You can download *Netcat* from *http://www.atstake.com/research/tools*.

When running, use the format:

```
nc -u  -w5  -p 2222 marcl-lnx 4443 < invMsg.SIP
```

For example, you can modify the appropriate fields in the SIP message file to test different portions of the RS. To use this file, type:

```
nc -u -w5 -p <any unused port> <machine running RS> <RS SIP port> < <SIP message
file>
```

# Index

We'd like to hear your suggestions for improving our indexes. Send email to *index@oreilly.com*.

forking
    characteristics, 161–163
    least-cost routing and, 258
    Level II transactions and, 185
    proxies and, 134
fractional T1 lines, 42
fraud, billing and toll, 399
FreeBSD operating system, 6
From field (SIP message header)
    Level I transactions, 185
    Redirect server processing, 255
    sample SIP call message flow, 151, 153
    syntax, 142
FsBuilder (base file), 294
FsConfigFile (base file), 294
FsConfigFile class (Feature server), 309
FsFeature (base file), 294
FsFeature class (Feature server), 309
FsProvisioning (base file), 294
FsStateMachnData (base file), 295
FsUserData (base file), 294
FsUsers (base file), 294
FsUsers class (Feature server), 309
FSVM (Voice Mail Feature server), 119–122,
    307–312
FwdCall operator (CPL), 281
FwdCallOp, 282
FXO (Foreign Exchange Office), 41
FXS (Foreign Exchange Status), 41

## G

G.711 codec, 31, 49
G.729 A codec, 31
garbage collection, 179, 181
Garbage Collector (SIP thread), 187
Gatekeeper (H.323 translator), 352, 359
gatekeepers
    H.323 translator, 352, 353
    siph323csgw, 359
Gateway Marshal servers
    calling phones and, 97
    features, 112
    heartbeats and, 121
    provisioning, 65
    PSTN Gateway Marshal servers, 223, 244
    reprovisioning, 59
    SIP call message flow, 154
    software configuration and, 50
    VOCAL type, 10
gateway-level authentication, 244

gateways
    endpoints as, 323
    firewalls and, 237, 238
    H.323 interoperability, 360
    links, 41, 324
    Marshal server, 114, 223, 224, 244
    MGCP gateways, 328, 333
    PEPs, 386
    provisioning for multiple, 65
    responsibilities of, 324
    SIP user agents and, 9
    VOCAL requirements for, 16
    (see also ATAs; PSTN gateways)
gcc compiler
    code optimization and, 22
    SIP stack size and, 190
    siph323csgw and, 361
GCF (Gatekeeper Confirm) message, 350,
    353
genPutUsers.pl (testing tool), 466
get command (SNMP), 364, 374
get method (HTTP), 409
get request, 371, 409
getCdrServers call, 405
getCurrentRecordsInDomain function, 423
getData( ) method, 436, 443
GetFields function, 419
getFieldsInDomain function, 423
getFieldValue function, 426
getFieldValueAsBool function, 426
getFieldValueAsDouble function, 426
getFieldValueAsInt function, 426
getFieldValueAsList function, 426
getFieldValueAsLong function, 426
getMailboxFromProvisioning( ), 312
getMarshalData call, 405
getnext request, 371
GetRecord function, 419
GetRecordCount function, 419
GetRecordsInDomain command, 410
getRecordsInDomain function, 423
GetSearchFilter function, 419
GetStackPortFromIndex, 375
getVmFeature( ), 310
glare, 339
"global village", 13
GNU C Library, 200
GNU public license, 5, 34
graphical user interface (see GUI)
Group class (RS), 246
Group field, 73
group option (Marshal field), 73

plus sign (+), 96
pointers, 176
points of failure, 7, 42
policy
    defined, 386
    as VoIP data type, 8
policy decision point (PDP), 386, 387
policy enforcement point (see PEP)
Policy servers
    features, 123
    function of, 386
    heartbeats, 121
    Marshal server, 230
    purpose of, 12
    QoS and, 385
    redundant systems and, 45
Portable Operating System Interface
            (POSIX), 200
ports
    dedicated SNMP ports, 363
    distinguishing numbers for, 107
    HeartLessProxy class, 198
    Provisioning servers, 37–38
    Redirect servers, 39
    SNMP and, 363
    synchronization port, 252
    UDP port, 371
POSIX (Portable Operating System
            Interface), 200
post method (HTTP), 409
POTS (plain old telephone system), 48, 339
potStateEngine module, 333
prepaid calls, 217
Pretty Good Privacy (PGP), 237
PRI (Primary Rate Interface), 42, 48
primary and backup, 43, 46
Primary Rate Interface (see PRI)
<priority> CPL tag, 275, 297
Priority operator (CPL), 281
Priority (SIP message header), 145
<priority-switch> CPL tag, 275, 297
Priority-Switch operator (CPL), 281
private branch exchanges (see PBXs)
process( ) function, 195, 198, 301
process( ) method, 228
process ( ) method (Location), 288
process controllers, 379
    SNMP, 379
process ID, 143
processes
    multicall processing, 206–212
    NetMgnt and, 364, 372–374
    SNMP GUI, 378

processVMRedirect( ) method, 312
PromptTable, 319
protocol stacks
    availability of open source, 13
    MGCP stack, 325, 332
    transport layer, 168
    (see also SIP stack)
provideDtmf, 316
Provide_Ringback tag (VOCAL UA), 459
provideRingStart method, 316
ProvisionableInstance class, 432, 435
ProvisionableType class, 432
Provisioning server
    Apache server and, 18
    authentication and, 240, 244
    booting up process, 270
    CPL limitations, 273
    direct command and, 37
    FSVM call processing, 312
    heartbeats and, 43, 122
    MGCP translator, 331
    navigating through GUI, 59
    Network Management System, 368
    Network Manager, 373
    old system, 400–405
    prerequisite for provisioning users, 67
    PSUTILIB directory and, 200
    Redirect servers and, 251
    redundant systems and, 44
    regenerating users and, 107
    syntax for starting, 38, 39
    (see also Mascarpone Provisioning server)
provisioning users (see user provisioning)
ProvisioningInterface class, 422–427
ProvisioningInterface::Instance( ), 331
ProvisionServer interface, 405
proxy
    adding UAs and, 32
    defined, 10
    ingress/egress, 137
<proxy> CPL tag, 276, 297, 306
Proxy operator (CPL), 281
<proxy ordering> CPL tag, 297, 306
Proxy servers (Proxy)
    B2BUA and, 217
    CFNA, 10, 130
    forking INVITEs, 161
    ingress/egress depicted, 137
    INVITE sample, 140
    message flows, 134–136
    Redirect servers, 10
    Registrar servers, 10
    registration and, 136

# About the Authors

**Luan Dang** is a Director of Software Development at Cisco Systems. Previously, Luan was Chief Technology Officer and Co-Founder of Vovida Networks. Luan is currently a member of the Technical Advisory Council for the International Softswitch Consortium and has previously filed telephony patents for voice-over-IP (1999) and Caller IP (1998). Luan was also granted a patent for the display screen management apparatus in 2000. Luan has a BS degree from UCSD and an MS in Computer Science from Stanford.

**Cullen Jennings** has management, consulting, and development experience for both technology-based companies and educational institutions. His industry experience includes writing software for semiconductor manufacturing, safety-critical air traffic control systems, biomedical engineering, energy, government, and academia. Cullen holds an MS in Computer Science and a BS in Applied Mathematics from the University of Calgary. He also has a Ph.D. in Computer Science from the University of British Columbia. Cullen is a member of the IEEE and ACM and has published more than 15 articles in the areas of computer vision and pattern recognition, vectorization, thinning algorithms and strategy, character recognition, multiprocessor programming, and data capture and conversion. At Vovida Networks, Cullen was Vice President of Engineering.

**David Kelly** joined Cisco as a technical writer as part of the Vovida Networks acquisition in 2000. Prior to moving to California, David owned and operated a successful technical writing firm in Vancouver, Canada. David holds a BA from the University of British Columbia and has traveled extensively: his globetrotting adventures include picking bananas on a kibbutz in Israel and learning Spanish in Guatemala.

# Colophon

Our look is the result of reader comments, our own experimentation, and feedback from distribution channels. Distinctive covers complement our distinctive approach to technical topics, breathing personality and life into potentially dry subjects.

The animals on the cover of *Practical VoIP Using VOCAL* are snipefish. There are about 10 or 12 species of snipefish, living mainly in tropical, subtropical, and temperate waters—just about all over the world, in other words—and eating small marine animals. Their distinguishing characteristic is a long, tubular snout, which starts out short and stubby but lengthens as the fish grows to its maximum size of up to 30 centimeters. Their tiny jaw, sans teeth, is at the very end of the snout. Some snipefish, such as the ones on the cover, also have a long, pointed spine stretching back from the dorsal fin.

Jane Ellin was the production editor, Norma Emory was the copyeditor, and Audrey Doyle was the proofreader for *Practical VoIP Using VOCAL*. Darren Kelly, Sue

Willing, and Phil Dangler provided production assistance. Lucie Haskins wrote the index.

Ellie Volckhausen designed the cover of this book, based on a series design by Edie Freedman. The cover image is a 19th-century engraving from the Dover Pictorial Archive. Emma Colby produced the cover layout with QuarkXPress 4.1 using Adobe's ITC Garamond font.

David Futato designed the interior layout. This book was converted to FrameMaker 5.5.6 with a format conversion tool created by Erik Ray, Jason McIntosh, Neil Walls, and Mike Sierra that uses Perl and XML technologies. The text font is Linotype Birka; the heading font is Adobe Myriad Condensed; and the code font is Lucas-Font's TheSans Mono Condensed. The illustrations that appear in the book were produced by Robert Romano and Jessamyn Read using Macromedia FreeHand 9 and Adobe Photoshop 6. The tip and warning icons were drawn by Christopher Bing. This colophon was written by Leanne Soylemez.

# More Titles from O'Reilly

## Network Administration

### DNS and BIND, 4th Edition

*By Paul Albitz & Cricket Liu*
*4th Edition April 2001*
*622 pages, ISBN 0-596-00158-4*

*DNS and BIND*, 4th Edition, covers the new 9.1.0 and 8.2.3 versions of BIND as well as the older 4.9 version. There's also more extensive coverage of NOTIFY, IPv6 forward and reverse mapping, transaction signatures and the new DNS Security Extensions; and a section on accommodating Windows 2000 clients, servers, and Domain Controllers.

### Internet Core Protocols: The Definitive Guide

*By Eric Hall*
*1st Edition February 2000*
*472 pages, Includes CD-ROM*
*ISBN 1-56592-572-6*

*Internet Core Protocols: The Definitive Guide* provides the nitty-gritty details of TCP, IP, and UDP. Many network problems can only be debugged by working at the lowest levels—looking at all the bits traveling back and forth on the wire. This guide explains what those bits are and how to interpret them. It's the only book on Internet protocols written with system and network administrators in mind.

### Network Troubleshooting Tools

*By Joseph D. Sloan*
*1st Edition August 2001*
*364 pages, ISBN 0-596-00186-X*

*Network Troubleshooting Tools* helps you sort through the thousands of tools that have been developed for debugging TCP/IP networks and choose the ones that are best for your needs. It also shows you how to approach network troubleshooting using these tools, how to document your network so you know how it behaves under normal conditions, and how to think about problems when they arise so you can solve them more effectively.

### TCP/IP Network Administration, 3rd Edition

*By Craig Hunt*
*3rd Edition April 2002*
*746 pages, ISBN 1-56592-322-7*

This is a complete guide to setting up and running a TCP/IP network for administrators of networks of systems, as well as for lone home systems that access the Internet. It starts with the fundamentals: what the protocols do and how they work, how addresses and routing are used to move data through the network, and how to set up a network connection.

### Managing NFS and NIS, 2nd Edition

*By Hal Stern, Mike Eisler & Ricardo Labiaga*
*2nd Edition July 2001*
*510 pages, ISBN 1-56592-510-6*

This long-awaited new edition of a classic, now updated for NFS Version 3 and based on Solaris 8, shows how to set up and manage a network filesystem installation. *Managing NFS and NIS* is the only practical book devoted entirely to NFS and the distributed database NIS; it's a "must-have" for anyone interested in Unix networking.

### sendmail, 2nd Edition

*By Bryan Costales & Eric Allman*
*2nd Edition January 1997*
*1050 pages, ISBN 1-56592-222-0*

*sendmail*, 2nd Edition, covers sendmail Version 8.8 from Berkeley and the standard versions available on most systems. This cross-referenced edition offers an expanded tutorial and solution-oriented examples, plus topics such as the #error delivery agent, sendmail's exit values, MIME headers, and how to set up and use the user database, mailertable, and smrsh.

# O'REILLY®

TO ORDER: **800-998-9938** • **order@oreilly.com** • **www.oreilly.com**
ONLINE EDITIONS OF MOST O'REILLY TITLES ARE AVAILABLE BY SUBSCRIPTION AT **safari.oreilly.com**
ALSO AVAILABLE AT MOST RETAIL AND ONLINE BOOKSTORES

# Network Administration

## Essential SNMP

By Douglas Mauro & Kevin Schmidt
1st Edition July 2001
326 pages, ISBN 0-596-00020-0

This practical guide for network and system administrators introduces SNMP along with the technical background to use it effectively. But the main focus is on practical network administration: how to configure SNMP agents and network management stations, how to use SNMP to retrieve and modify variables on network devices, how to configure management software to react to traps sent by managed devices. Covers all SNMP versions through SNMPv3.

## Unix Backup & Recovery

By W. Curtis Preston
1st Edition November 1999
734 pages, Includes CD-ROM
ISBN 1-56592-642-0

This guide provides a complete overview of all facets of Unix backup and recovery and offers practical, affordable backup and recovery solutions for environments of all sizes and budgets. It explains everything from freely available backup systems to large-scale commercial utilities.

## T1: A Survival Guide

By Matthew Gast
1st Edition August 2001
304 pages, ISBN 0-596-00127-4

This practical, applied reference to T1 for system and network administrators brings together in one place the information you need to set up, test, and troubleshoot T1. You'll learn what components you need to build a T1 line; how the components interact to transmit data; how to adapt the T1 to work with data networks using standardized link layer protocols; troubleshooting strategies; and working with vendors.

## Cisco IOS in a Nutshell

By James Boney
1st Edition December 2001
608 pages, ISBN 1-56592-942-X

This two-part reference covers IOS configuration for the TCP/IP protocol family. The first part includes chapters on the user interface, configuring lines and interfaces, access lists, routing protocols, and dial-on-demand routing and security. The second part is a classic O'Reilly-style quick reference to all the commands you need to work with TCP/IP and the lower-level protocols on which it relies, with lots of examples of the most common configuration steps for the routers themselves.

## Cisco IOS Access Lists

By Jeff Sedayao
1st Edition June 2001
272 pages, ISBN 1-56592-385-5

This book focuses on a critical aspect of the Cisco IOS—access lists, which are central to securing routers and networks. Administrators cannot implement access control or traffic routing policies without them. The book covers intranets, firewalls, and the Internet. Unlike other Cisco router titles, it focuses on practical instructions for setting router access policies rather than the details of interfaces and routing protocol settings.

## Hardening Cisco Routers

By Thomas Akin
1st Edition February 2002
192 pages, ISBN 0-596-00166-5

This small, handy reference helps system and network administrators make sure their Cisco routers are secure. Because it's about securing the routers themselves, and not the entire network, it's highly practical. The book includes Cisco Router Security Checklists for quick reference, not to mention value-added topics that incorporate the most current thinking about security: DoS attack mitigation, router auditing, and FBI recommendations on incident response.

# Network Administration

## Server Load Balancing

By Tony Bourke
1st Edition August 2001
192 pages, ISBN 0-596-00050-2

Load balancing distributes traffic efficiently among network servers so that no individual server is overburdened. This vendor-neutral guide to the concepts and terminology of load balancing offers practical guidance to planning and implementing the technology in most environments. It includes a configuration guide with diagrams and sample configurations for installing, configuring, and maintaining products from the four major server load balancing vendors.

## Solaris 8 Administrator's Guide

By Paul Watters
1st Edition, January 2002
400 pages, ISBN 0-596-00073-1

This guide covers all aspects of deploying Solaris as an enterprise-level network operating system, with a focus on e-commerce. Written for experienced network administrators who want an objective guide to networking with Solaris, the book covers installation on the Intel and Sparc platforms, and instructs you how to setup Solaris as a file server, application server, and database server.

## Designing Large-Scale LANs

By Kevin Dooley
1st Edition, January 2002
400 pages, ISBN 0-596-00150-9

This unique book outlines the advantages of a top-down, vendor-neutral approach to network design. Everything from network reliability, network topologies, routing and switching, wireless, virtual LANs, firewalls and gateways to security, Internet protocols, bandwidth, and multicast services are covered from the perspective of an organization's specific needs, rather than from product requirements. The book also discusses proprietary technologies that are ubiquitous, such as Cisco's IOS and Novell's IPX.

## IP Routing

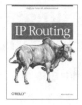

By Ravi Malhotra
1st Edition, January 2002
240 pages, ISBN 0-596-00275-0

This concise guide offers the basic concepts of IP routing, free of hype and jargon. It begins with the simplest routing protocol, RIP, and then proceeds in successive chapters to IGRP, EIGRP, RIP2, OSPF, and finally to the most complex, BGP. By the end, you will have mastered not only the fundamentals of all the major routing protocols, but also the underlying principles on which they are based.

## Using SANs and NAS

By W. Curtis Preston
1st Edition, February 2002
218 pages, ISBN 0-596-00153-3

Storage Area Networks (SANs) and Network Attached Storage (NAS) allow organizations to manage and back up huge file systems quickly. W. Curtis Preston's insightful book takes you through the ins and outs of building and managing large data centers using SANs and NAS. Whether you're a seasoned storage administrator or a network administrator charged with taking on this role, you'll find all the information you need to make informed architecture and data management decisions.

## Building Wireless Community Networks

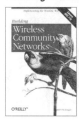

By Rob Flickenger
1st Edition, November 2001
138 pages, ISBN 0-596-00204-1

*Building Wireless Community Networks* offers a compelling case for building wireless networks on a local level: They are inexpensive, and they can be implemented and managed by the community using them, whether it's a school, a neighborhood, or a small business. This book also provides all the necessary information for planning a network, getting the necessary components, and understanding protocols that you need to design and implement your network.

# How to stay in touch with O'Reilly

## 1. Visit our award-winning web site

*http://www.oreilly.com/*

★ "Top 100 Sites on the Web"—PC Magazine
★ CIO Magazine's Web Business 50 Awards

Our web site contains a library of comprehensive product information (including book excerpts and tables of contents), downloadable software, background articles, interviews with technology leaders, links to relevant sites, book cover art, and more. File us in your bookmarks or favorites!

## 2. Join our email mailing lists

Sign up to get email announcements of new books and conferences, special offers, and O'Reilly Network technology newsletters at:

*http://www.elists.oreilly.com*

It's easy to customize your free elists subscription so you'll get exactly the O'Reilly news you want.

## 3. Get examples from our books

To find example files for a book, go to:

*http://www.oreilly.com/catalog*

select the book, and follow the "Examples" link.

## 4. Work with us

Check out our web site for current employment opportunites:

*http://jobs.oreilly.com/*

## 5. Register your book

Register your book at:

*http://register.oreilly.com*

## 6. Contact us

**O'Reilly & Associates, Inc.**
1005 Gravenstein Hwy North
Sebastopol, CA 95472  USA
TEL:     707-827-7000 or 800-998-9938
             (6am to 5pm PST)
FAX:     707-829-0104

**order@oreilly.com**
For answers to problems regarding your order or our products. To place a book order online visit:

*http://www.oreilly.com/order_new/*

**catalog@oreilly.com**
To request a copy of our latest catalog.

**booktech@oreilly.com**
For book content technical questions or corrections.

**proposals@oreilly.com**
To submit new book proposals to our editors and product managers.

**international@oreilly.com**
For information about our international distributors or translation queries. For a list of our distributors outside of North America check out:

*http://international.oreilly.com/distributors.html*

# Notes